D1238363

CHLOROPLAST METABOLISM

The structure and function of chloroplasts in green leaf cells

CHLOROPLAST METABOLISM

The structure and function of chloroplasts in green leaf cells

BARRY HALLIWELL

Department of Biochemistry,
King's College, London

CLARENDON PRESS · OXFORD 1981

Oxford University Press, Walton Street, Oxford OX2 6DP
OXFORD LONDON GLASGOW
NEW YORK TORONTO MELBOURNE WELLINGTON
KUALALUMPUR SINGAPORE JAKARTA HONGKONG TOKYO
DELHI BOMBAY CALCUTTA MADRAS KARACHI
NAIROBI DARESSALAAM CAPE TOWN

Published in the United States
by Oxford University Press, New York

British Library Cataloguing in Publication Data

Halliwell, Brian
Chloroplast metabolism.
1. Chloroplasts 2. Biological chemistry
I. Title
581.87'33 QK882

ISBN 0-19-854549-5

Typeset by Anne Joshua Associates, Oxford
Printed in Great Britain
by Eric Buckley
Printer to the University
Oxford University Press

PREFACE

Standard textbooks of biochemistry pay scant attention to photosynthesis, although a study of this process at an advanced level is becoming an increasingly important part of degree courses in biology, biochemistry, and plant sciences. To some extent this defect can be remedied by a study of Gregory's excellent book (R. P. F. Gregory (1977). *Biochemistry of photosynthesis.* Wiley, London). However, this book is biased towards physical studies of the light reactions of photosynthesis and it does not cover the recent great advances in our understanding of chloroplast carbon, nitrogen, and sulphur metabolism. The present volume aims to deal with such topics in an integrated fashion. In order to do so, it is necessary to discuss the 'light reactions' to some extent (Chapter 2) but I have tried to avoid extensive repetition of work already covered by Gregory. I have taken the opportunity to include a chapter on 'toxic oxygen effects', which are becoming of increasing importance to our understanding of the metabolism of all aerobes, including green plants.

There is extensive metabolic co-operation between chloroplasts, other leaf organelles, and cytoplasmic enzymes. Sometimes, because of the use of inadequately-characterized chloroplast preparations, it is not clear exactly which reactions in a given pathway occur in chloroplasts and which do not. I have tried to indicate where such problems exist and to point the way to experiments which should resolve them.

I most gratefully acknowledge the help given to me in the typing of this manuscript by secretaries within King's College (Ms Yvonne Chmielowiec) and outside (Ms Bernice Charles), in the preparation of the figures by Ms Frances Hodsman and Mr Colin Chinnery and in proof-reading and expert comments by Dr Stephen Charles. I thank also the following for the gift of figures or for permission to reproduce figures from their published works: Professor D. O. Hall (Fig. 1.1); Professor E. Beck and Elsevier/North Holland Biomedical Press (Fig. 7.1); Professor L. E. Anderson and the American Society of Plant Physiologists (Fig. 3.4); Dr M. C. W. Evans and Elsevier/North Holland Biomedical Press (Fig. 2.7); Professor D. A. Walker and Elsevier/North Holland Biomedical Press (Fig. 4.1); Professor I. P. Ting and Academic Press (Fig. 1.10); Professor C. B. Osmond (Fig. 5.9); Dr B. M. R. Harvey and Blackwell Scientific Publications (Fig. 8.10); Dr C. K. M. Rathnam and Blackwell Scientific Publications (Table 5.3); Professor D. A. Walker and Dr C. H. Foyer (Fig. 1.9); Professor J. Barber and Elsevier/North Holland Biomedical Press (Figs. 2.4 and 1.5); Professor M. Kluge and Springer-Verlag (Figs 5.6, 5.7 and 5.8); Professor C. C. Black and Springer-Verlag (Figs 5.1 and 5.2).

Finally, I would like to thank Sean for support, tolerance, and love during the hectic period in which the manuscript was completed.

London
July 1980

BH

CONTENTS

9 SYNTHESIS OF PHENOLIC COMPOUNDS, MEMBRANE LIPIDS, CHLOROPHYLLS, AND CAROTENOIDS IN CHLOROPLASTS

10 NITROGEN AND SULPHUR METABOLISM IN PHOTOSYNTHETIC TISSUES

1 STRUCTURE, FUNCTION, AND ISOLATION OF CHLOROPLASTS

1.1 The nature of photosynthesis

Photosynthesis may be defined as the process by which plants, algae, and a few species of bacteria use the energy of light to generate stored chemical energy which is used not only by these organisms but also by all other living cells.

Light energy trapped by photosynthetic pigments is used to remove electrons from a donor and transfer them on to an acceptor. For higher plants, with which this book is largely concerned, and algae, the electron donor is water. Indeed, all the oxygen evolved during their photosynthesis comes from the splitting of water (Radmer and Ollinger 1980). By contrast, photosynthetic bacteria utilize other substances, such as hydrogen sulphide, as electron donors. In this case, elemental sulphur is generated as a by-product.

Electrons obtained from the splitting of water pass through a series of acceptors. Most of the electrons are eventually used to produce NADPH, which is required for the reduction of carbon dioxide into carbohydrate (the overall process may be represented as in eqn (1.1)) but small numbers are diverted onto other terminal electron acceptors, such as nitrate or sulphate.

$$n\mathrm{H_2O} + n\mathrm{CO_2} \xrightarrow{h\nu} \underset{\text{carbohydrate}}{(\mathrm{CH_2O})_n} + n\mathrm{O_2} \qquad (1.1)$$

In higher plants the process of photosynthesis occurs within chloroplasts. Under the electron microscope these organelles are seen to be bounded by an outer envelope consisting of two membranes separated by an electron-translucent space of about 10 nm (Fig. 1.1). The envelope encloses the stroma of the chloroplast, in which floats a complex internal membrane structure. The stroma is an aqueous solution containing various low-molecular-weight compounds plus a high concentration of proteins, the major protein being the enzyme ribulose diphosphate carboxylase (Chapter 3). It contains the enzymes necessary to convert carbon dioxide into carbohydrate, by a cyclic metabolic pathway known as the *Calvin cycle*. Starch grains and other particulate matter are often present. For example, many small spherical electron-opaque bodies, known as plastoglobuli, are seen in almost all chloroplasts and may function as stores of lipid material such as plastoquinone and tocopheryl-quinones (Coombs and Greenwood 1976). Ribosomes are also present in the stroma, since chloroplasts have the capacity to synthesize some proteins (Chapter 2).

The internal membrane structure of the chloroplast is extremely complex. In

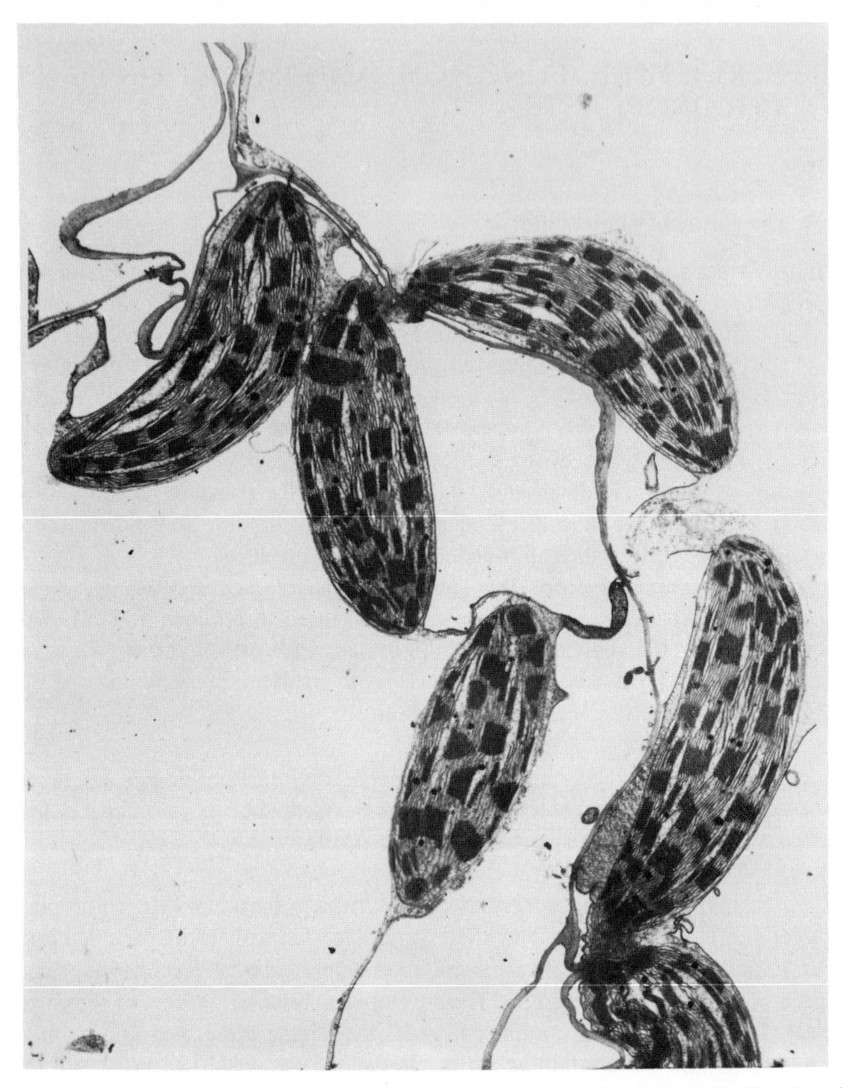

Fig. 1.1. Electron micrograph of a thin section of the leaf mesophyll cells of a higher plant to show the chloroplasts.

the electron micrograph shown in Fig. 1.1 two distinct features may be recognized, viz. regions of closely stacked membranes known as *grana* that are interconnected by a three-dimensional network of membranes known as the *stroma thylakoids*. The grana were at one time regarded as being made up of piles of essentially separate discs (*thylakoids*) interconnected by stromal thylakoids, but in fact the network of interconnections is so extensive and complex that the internal spaces of the granal and stroma thylakoids effectively form a single large space separated

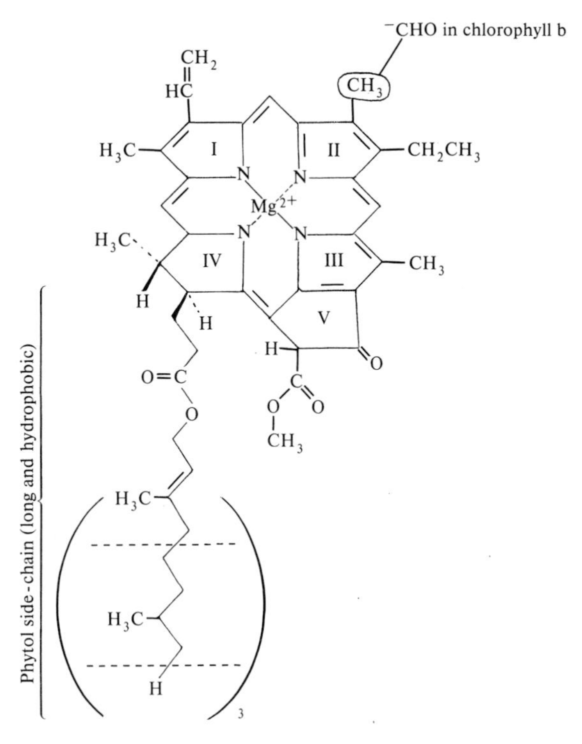

Fig. 1.2. Structure of chlorophylls a and b.

from the stroma by the thylakoid membranes (Anderson 1975; Coombs and Greenwood 1976). The membranes of this complex internal chloroplast structure contain the photosynthetic pigments and generate the NADPH and, by the process of photophosphorylation, the ATP needed to drive carbon dioxide fixation by the Calvin cycle in the stroma. The pigments of higher plant chloroplasts comprise chlorophylls a and b (Fig. 1.2) plus carotenes and xanthophylls, yellow pigments absorbing blue light. Chlorophyll a absorbs light in the blue (420–435 nm) and red (660–680 nm) parts of the spectrum (Fig. 1.3) although the exact shape of the absorption spectrum depends on the polarity of the environment in which the chlorophyll molecule is placed. The electron-dense tetrapyrrole 'head' of the chlorophyll molecule is responsible for the light absorption, whilst the long hydrophobic phytol 'tail' probably helps to anchor the molecule into the thylakoid membrane. Different forms of chlorophylls a and b, with modified side chains, have been detected in higher plants (Chapter 9).

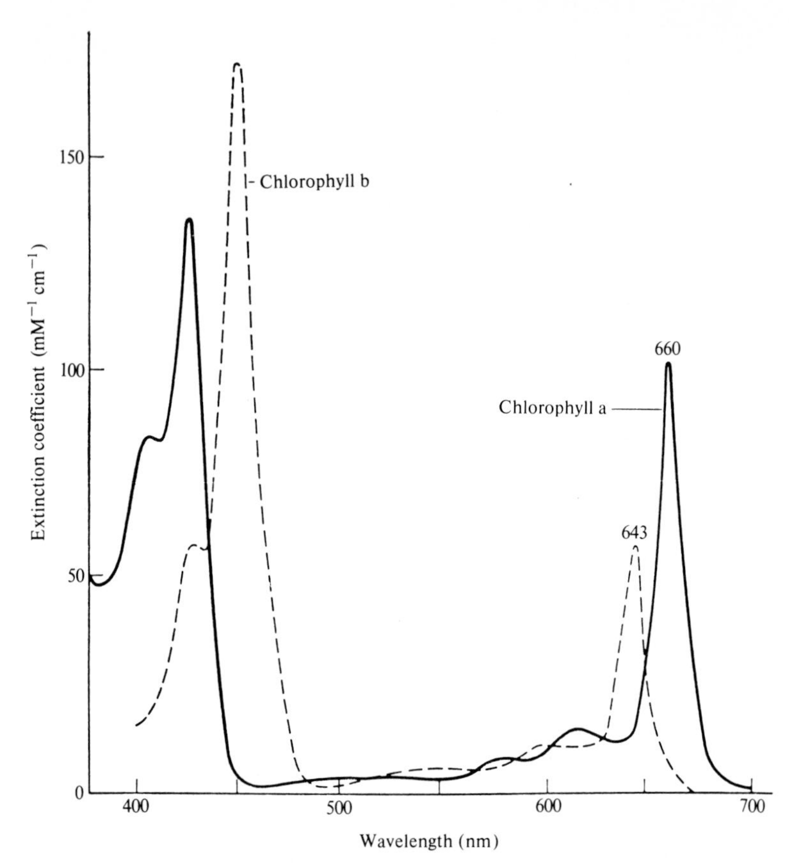

Fig. 1.3. The absorption spectra of chlorophylls a and b in acetone solution. Chlorophylls are the pigments that give plants their characteristic green colour. They are insoluble in water but soluble in organic solvents.

1.2 Function of the photosynthetic pigments

Absorption of light energy by chlorophyll or other pigment molecules results in the formation of higher electronic excitation states (*excited singlet states*). Absorption of a quantum of blue light by chlorophyll results in the formation of a more excited singlet state (*second excited state*) than does absorption of a quantum of red light (*first excited state*), but the former excited state loses its excess energy as heat extremely quickly, so that the absorption of either red or blue quanta effectively produces the same first excited state of the chlorophyll molecule (Fig. 1.4). Energy so absorbed can be lost by re-emission of a photon to give *fluorescence*, it can be lost as heat by various processes, or it can be transferred to another molecule, e.g. an adjacent chlorophyll, the first molecule

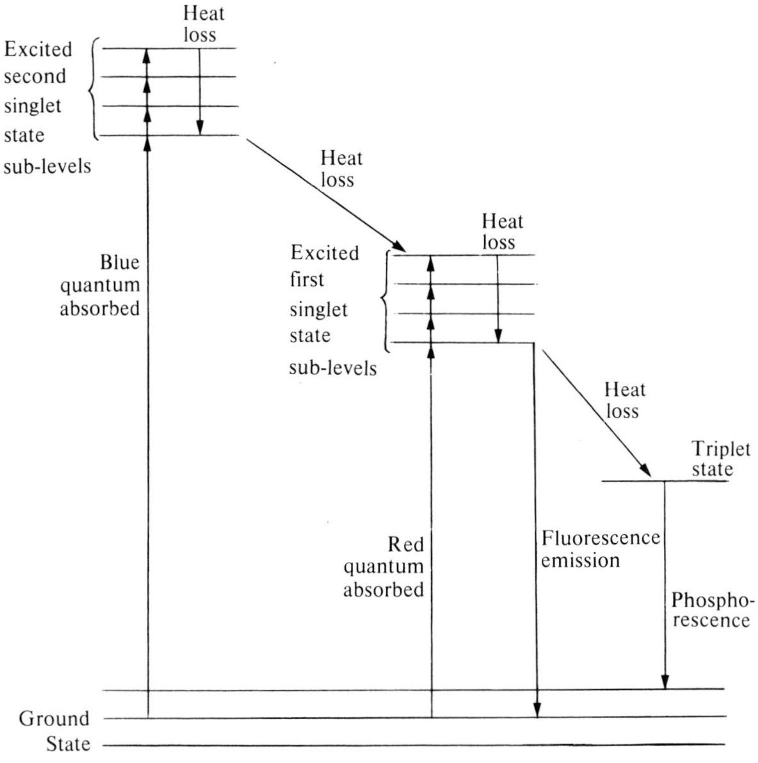

Fig. 1.4. Energy changes in the chlorophyll molecule. Absorption of light energy by the chlorophyll molecule results in the formation of higher electronic excitation states. Although the energy stored by the second singlet state is much greater than that in the first singlet state, the excess is lost as heat and does not contribute to photosynthesis. Both ground and excited states possess a series of sub energy-levels. Excitation can proceed from any sub-level of the ground state to any sub-level of an excited state, but heat loss occurs to give the lowest sub-energy level of the excited state. Hence any energy omitted by fluorescence is usually less than that absorbed during excitation and so the fluorescence spectrum is shifted to the red end of the spectrum as compared with the spectrum of the exciting light (*Stokes shift*). A triplet state can also be formed from the excited singlet states. This arises because an orbital can contain a maximum of two electrons, each of which must have a different value of the spin quantum number ($+\frac{1}{2}$ or $-\frac{1}{2}$). When one of the electrons is promoted to a higher orbital by absorption of energy or by other means, the value of its spin quantum number is not normally changed. However, spin reversal can take place infrequently during the lifetime of an excited state, to form a triplet state in which the excited electron and the remaining electron in the orbital in the ground state both have the same spin quantum number. Direct light emission from the triplet state (*phosphorescence*) has to be accompanied by spin reversal and has a longer wavelength than normal fluorescence. Alternatively, the triplet can be converted back to the singlet, a process with low probability. Nevertheless, this requires only a small amount of energy and can result in a 'delayed' fluorescence with the normal fluorescence spectrum. Triplet chlorophyll states are not thought to play a role in photosynthesis, although their formation has been observed *in vivo* (Gregory 1977).

returning to the ground state whilst the second becomes excited. The probability of efficient transfer of excitation energy between chlorophyll molecules is inversely proportional to the sixth power of the distance between them (Gregory 1977; Barber 1978). Energy absorbed by carotenoids can be directly transferred to chlorophyll molecules in this way.

However, certain special molecules of chlorophyll a in the thylakoid membrane (*reaction centre chlorophylls*) can, when excited, lose electrons to a neighbouring electron acceptor (A). An asterisk is used to represent the excited state of chlorophyll in the equation below:

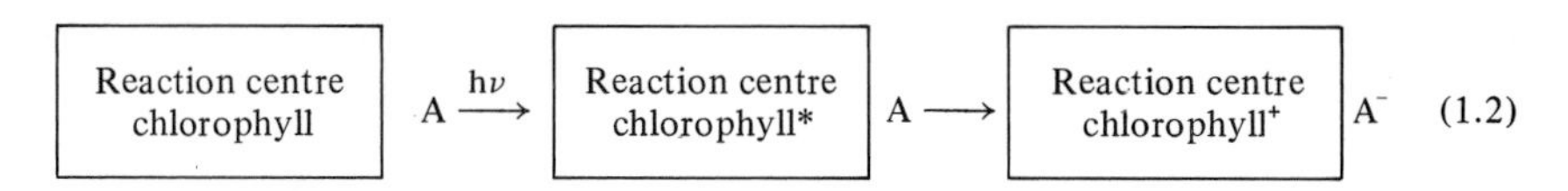

and this initial charge separation is the basic reaction of photosynthesis. The fate of the ionized species is discussed further in Chapter 2.

Light energy absorbed by other chlorophyll and carotenoid molecules can be transferred quite efficiently in the thylakoid membrane to the reaction centre chlorophylls. Indeed, each reaction centre chlorophyll seems to be associated with a 'light-harvesting array' of other pigment molecules that can funnel energy to it, the whole complex often being referred to as a 'photosynthetic unit'. Such an organization enables light of widely varied wavelengths to be utilized for photosynthesis and it also allows photosynthesis to occur at low light intensities, at which the probability of direct excitation by light of a reaction centre chlorophyll would be very low. Transfer through the 'light harvesting antenna system' has to be rapid in order to avoid loss of energy by processes such as fluorescence. Such efficient transfer between chlorophyll molecules requires correct orientation of them and proceeds most rapidly when the molecules are about 1 nm apart and the wavelength of maximum absorption of the acceptor chlorophyll is closer to the red end of the spectrum than that of the donor (Barber 1978).

Two different types of reaction centre, known as P700 and P680, can be detected in the chloroplasts of higher plants by observing their light-induced absorption changes. P700 seems to be a dimer of chlorophyll a molecules located in a special environment. The positive charge in the excited form (eqn 1.2) is delocalized over the rings of both chlorophyll molecules. The P700 spectrum can be observed during steady-state illumination and it also produces an electron spin resonance signal.

Rather less is known about P680 because its absorption cannot be detected during steady-state illumination of thylakoids, since the oxidized (chl^+) form rapidly regains electrons (Chapter 2). It seems to be chlorophyll a, possibly as a dimer, in a special environment. The two reaction centres P680 and P700 operate in series so that two photons are absorbed for every electron that is extracted from water and transferred to the redox level necessary to operate the

Calvin cycle. Each reaction centre is served by its own light-harvesting antenna system: the complexes containing P700 may be referred to as photosystems I (PSI) and those containing P680 as photosystems II (PSII). Chlorophyll b preferentially donates its absorbed energy to PSII, whereas carotenoids are present in both light-harvesting complexes. Above 700 nm only PSI absorbs light energy (Fig. 1.5). Nevertheless, there is some 'spillover' of the energy absorbed by PSII into PSI, which is necessary to equalize the quantal intakes of PSI and II under normal light conditions (Fig. 1.5). An increase in the degree of 'spillover' is brought about by shining light which preferentially excites PSII (below 680 nm). This apparently causes slow conformational changes of the thylakoid membrane that alter the configuration of the pigment molecules so as to balance the distribution of light absorbed by the 'bulk' pigments to the light-harvesting complexes of PSI and PSII. The regulation of spillover is closely associated with movement of cations such as Mg^{2+} between stroma and thylakoids (Barber 1976) (Chapter 2).

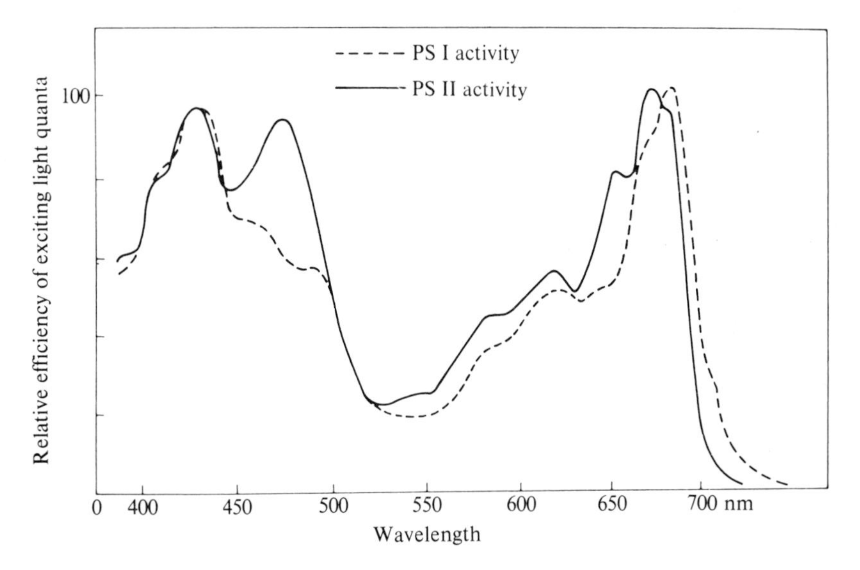

Fig. 1.5. Action spectra for PSI and PSII. Data were taken from Thornber and Barber (1979). PSII does not utilize light energy above 700 nm, whereas below about 680 nm light is absorbed by both systems, PSII absorbing slightly more. Normal photosynthesis requires the simultaneous operation of both photosystems, and so above 680 nm the light absorbed is utilized inefficiently, i.e. quanta absorbed by chlorophyll above 680 nm are less effective in allowing photosynthesis than those of shorter wavelength. This phenomenon is sometimes called the *red drop*. The red drop can be obviated if a weak supplementary beam of light of a shorter wavelength is supplied (*the Emerson enhancement effect*).
Further discussion of this phenomenon may be found in Gregory (1977).

1.3 Structure of the thylakoid membrane

In view of current knowledge of membrane structure as formulated in the fluid mosaic model (Singer and Nicolson 1972), it seems reasonable to regard the chloroplast membranes as consisting essentially of a lipid bilayer into which various proteins are embedded. The lipid bilayers should be essentially fluid under physiological conditions, allowing a rapid exchange of places between lipid molecules and their neighbours in any one half of the bilayer, and a fairly-free translational mobility of any proteins in the membrane. It should be noted, however, that only about half of the dry weight of thylakoids is lipid: there is a considerable protein content. Protein molecules embedded in a membrane often decrease the mobility of adjacent lipid molecules by hydrophobic interactions, so that there may be a considerable degree of ordered lipid in areas of the thylakoid membrane that contain large numbers of protein complexes (Hiller and Raison 1980). About 26 per cent of the total lipids present represent the photosynthetic pigments, which must be held in correct orientation to allow energy transfer to occur in the light-harvesting complexes. The other major lipids present are glycolipids (monogalactosyl and digalactosyldiglycerides – Fig. 1.6) together with some sulphoquinovosyl diglyceride, often called 'the plant sulpholipid' (Table 1.1) (Quinn and Williams 1978). Presumably these form the basic lipid bilayer of the membrane. Most of the fatty acids present in chloroplast lipids are C_{18} acids containing one or more double bonds, a high degree of unsaturation promoting membrane fluidity. In some species α-linolenic acid (Fig. 1.7) can account for over 90 per cent of the fatty acids present in chloroplast lipids (Leech and Murphy 1976; Poincelot 1976). An unsaturated fatty acid unique to photosynthetic tissues is *trans*-3-hexadecenoic acid ($C_{16:1}$) which occurs acylated to phosphatidylglycerol (Leech and Murphy 1976).

Some of the thylakoid proteins might be loosely attached to the membranes, held by ionic forces to the polar groupings of membrane lipids (*extrinsic proteins*), others might be located in the hydrophobic membrane interior, so requiring detergents to remove them from the membranes (*intrinsic proteins*), and some proteins might span the membrane completely. For example, the coupling factor (CF_1) involved in photophosphorylation (Chapter 2) is an extrinsic protein complex, being easily removed by treatment of thylakoid membranes with the divalent cation chelator ethylene-diamine tetraacetic acid (EDTA). In contrast, most, if not all, chlorophyll is associated with proteins that can only be released from the thylakoids after treatment with detergents such as sodium dodecyl sulphate (SDS). Such an association with proteins might perhaps be predicted as being necessary to give that correct orientation of pigment molecules needed for efficient transfer of absorbed energy between them. When chloroplast membranes are solubilized in SDS the green extract so obtained may be separated by electrophoresis on polyacrylamide gels. A series of bands are resolved. Most of the chlorophyll is found to be associated with

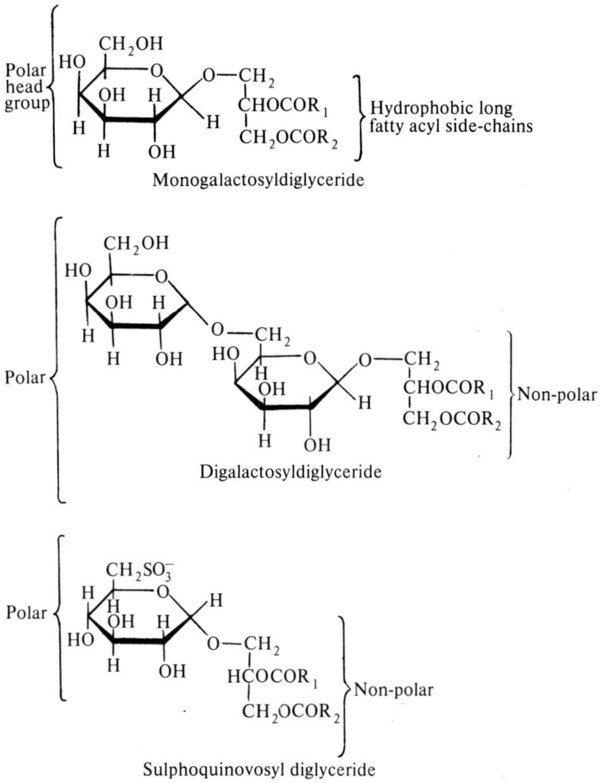

Polar head group

Hydrophobic long fatty acyl side-chains

Monogalactosyldiglyceride

Polar

Non-polar

Digalactosyldiglyceride

Polar

Non-polar

Sulphoquinovosyl diglyceride

Fig. 1.6. Structure of the chloroplast galactolipids and of sulphoquinovosyl diglyceride.

three distinct bands (Anderson 1975, 1980; Argyroudi-Akoyunoglou and Castorinis 1980), although many others can be detected. One major band, referred to as a P700-chlorophyll a-protein complex, contains about 28 per cent of the membrane protein, has a high ratio of chlorophyll a to chlorophyll b (at least 12:1) and contains the P700 reaction centre. Some β-carotene is also present. The complex is detected in gels made from thylakoid extracts of all higher plant species and it presumably represents the P700 reaction centre with part of its antenna light-harvesting complex. The proteins present contain a high proportion of hydrophobic amino acid residues, as one would expect for intrinsic membrane proteins that interact with the fatty acyl side chains of lipids.

The second major electrophoretic band has a much lower ratio of chlorophyll a to chlorophyll b (1:1), and contains about 50 per cent of the membrane protein. Some carotene is present, and, again, the protein present is extremely hydrophobic. This band is not found in mutant plants that lack chlorophyll b.

Table 1.1. Lipid composition of subcellular fractions from leaves of *Vicia faba*

Fraction	Lipid present as % of total lipids analysed			
	Monogalactosyl diglyceride	Digalactosyl diglyceride	Phosphatidylcholine	Phosphatidylglycerol
Chloroplast envelope	29	32	29	9
Chloroplast thylakoids	65	26	3	6

Data from Leech and Murphy (1976).

Only the lipids listed above were analysed. When all lipids present are taken into account, monogalactosyl diglyceride accounts for ca. 30 per cent of total lipids in chloroplasts, digalactosyl diglyceride 15 per cent, sulphoquinovosyl diglyceride 5 per cent, phospholipids 10 per cent, neutral lipids 1 per cent, pigments 26 per cent and sterol glycosides 2 per cent (Leech 1976). The envelope contains no chlorophyll. It may be seen that the lipid compositions of envelope and thylakoid membranes are very different. It has recently been suggested, however, that there may be some interconversion of envelope galactolipids during isolation (Wintermans, Van Besouw and Bogemann 1981).

Major component (R_1R_2)—α linolenic acid

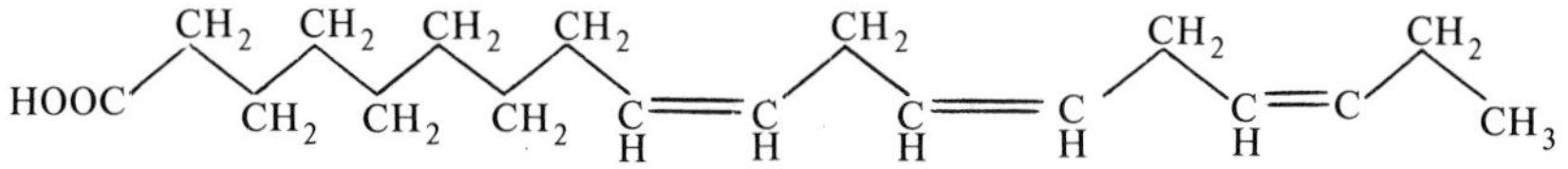

CH_2 CH_2 CH_2 CH_2 CH_2 CH_2 CH_2
HOOC CH_2 CH_2 CH_2 C=C C=C C=C CH_3
H H H H H H

Minor component (R_1R_2) *Trans*-3-hexadecenoic acid

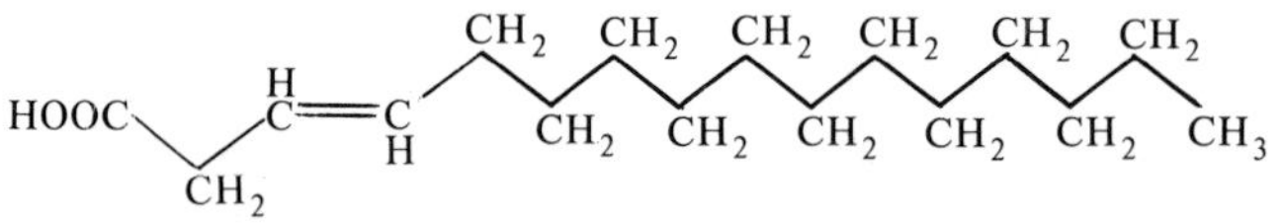

CH_2 CH_2 CH_2 CH_2 CH_2 CH_2
H
HOOC C=C CH_2 CH_2 CH_2 CH_2 CH_2 CH_3
H
CH_2

Fig. 1.7. Fatty acids found in chloroplast lipids. α-Linolenic acid is a major component of lipids in chloroplasts of all higher plants and in some species comprises over 90 per cent of all esterifed fatty acids present. Note the high degree of unsaturation and the *cis* configuration of the double bonds.

Energy transfer from chlorophyll b to chlorophyll a is highly efficient in the isolated complex. No reaction centre chlorophyll, either P680 or P700, is present and it is thought that this complex functions *in vivo* as a light-harvesting system donating mainly to the photosystem II reaction centre, P680. It is therefore often referred to as a 'light harvesting chlorophyll a/b protein complex'. Illumination of isolated chloroplasts has been shown to cause phosphorylation of several thylakoid peptides, the major ones phosphorylated being components of this complex. Phosphorylation is achieved by the light-dependent activation of a kinase enzyme bound to the thylakoids (Alfonzo, Nelson, and Racker 1980). The phosphate groups are removed in the dark by the action of a phosphatase enzyme (Bennett 1979, 1980). The physiological role of this phosphorylation/dephosphorylation cycle in the light-harvesting chlorophyll a/b protein complex remains to be established, however, but it could play some role in the conformational changes that occur in thylakoid membranes on illumination, e.g. in the 'spillover' phenomenon.

The third major band detected on electrophoresis is a pigment-lipid-SDS complex that runs at the gel front. The amount of this depends on the time of exposure of the thylakoid membranes to detergent. It is possible, with low amounts of detergent and quick electrophoretic runs, to retain as much as 80 per cent of the chlorophyll in the form of protein complexes (Anderson 1975, 1980). Longer exposure to detergent strips off more chlorophyll.

Some information about the arrangement of protein complexes within the thylakoid membrane has been provided by the technique of freeze-fracture and etching. In the freeze-fracture technique the specimen is frozen and then fractured with a knife edge to yield cleavage surfaces. They are then metal-shadowed and the replica so obtained is examined under the electron microscope. This technique is thought to split membranes through their most hydrophobic regions, viz. in the region between the two lipid bilayers and around the outer lipid shells surrounding intrinsic proteins. Freeze-etching, a modification of the above technique, reveals some of the outer membrane surfaces as well. Whereas artificial membranes containing only lipids reveal smooth surfaces after freeze-fracture, the thylakoid membrane with its high protein content shows very great complexity. Particles of about 10-11 nm in diameter are seen to be attached to the external membrane surfaces of the thylakoids in those regions that are not stacked to form grana. Since they can be removed by washing the membranes with EDTA before freeze-etching, the particles presumably represent CF_1, an extrinsic protein complex involved in ATP synthesis (Chapter 2).

The two interior membrane surfaces revealed by freeze-etching are strikingly different in appearance, as is shown diagrammatically in Fig. 1.8. In the regions where membranes are stacked together to form grana, one interior membrane face (known as the EF_s face – s for 'stacked') is densely studded with large particles, 13-17 nm in diameter and 8-9 nm in height (Coombs and Greenwood 1976). They are sometimes observed to be in a crystalline array. Particles are less numerous and smaller in the unstacked thylakoid EF faces (EF_u). When thylakoid membranes are suspended in media of low salt concentration the stacked regions disappear and the membranes appear freely unfolded. This unstacking causes no loss of photochemical activities and is reversible by addition of monovalent cations (50-100 mM) or much lower concentrations (3-5 mM) of divalent cations. When such unstacked membranes are studied by freeze-fracturing, internal surfaces that have dense populations of large particles are no longer seen

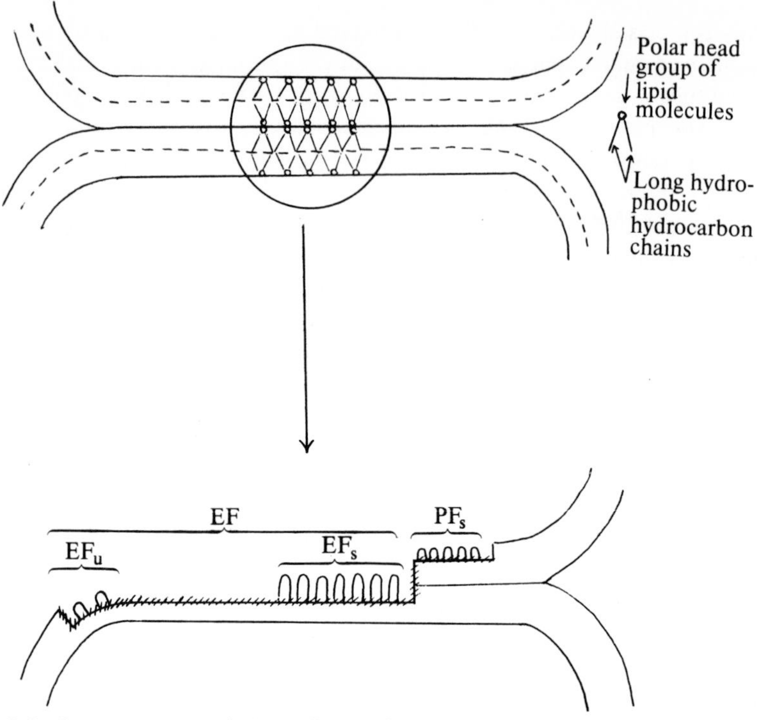

Fig. 1.8. Structures revealed on freeze–fracture of thylakoid membranes. The diagram represents two thylakoid membranes stacked together. Each membrane consists of a lipid bilayer, part of which is shown (circled). A fracture path has been shown passing through both bilayers so as to expose two different types of membrane face (hatched line). The largest particles are found in the region of the EF face where the membrane was in contact with other membrane (stacked). Where the two membranes are not in contact the particles are reduced in size and number (EF_u). The PF face contains many smaller particles in both stacked (PF_s) and unstacked membrane regions.

(Anderson 1975). In contrast to the EF face, the opposing (PF) face contains particles of about 8 nm in diameter which are found in both granal and stromal thylakoids. Whether the particles seen on fracture faces are located in only one half of a lipid bilayer, extend completely across one membrane or penetrate into adjacent bilayers of two different membranes in stacked regions is still a matter of controversy. An intrinsic membrane protein spanning the thylakoid membrane might appear to be associated with only one half of the bilayer on freeze-fracture simply because it interacts more strongly with other components of that half and therefore pulls away from the other half of the bilayer on freeze–fracture. The chlorophyll a/b protein complex has been suggested to play some role in the stacking of the thylakoids.

The functional significance of the EF and PF particles is also controversial (for a review see Arntzen and Burke 1980). It has been suggested that the large EF particle contains the PSII reaction-centre chlorophyll surrounded by the chlorophyll a/b light-harvesting complex, which would explain why the size of this particle differs between higher plants and mutant strains that contain different amounts of the a/b complex. The 8-nm particles on the PF face have been suggested to represent PSI. In considering the static picture of membrane structure provided by freeze-etching, it must be borne in mind that illumination of thylakoid membranes causes extensive conformational changes within them (Wang and Packer 1973) and also that the movement of ions from thylakoids to stroma accompanied by a flow of protons in the opposite direction (Chapter 2) will exert strong influences on membrane stacking (for reviews see Anderson 1975; Barber 1976; Thornber and Barber 1979). Stroma thylakoids are enriched in PSI relative to the stacked granal thylakoids, but the reverse is true for PSII.

1.4 The isolation of chloroplasts

We have seen that chloroplasts are extremely complex structures. Most of our information about the metabolic processes occurring within them has come from the study of isolated chloroplasts. Before discussing such processes further, it is worth reviewing exactly what is meant by 'isolated chloroplast preparations'. The methods of isolation that are frequently used to obtain chloroplasts for the study of electron transport and photophosphorylation and sometimes for enzyme studies involve the grinding of a suitable leaf tissue (most often spinach, *Spinacia oleracea*) in ice-cold buffer containing osmoticum (sodium chloride or sucrose) in a blender for 10–15 s, followed by filtration through muslin. The filtrate is usually subjected to two cycles of differential centrifugation: a low speed spin (e.g. 500 *g* for 2 min) to remove nuclei and 'cell-debris', followed by a higher-speed spin (e.g. 6000 *g* for 15 min) to pellet the actual chloroplast preparation. Table 1.2 outlines some procedures that have been used.

However, the envelope of chloroplasts isolated from leaf tissues is very fragile and easily becomes torn or lost altogether during such prolonged preparation procedures. Hence the isolated chloroplast pellet can contain organelles of variable integrity: some chloroplasts might remain completely intact, but most will have leaked metabolites and/or enzymes through damaged envelopes, and in others the envelopes and stromal contents will have been lost completely, leaving a more-or-less naked thylakoid preparation. Such damaged organelles are usually still capable of oxygen evolution, photophosphorylation, and the reduction of added electron acceptors under appropriate assay conditions, but they do not fix carbon dioxide at significant rates because of this loss of stromal metabolites and/or enzymes. Table 1.2 illustrates some typical rates of carbon dioxide fixation and enzyme activities in 'broken chloroplasts'. Even when the isolated chloroplast preparation is further separated, e.g. by density-gradient techniques, to obtain bands of organelles that appear intact under the light microscope, the

Table 1.2. Carbon dioxide fixation and specific activity of enzymes in spinach chloroplasts isolated by different methods

Method of isolation	Rate of CO_2 fixation (μmol h^{-1} mg chlorophyll^{-1})	Activity of enzymes (μmol h^{-1} mg chlorophyll^{-1})		
		Ribulose diphosphate carboxylase	$NADP^+$-glyceraldehyde 3-phosphate dehydrogenase	Alkaline fructose diphosphatase
A	5	7.8	104	43
B	7	8.5	210	46
C	123	28	505	101

Data from Table 3 of Latzko and Gibbs (1968). Much higher activities of ribulose diphosphate carboxylase have subsequently been obtained using improved assay methods.

Method A: As for B but using sucrose as osmoticum.

Method B: Homogenization in 0.35 M NaCl buffered with Tris-HCl, pH 8. Filter and centrifuge 4 min at 2000 *g*. Resuspend in 0.35 M NaCl (2–3 ml) and dilute to 40 ml with 0.35 M NaCl. Spin 4 min at 2000 *g*. Resuspend the pellet again in 0.35 M NaCl (essentially the procedure of Whatley and Arnon 1963).

Method C: Method of Jensen and Bassham (1966) as outlined in Table 1.3.

chloroplasts obtained still cannot fix carbon dioxide at high rates because of the loss of soluble metabolities.

Many plant tissues contain high activities of enzymes which catalyse oxidation of *o*-diphenols into quinones (see Chapter 9) and they often accumulate large quantities of diphenols, e.g. caffeic and chlorogenic acids, in the central vacuoles of the cells. When the tissue is homogenized the diphenols are released and come into contact with the diphenol oxidases in the cytoplasm. The quinones so produced polymerize to give melanin-like products (the 'browning reaction') and also react with proteins, causing severe damage. It is usually impossible to extract functional organelles, or even active enzymes, from tissues with a high content of phenols unless inhibitors of diphenoloxidases, such as diethyldithiocarbamate or 2-mercaptoethanol, are included in the homogenizing medium. Alternatively, reducing agents such as ascorbate may be added, which serve to reduce the quinones back to diphenols (Anderson 1968; Mayer and Harel 1979). One advantage of spinach is that it rarely accumulates phenolic compounds to an extent that interferes with organelle isolation.

1.5 Preparation and properties of intact chloroplasts

Nevertheless, it is possible to obtain chloroplast preparations in which a significant percentage of the organelles retain intact envelopes. The preparation methods usually involve a combination of special isolation media, brief grinding of the leaves in an efficient homogenizer, such as a Polytron, and rapid centrifugation to separate the organelles from damaging components in the homogenate. To speed up the preparation, the preliminary centrifugation to remove 'cell debris' is usually replaced by filtration through several layers of muslin or cotton wool. A useful product for achieving this is gamgee tissue, which consists of a layer of cotton wool sandwiched between two layers of gauze. Table 1.3 illustrates some of the different procedures that have been used and Table 1.2 shows that such procedures can yield chloroplasts capable of fixing carbon dioxide at high rates (between 50 and 150 μmol carbon dioxide fixed h^{-1} mg chlorophyll $^{-1}$). The isolated chloroplast preparation still contains some broken chloroplasts, however, and integrity may be further improved by resuspending the organelles in homogenizing medium and rapidly pelleting for a second time, when some of the damaged chloroplasts remain in the supernatant. The highest figures reported for carbon dioxide fixation (e.g. 350 μmol h^{-1} mg chlorophyll^{-1}) have been obtained using chloroplasts washed in this way. Each washing step, however, causes loss of a considerable proportion of the chloroplast fraction.

The quality of the 'intact' chloroplast fraction obtained by the above procedures (i.e. the percentage of totally intact organelles fixing carbon dioxide at high rates that it contains) depends on many factors, not least the experience of the isolator. The nature of the plant material is also important. Chloroplasts fixing carbon dioxide at high rates have been obtained only from a small number

Table 1.3. Methods commonly used for isolating chloroplasts capable of fixing carbon dioxide at high rates

Method of Jensen and Bassham (1966)
De-ribbed spinach leaves (10 g) are homogenized at high speed for 5 s only in 0.33 M sorbitol, 0.002 M $NaNO_3$, 0.002 M EDTA, 0.002 M sodium isoascorbate, 0.02 M NaCl, 0.001 M $MgCl_2$, 0.0005 M K_2HPO_4, 0.05 M MES adjusted to pH 6.1 with NaOH. The homogenate is squeezed through six layers of muslin and spun at 2000 *g* for 50 s. The pellet is resuspended in a similar medium, but with HEPES buffer pH 6.7 instead of MES.
Method of Walker (1971)
Leaves are homogenized briefly in a Polytron homogenizer in a partially frozen slurry of 0.33 M sorbitol, 0.002 M EDTA, 0.001 M $MgCl_2$, 0.002 M sodium isoascorbate, 0.0005 M K_2HPO_4, 0.005 M $Na_4P_2O_7$, 0.05 M HEPES adjusted to pH 7.6 with NaOH. The preparation is then carried out essentially as above.

of plants, such as spinach (*Spinacea oleracea*), wheat (*Triticum aestivum*), sunflower (*Helianthus anuus*), and pea (*Pisum sativum*). Other plants, such as spinach beet (*Beta vulgaris*, often sold in shops in England as 'spinach') give much poorer results. It is usually impossible to extract intact chloroplasts from the leaves of plants such as grasses because the prolonged grinding required to rupture the exceptionally-hard cell walls destroys the integrity of the organelles. However, promising results have been achieved in the use of protoplasts as starting material for the isolation of chloroplasts capable of fixing carbon dioxide from the leaves of certain grasses (Huber and Edwards 1975; Rathnam and Edwards 1976). Protoplasts may also be used to obtain intact chloroplasts from other plants (e.g. Robinson and Walker 1979; Hampp and Ziegler 1980).

Even for plants such as spinach there is a seasonal variation in the quality of the commercially-available materials as far as chloroplast isolation is concerned. In our own laboratory, it has been found that field-grown plants often give poor-quality organelles in summer. Of course, such variations can be minimized by the growth of plants under controlled environmental conditions (Walker 1980), especially in water culture. Nevertheless, spinach is the most commonly used plant because of the softness of the leaf and its low content of phenolic compounds. One problem that may occur is that if the leaves are allowed to accumulate large quantities of starch, the acceleration during centrifugation can drag the starch grains within the chloroplasts through the envelopes, so causing rupture. This problem can be overcome by harvesting the leaves at the beginning of a light period rather than after prolonged illumination.

The nature of the isolation medium used must also be considered. Buffered sorbitol media are now the most common (Table 1.3): often the medium is used as a partially-frozen slurry so that the ice particles can minimize any localized temperature changes during homogenization. However, the major osmoticum present in spinach leaves themselves is the compound tri-*N*-methylglycine (often called 'betaine' or 'glycine-betaine'), which has the structure of $(CH_3)_3 N^+CH_2{\cdot}COO^-$. Larkum and Wyn Jones (1979) found that intact spinach chloroplasts isolated in a medium containing betaine in place of sorbitol maintained high rates of carbon dioxide fixation *in vitro* for longer periods than did sorbitol-isolated chloroplasts. Some other plants, such as *Plantago maritima* (Ahmad, Larher, and Steward 1979), are more respectful of plant biochemistry and accumulate sorbitol as an osmoticum, however!

Pyrophosphate buffer is sometimes used in chloroplast isolation media, but care should be taken to see that its content of inorganic phosphate does not reach a level that is inhibitory to photosynthesis (Walker 1980). Often Good's buffers such as MES, HEPES, or TRICINE are used, but it should not be assumed that these are always metabolically inert. For example, they can act as electron donors to flavins and flavoproteins in chloroplast systems (Nelson, Nelson and Racker 1972; Halliwell and Butt 1972). Nevertheless, they usually do not seem to interfere with carbon dioxide fixation reactions and are widely used. Ascorbate (or *iso*-ascorbate, which is supposed to be more stable) is also often added. As well as protecting against phenolic compounds (Section 1.4) it appears to protect chloroplasts against the accumulation of hydrogen peroxide during photosynthesis, a point that will be discussed further in Chapter 8. Its effects are variable; carbon dioxide fixation by some preparations is markedly improved by ascorbate addition, whereas fixation by others is scarcely affected. Perhaps this is related to variations in the endogenous ascorbate content of the chloroplasts, as suggested by Walker (1971). Magnesium salts are frequently added to isolation media (Table 1.3). Often, however, EDTA is also present, so that some or all of the Mg^{2+} is chelated! (See Table 1.3.) Mg^{2+} appears to be essential for the success of photophosphorylation assays on broken chloroplasts, but it is not required for carbon dioxide fixation by intact spinach chloroplasts (Avron and Gibbs 1974) and when present at concentrations of 1 mM or greater severely inhibits carbon dioxide fixation (Huber 1979). Perhaps it is best omitted: the internal concentration of Mg^{2+} in intact chloroplasts is high enough to allow them to photosynthesize rapidly and the envelope is impermeable to Mg^{2+}, so that it cannot leak away unless the membrane is damaged, which will inhibit fixation in any case. Indeed, Nakatani and Barber (1977) found that the percentage of intact chloroplasts obtained by a conventional rapid preparation could be greatly improved by washing the pellet in a medium containing no added metallic cations (0.33 M sorbitol buffered slightly with 0.5 mM Tris, pH 7.5). The isolated intact chloroplasts were more *stable* if 1 mM KCl plus 1 mM $MgCl_2$ were added to the suspension medium (Nakatani, personal communication). Inclusion of EDTA in the homogenizing medium is necessary to obtain intact chloroplasts from *Avena sativa* and sunflower protoplasts. (Hampp and Ziegler 1980; Edwards, Robinson, Tyler, and Walker 1978).

1.6 A classification of chloroplasts

So far in this chapter, 'intact' has been used to describe chloroplasts with whole envelopes that are capable of fixing carbon dioxide at high rates, and 'broken' to describe other chloroplasts, whatever their degree of integrity. Obviously, a better classification is required and such a scheme was introduced by Hall (1972). It is reproduced, with some modifications, in Table 1.4. The envelope is impermeable to ferricyanide ($Fe(CN)_6^{3-}$) and so this compound cannot be reduced by illuminated chloroplasts that have intact envelopes (Type A). Sometimes the envelope is partially damaged, so that although the chloroplasts may look whole under the microscope, they have lost essential metabolites (Type B). $Fe(CN)_6^{3-}$ can penetrate such leaky envelopes and reach the electron-transport chain, so that ferricyanide-dependent oxygen uptake can be demonstrated in illuminated type B preparations. Most chloroplast enzymes will still be retained, however, and carbon dioxide fixation can often be partially restored by adding Calvin cycle intermediates (Walker 1964). More extensive envelope damage causes loss of a greater percentage of the stromal components, to produce type C chloroplasts. Thylakoid preparations are often required for studies of electron-transport. It would seem sensible to 'standardize' such preparations by first making type A chloroplasts, disrupting them by osmotic shock and washing the thylakoid fraction (Type D chloroplasts – see Table 1.4) rather than relying on older preparative methods (Section 1.4) which might allow retention of variable proportions of the stromal contents.

Isolated type A chloroplast fractions will always contain some type B and C chloroplasts, because of damage during the isolation procedure. In addition to the ferricyanide method, the percentage of such damaged chloroplasts may be assessed by phase contrast microscopy (type A chloroplasts give a bright, opaque appearance surrounded by a halo, whereas types B and C show distinct grana and are dull in appearance (Fig. 1.9)) or by measurement of the rate of photophosphorylation dependent on externally-added ADP, since the rate of penetration of ADP across the intact envelope of chloroplasts isolated from mature spinach leaves is insignificant (see Chapter 6). All these techniques give comparable results (Lilley, Fitzgerald, Rienits, and Walker 1975), but the ferricyanide method is most popular. The presence of types B and C chloroplasts in a type A preparation must always be borne in mind, particularly during the studies of the permeability properties of the envelope of type A chloroplasts. The washing technique described by Nakatani and Barber (1977) appears to produce fractions containing 90 per cent or more of type A chloroplasts and it is recommended for use in such experiements.

Lilley *et al.* (1975) have suggested that damaged envelopes can sometimes reseal after breakage, so that one obtains chloroplasts impermeable to $Fe(CN)_6^{3-}$ that have lost some of their stromal contents and will therefore be unable to fix carbon dioxide at the highest rates. It cannot be ruled out that

Table 1.4. Types of chloroplast preparation

Type	Other descriptions often found in the literature	Preparation method	Envelope	Rate of CO_2 fixation (μmol h^{-1} mg chlorophyll^{-1})	Permeability properties
A	Whole, intact chloroplasts	Rapid grinding and centrifugation in special media	Intact	50–200	$NAD(P)^+$ or ferricyanide do not penetrate
A_2	Resealed chloroplasts	As type A: envelope appears to rupture and then reseal, causing loss of some stromal components	Intact	<50, presumably depending on how much stroma is lost	As type A
B	Unbroken chloroplasts (sometimes called 'intact chloroplasts', especially in older papers)	Usually in isotonic sugar or salt, prolonged centrifugation (often two or three steps)	Often seems intact by microscopy, but has been damaged	$\leqslant 5$	$NADP^+$, ferricyanide and other small molecules penetrate
C	Broken chloroplasts	Vigorous grinding and prolonged centrifugation	Disrupted	O	As for B
D	Free-lamellar chloroplasts	Osmotic shock of type A followed by return to isotonic medium	Removed	O	As for B
E	Chloroplast fragments	Resuspension of chloroplasts in hypotonic medium	Lost	O	–
F	Sub-chloroplast particles	Sonication, detergent treatment, French press chaotropic agents	–	O	–

The classification is a modification of that introduced by Hall (1972). 'Intact' chloroplast preparations obtained by the methods in Table 1.3 may contain types A and A_2 and a variable percentage of types B and C.

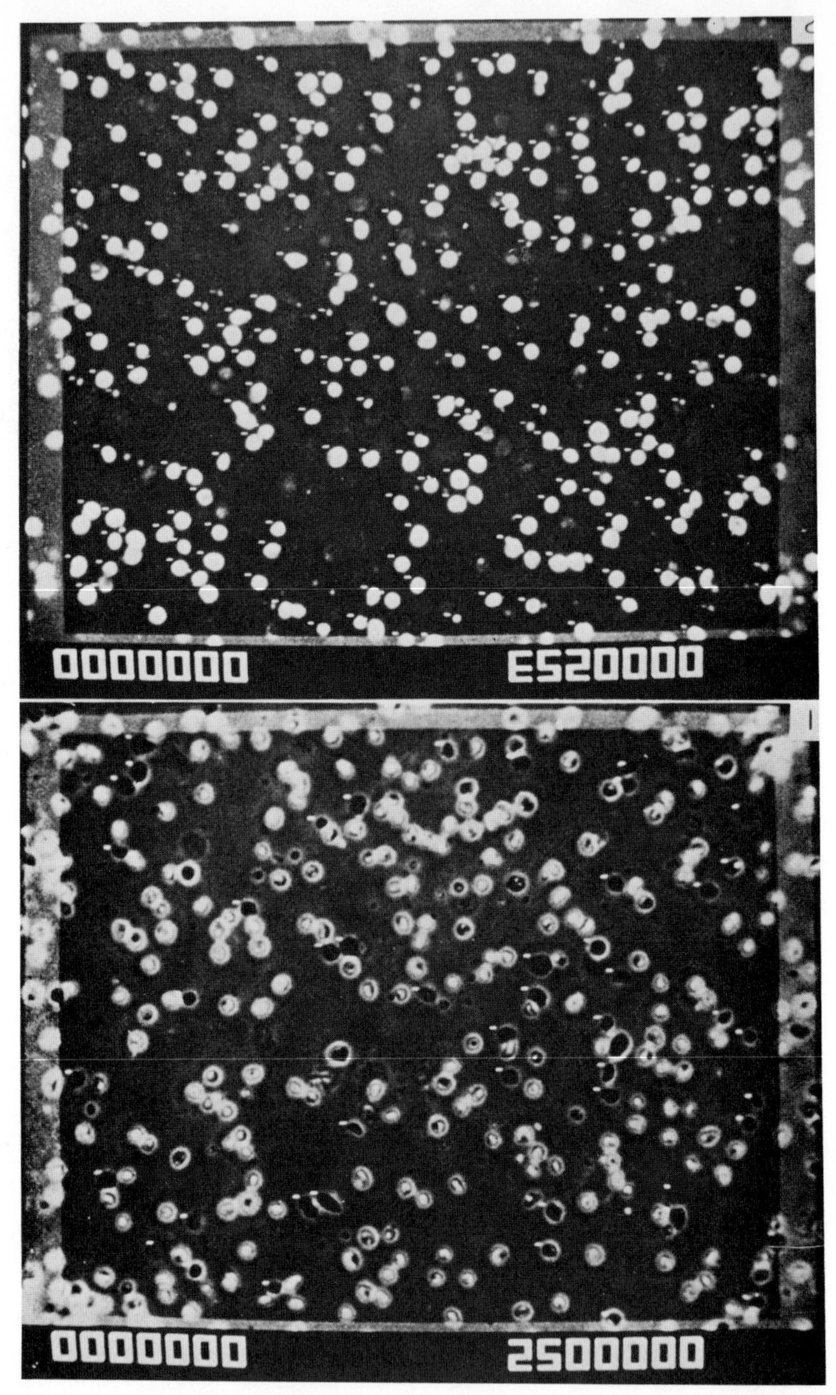

Fig. 1.9. Preparations of type A and type C chloroplasts were examined using phase contrast microscopy. Chloroplasts at a concentration of approximately 50 μg chlorophyll ml^{-1} were viewed with a × 40 phase objective. The upper plate uses positive phase contrast. Intact chloroplasts have a highly reflective, bright, opaque appearance and the presence of a halo. The lower plate shows a negative phase contrast. (Photographs by courtesy of Professor D. A. Walker of the University of Sheffield.)

type A preparations could contain a substantial percentage of these resealed ('type A_2') chloroplasts, but there is no direct evidence for it.

1.7 The problem of cytoplasmic contamination

As well as containing some 'broken' chloroplasts, type A chloroplast fractions will also be contaminated with cytoplasmic components. Mitochondrial (e.g. cytochrome oxidase), peroxisomal (e.g. catalase, glycollate oxidase), and soluble (e.g. dehydroascorbate reductase) enzymes can readily be detected in such fractions. Hence the presence of an enzyme in a type A chloroplast preparation does not mean that this enzyme is located within chloroplasts: it could be in the other organelles present or else loosely adsorbed on to the chloroplast envelope. Further, the observation of light-dependent synthesis of metabolites by type A chloroplast fractions does not mean that *all* the enzymes needed for their production are located within chloroplasts. This can be illustrated by using results obtained by Larsson and Albertsson (1974), who employed a two-phase separation procedure to obtain spinach chloroplasts. The chloroplasts that had low contamination by other organelles fixed carbon dioxide into glycollic acid and starch as major end-products. Another chloroplast fraction, which had a substantial degree of cytoplasmic contamination, fixed carbon dioxide into a wider range of products, including sucrose and alanine. In fact, it is now known that the enzymes of sucrose synthesis are located in the cytosol of the cell and use metabolites exported by the chloroplast (see Chapter 4), and that the capacity of chloroplasts to synthesize amino acids is very restricted (Buchholz, Reupke, Bickel, and Schultz 1979; Chapter 10).

Proof that an enzyme is located within the chloroplast requires removal of contaminating organelles. This can be done partially by 'washing' the preparation, i.e. resuspending in fresh homogenizing medium and recentrifuging, although each washing procedure causes loss of some of the chloroplasts. The enzyme activity of the contaminating organelles must be measured, and the number of these organelles in the final chloroplast preparation determined, either by microscopy or more conveniently, since quantitative microscopy is difficult and time-consuming, by the use of marker enzymes specific for each organelle. The use of marker enzymes themselves is subject to certain qualifications, as summarized in Table 1.5.

A related problem arises during experiments designed to localize enzymes generally within leaf cells. How have researchers set about determining whether or not a leaf enzyme is located in chloroplasts? Often the leaf is ground up in an aqueous solution containing buffer plus osmoticum and the homogenate is subjected to several stages of differential centrifugation that provide pellets enriched in different subcellular fractions. Obviously, while this slow procedure is being carried out the envelopes of many chloroplasts will rupture and their stromal enzymes will either appear in the final supernatant fraction as 'soluble'

Table 1.5. Precautions to be taken in subcellular localization of enzymes (although many of these precautions are self-evident, it is surprising how often they are neglected in published work!)

(1) *The assay used must accurately measure the activity of the enzyme in all fractions tested*
The presence of inhibitors or activators may confuse the results obtained. Some fractions from density-gradients contain high concentrations of gradient materials, which may affect the assay (Hartman, Black, Sinclair, and Hinton 1974). This is true of sucrose. Some enzymes are 'latent' and require rupture of particles by sonication or detergents before they can be assayed accurately. It is essential to ensure that the envelopes of type A chloroplasts are ruptured by osmotic shock before enzyme assays are performed on them

(2) *Account must be taken of the stability of the enzyme in the various fractions tested*
If activity is lost rapidly in one fraction and assays are delayed, then that fraction will appear to contain only a low percentage of the activity present. For example, glutamate dehydrogenase in leaf mitochondria seems to be less stable than the enzyme in chloroplasts (Lea and Thurman 1972)

(3) *A balance sheet should be constructed and percentage recovery determined*
This is a way of checking the possibilities listed under (1) and (2). The enzyme activity in the starting fraction should be measured as soon as it is available and then measured again when fractionation is complete. The sum of the activity in the fractions obtained should be equal to the starting activity. If there is a gross discrepancy, the fractionation cannot be regarded as rigorous proof of enzyme localization. For example, if only 50 per cent of the enzyme activity applied to a sucrose gradient is recovered and activity is only found in a chloroplast fraction, then it cannot be concluded that the enzyme is present only in chloroplasts

(4) *The use of marker enzymes rests on two assumptions: that they belong to a single class of organelle and are uniformly distributed among all members of that class*
Heterogeneity of isolated organelles has been demonstrated for lysosomes and mitochondria in animal tissues, but its occurrence in chloroplasts (apart from the special case of C_4 plants, see Chapter 5) has not been reported. Enzymes must not redistribute on homogenization, e.g. a cytosolic marker should not adsorb on to the outside of organelles (Quail 1979)

enzymes or, even worse, will adsorb on to the outside of other organelles. This can best be illustrated by taking an example. When de-ribbed spinach leaves are homogenized in 0.5 M sucrose buffered with 50 mM Tris HCl at pH 7.5 and the homogenate is subjected to differential centrifugation, over 90 per cent of the $NADP^+$-dependent glutathione reductase activity is found in the final supernatant after high speed centrifugation, together with the enzymes of the Calvin cycle (Foyer and Halliwell 1976). The pelleted intact microsomes, mitochondria, and peroxisomes contain no glutathione reductase and so the enzyme cannot be located within these organelles. Hence the enzyme is either cytosolic or could have been released from chloroplasts during fractionation of the homogenate. In agreement with the latter possibility, type A chloroplasts isolated by rapid centrifugation and then 'washed' contain substantial glutathione reductase activity (Table 1.6). This enzyme must either be located within the chloroplasts or might perhaps have been adsorbed on to the outside of them. The latter

Table 1.6. Evidence for enzyme latency in spinach chloroplast fractions

<table>
<tr><th rowspan="2">Chloroplasts</th><th colspan="3">Enzyme assayed (units mg chlorophyll^{-1})</th><th colspan="3">Latency factor (activity in broken/activity in intact)</th></tr>
<tr><th>NADP$^+$-TPDH</th><th>Glutathione reductase</th><th>CN$^-$-insensitive SOD</th><th>NADP$^+$-TPDH</th><th>Glutathione reductase</th><th>CN$^-$-insensitive SOD</th></tr>
<tr><td>Intact</td><td>0.33</td><td>0.06</td><td>29</td><td rowspan="2">14.7</td><td rowspan="2">14.0</td><td rowspan="2">1.0</td></tr>
<tr><td>Disrupted before assay</td><td>4.87</td><td>0.84</td><td>29</td></tr>
</table>

Details of assays and chloroplast preparations may be found in Foyer and Halliwell (1976) and in Jackson *et al.* (1978). Assays of spinach chloroplasts for NADP$^+$-glyceraldehyde-3-phosphate dehydrogenase (TPDH), cyanide-insensitive superoxide dismutase (SOD), and glutathione reductase were carried out using reaction media to which 0.33 M sorbitol had been added to keep the chloroplasts intact. Washed Type A preparations were either added directly to the assay system or were ruptured by osmotic shock and then assayed.

possibility can be ruled out by determining the 'latency' of the enzyme. Normally the chloroplast envelope is ruptured before assaying glutathione reductase, but if this is not done and sorbitol is added to the assay mixture to ensure that the envelope remains intact, little glutathione reductase activity can be detected. The latency factor (see Table 1.6) is the same for both glutathione reductase activity and for the Calvin cycle enzymes. Hence glutathione reductase must be located inside the chloroplast.

Washed type A chloroplast fractions also contain a superoxide dismutase activity that is not inhibited by cyanide (see Chapter 8 for further discussion of this enzyme). However, this activity is the same whether or not the envelope is ruptured during the assay and it presumably represents an activity adsorbed on to the surface of the chloroplast, although a true location on the outside of the envelope obviously cannot be ruled out (Jackson, Dench, Moore, Halliwell, Foyer, and Hall 1978). A similar phenomenon appears to account for the apparent catalase activity of chloroplasts (Allen 1977). Citrate synthetase can also adsorb on to the outside of the chloroplast envelope (Elias and Givan 1978). Hence latency experiments should be an essential part of enzyme localization work. Inclusion of bovine serum albumin in homogenizing media has been claimed to reduce non-specific enzyme binding, but it cannot be relied upon in all cases, since it was present in the media used by Jackson *et al.* (1978) and by Elias and Givan (1978).

The latency experiments described in Table 1.6 were all performed on washed type A chloroplast preparations. If *unwashed* pellets were resuspended, some latency of enzymes such as catalase and superoxide dismutase could be seen

Table 1.7. Evidence for enzyme latency in Type A spinach chloroplast fractions

Enzyme assayed	No. of washings of chloroplast preparation	Latency factor (activity in disrupted fraction/activity intact chloroplast fraction)
Cyanide-insensitive	0	1.9
Superoxide dismutase	1	1.2
	2	1.0
	3	1.0
Catalase	0	1.6
	1	1.3
	2	1.1
	3	1.0

Type A chloroplast fractions were prepared by conventional methods (see Allen 1977 and Jackson *et al.* 1978). The pellet was either resuspended in fresh homogenizing medium and used as such, or it was resuspended, repelleted by rapid centrifugation, and resuspended again to give a once-washed preparation. This procedure was repeated to obtain preparations washed two, three, or four times. Data for catalase are taken from Allen (1977): data for cyanide-insensitive superoxide dismutase were kindly provided by Dr C. H. Foyer (Sheffield University). The numbers of type A chloroplasts present in the pellets decreased by less than 3 per cent in each washing procedure.

(Table 1.7) although not to the same degree as for Calvin cycle enzymes (compare Tables 1.6 and 1.7). The reason for this is unknown, but some typical results are included in Table 1.7 to illustrate the sort of errors that might be obtained by the use of unwashed preparations. Type A preparations must be washed *at least once* before latency experiments are attempted and the 'latency' factor must be similar to that of known chloroplast enzymes such as ribulose diphosphate carboxylase or $NADP^+$-dependent glyceraldehyde-3-phosphate dehydrogenase. Elias and Givan (1978) have reached similar conclusions.

1.8 Density-gradient centrifugation

Density-gradient procedures are often used to supplement the information obtained about enzyme localization by differential centrifugation techniques. One of the pelleted fractions so obtained, or even the crude homogenate, may be resolved into its various components by density-gradient centrifugation. There are essentially two forms of this technique, depending on whether the separation is based on the rate at which organelles sediment in the gradient (kinetic-gradient or rate-gradient separation) or upon the final position achieved by the organelles when centrifugation has been carried out long enough for them to reach a density equal to their own buoyant density (isopycnic or equilibrium-gradient centrifugation). Leaf homogenates have been resolved into chloroplast, mitochondrial, and peroxisomal bands by equilibrium centrifugation on sucrose gradients. Type A chloroplasts do not survive the prolonged centrifugation procedures and the type B and C chloroplasts produced by their degradation behave differently on sucrose-gradients. This is seen clearly in Fig. 1.10. A spinach-leaf homogenate was centrifuged at 1000 *g* for 5 min. The pellet was resuspended and applied to a sucrose density-gradient, which was spun to equilibrium. Two peaks of chloroplasts were obtained, however, one of which had retained Calvin cycle enzymes (type B chloroplasts) and had a density similar to that of mitochondria whilst the other represented type C chloroplasts at a lower density. This might cause problems during localization work if only chlorophyll were used as a chloroplast marker. A stromal enzyme, present in type B but not in the type C band, might then be thought to be mitochondrial.

Centrifugation to equilibrium in sucrose gradients is time consuming and not only chloroplasts but also mitochondria are damaged (Jackson, Dench, Hall, and Moore 1979). Rate separation in sucrose gradients (Miflin and Beevers 1974) which relies on the fact that chloroplasts sediment to their equilibrium position in sucrose gradients more rapidly than other organelles and so form a band ahead of them, can achieve an equivalent fractionation in much less time (ca. 20–30 min). Nevertheless, type A chloroplasts still do not survive, so sucrose gradients cannot be used to remove contaminating organelles from them. It should also be noted that even if an enzyme appears to be associated with type B chloroplasts on density-gradients, a 'latency' test should be carried out to check the possibility of non-specific adsorption on to the envelope.

Equilibrium density-gradient centrifugation in sucrose gradients is slow mainly because of the high viscosity of sucrose solutions. Silica sol gradients have considerably lower viscosity and so separation is much more rapid. Problems initially arose in such experiments because of deleterious effects of the silica on enzymes and organelles, but these have been overcome by the introduction of a gradient material (Percoll, Pharmacia Fine Chemicals Ltd., Uppsala, Sweden) in which the silica sol particles are coated with an inert substance, polyvinylpyrrolidone. Gradients can be formed during the centrifugation procedure itself as well as by conventional methods (Pertoft, Laurent, Laas, and Kagedal 1978). Some experiments with silica sol gradients have been carried out which indicate that they may be useful in the removal of contaminating organelles from type A chloroplast fractions. For example, Morgenthaler, Price, Robinson, and Gibbs (1974) showed that type A chloroplasts could survive centrifugation on such gradients and both Takabe, Nishimura, and Akazawa (1979) and Mills and Joy (1980) showed that they were effective in decreasing contamination by other organelles. Nevertheless, the gradient techniques have not yet generally replaced the 'washing' of type A chloroplast fractions as a means of decreasing cytoplasmic contamination, especially as the washing procedure of Nakatani and Barber (1977) appears to produce pellets with a very high percentage of type A chloroplasts.

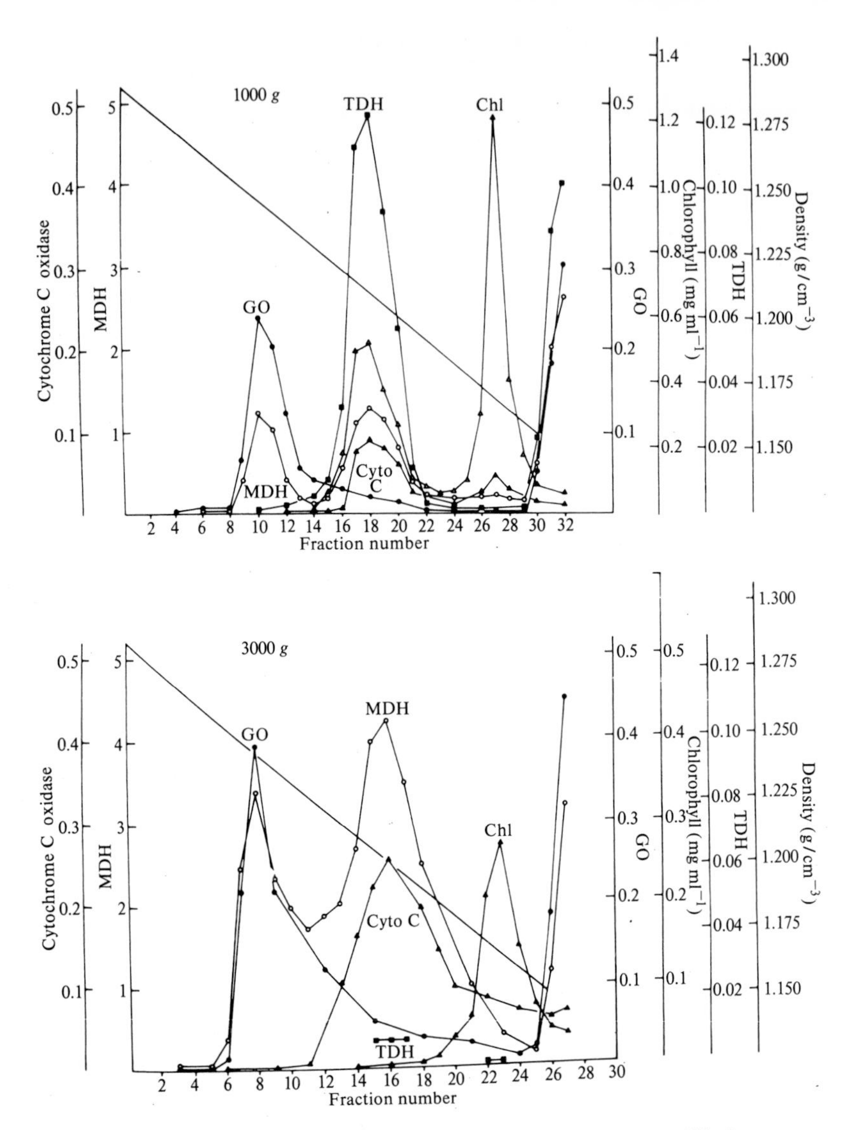

Fig 1.10. Resolution of organelles from spinach leaves on equilibrium sucrose density gradients (from Rocha and Ting 1970, with permission). A spinach leaf homogenate was centrifuged at 1000 *g* for 5 min and the supernatant then centrifuged at 3000 *g* for 15 min. The pellets were gently resuspended and then separately applied to 40-80% linear sucrose gradients and centrifuged to equilibrium. The '3000 *g* pellet' gradient shows a clear separation of the chlorophyll and mitochondrial peaks, whereas that obtained for the 1000 *g* pellet shows two bands of chloroplasts, one overlapping into the mitochondrial band and containing the Calvin cycle enzyme TDH. GO = glycollate oxidase; MDH = malate dehydrogenase; cyt c = cytochrome c oxidase; Chl - chlorophyll; TDH = $NADP^+$-glyceraldehyde 3-phosphate dehydrogenase. Enzyme activities are expressed as percentage of activity recovered.

1.9 Algal chloroplasts

Obtaining type A chloroplasts would be less troublesome if their envelopes were more stable. Chloroplasts from the green algae *Codium fragile*, *Codium vermilara* (Schonfeld, Rahat, and Neumann 1973), *Acetabularia* sp (Shephard and Bidwell 1973), and *Caulerpa* sp (Giles and Sarafis 1974)) possess very tough envelope membranes which usually require strong osmotic shock, detergent treatment, or even sonication to break them. For example, isolated *Codium* chloroplasts fix carbon dioxide at high rates and are stable for days. Indeed, when this seaweed is eaten by the mollusc *Elysia atroviridis*, the chloroplasts remain capable of fixing carbon dioxide within the digestive tract if the animal is illuminated! (Trench, Boyle, and Smith 1973 *a,b*, 1979; Gallop, Bartrop, and Smith 1980). Such algal chloroplasts would be expected to survive prolonged density-gradient or washing techniques and so might be expected to be valuable research tools. Unfortunately, only a small amount of work has yet been done on the enzyme activities of these algal chloroplasts. Indeed, it has sometimes proved difficult to demonstrate Hill reaction activity or reasonable activity of Calvin cycle enzymes in chloroplast extracts prepared by the rigorous techniques needed to rupture the envelope (D. O. Hall and P. Morris, personal communications). Some studies of the ability of algal chloroplasts to synthesise enzymes have been made however (Trench and Ohlhorst 1976) and the ribulose diphosphate carboxylase activity of *Codium* chloroplasts resembles that found in higher plants (Cobb and Rott 1978).

1.10 Composition of the chloroplast envelope

The envelope membranes of chloroplasts contain certain enzymes, as well as being the site of some specific systems for the transport of substances into and out of chloroplasts (see Chapter 6). The lipid composition of the envelope is very different from that of the thylakoids (Table 1.1). Envelope membrane fractions have been isolated by density-gradient centrifugation of disrupted type A chloroplasts (Douce, Holtz, and Benson 1973; Poincelot 1973; Joyard and Douce 1976). They contain a Mg^{2+}-dependent ATPase activity with a broad pH optimum, the activity of which is low in winter and summer grown plants, and higher in spring and autumn (2–4 and 8–12 μmol phosphate released h^{-1} mg protein^{-1} respectively) (Douce and Joyard 1979). The envelope also contains galactosyltransferase enzymes (Douce and Joyard 1979; Williams, Simpson and Chapman 1979), adenylate kinase (Murakami and Strotman 1978), enzymes involved in the synthesis and degradation of phosphatidic acid (Joyard and Douce 1979), and enzymes for the interconversion of xanthophyll pigments (Chapter 8). Since the envelope membranes have a net negative charge in most buffer systems, they can easily adsorb positively charged proteins from a leaf homogenate (Neurburger, Joyard, and Douce 1977). Further, it should be remembered that vesicles formed from disrupted chloroplast envelopes will contaminate other fractions obtained during subcellular fractionation procedures. Indeed, this may account for the reported ability of leaf microsomal fractions to synthesize galactolipids, an activity that apparently occurs exclusively in the chloroplast envelope (Chapter 9).

1.11 Non-aqueous chloroplast isolation methods

Non-aqueous chloroplast isolation methods, developed originally by Thalacker and Behrens (1959), Heber (1957), and Stocking (1959), use organic solvents rather than aqueous media for homogenization. Leaves are freeze-dried and then ground up in an organic solvent, often a mixture of hexane and tetrachloromethane (carbon tetrachloride). Chloroplast fractions may then be separated on density gradients prepared using mixtures of these solvents. Now that methods exist for the aqueous isolation of type A chloroplasts, non-aqueous techniques are no longer used for the localization of enzymes within chloroplasts. However, since many metabolities and ions easily cross chloroplast envelopes, a direct determination of their localization within the cell by aqueous methods is difficult because of the likelihood of redistribution during isolation. Although non-aqueous methods are potentially useful here (Stocking 1971), considerable doubt has been expressed as to the purity of 'chloroplast' fractions isolated by non-aqueous techniques (Bird, Cornelius, Dyer, and Keys 1973). On the whole, non-aqueous isolation techniques are not widely used today.

References

Ahmad, I., Larher, F., and Steward, G. R. (1979). *New Phytol.* **82**, 671–8.
Alfonzo, R., Nelson, N., and Racker, E. (1980) *Pl. Physiol., Lancaster*, 730–4
Allen, J. F. (1977). *FEBS Lett.* **84**, 221–224.
Anderson, J. M. (1975). *Biochim. biophys. Acta* **416**, 191–235.
— (1980). *FEBS Lett.* **117**, 327–31.
Anderson, J. W. (1968). *Phytochemistry.* **7**, 1973–88.
Argyroudi-Akoyunoglou, J. H. and Castorinis, A. (1980). *Archs Biochem. Biophys.* **200**, 326–35.
Arntzen, C. J. and Burke, J. J. (1980). *Meth. Enzymol.* **69**, 520–38.
Avron, M. and Gibbs, M. (1974). *Pl. Physiol.*, Lancaster, **53**, 136–9.
Barber, J. (1976). in *The intact chloroplast, Topics in photosynthesis* Vol. 1., (ed. J. Barber) pp. 89–134. Elsevier, Amsterdam.
— (1978). *Rep. Prog. Phys.* **41**, 1157–99.
Bennett, J. (1979). *Trends Biochem. Sci.* December, pp. 268–71.
— (1980). *Eur. J. Biochem.* **104**, 85–9.
Bird, I. F., Cornelius, M. J., Dyer, T. A., and Keys, A. J. (1973). *J. exp. Bot.* **24**, 211–15.
Buchholz, B., Reupke, B., Bickel, H., and Schultz, G. (1979). *Phytochemistry* **18**, 1109–11.
Cobb, A. H. and Rott, J. (1978) *New Phytol.* **81**, 527–41.

Coombs, J. and Greenwood, A. D. (1976) in *The intact chloroplast, Topics in photosynthesis,* Vol. 1 (ed. J. Barber) pp. 1–51. Elsevier, Amsterdam.

Douce, R. and Joyard, J. (1979). in *Plant organelles* (ed. E. Reid) pp. 47–59. Ellis Horwood, Chichester.

— Holtz, R. B., and Benson, A. A. (1973). *J. biol. Chem.* **248**, 7215–22.

Edwards, G. E., Robinson, S. P., Tyler, N. J. C., and Walker, D. A. (1978) *Archs Biochem. Biophys.* **190**, 421–33.

Elias, B. A. and Givan, C. V. (1978). *Planta* **142**, 317–20.

Foyer, C. H. and Halliwell, B. (1976). *Planta* **133**, 21–5.

Gallop, A., Bartrop, J., and Smith, D. C. (1980). *Proc. R. Soc.* **B207** 335–49.

Giles, K. L. and Sarafis, V. (1974). *Nature, Lond.* **248**, 512–513.

Gregory, R. P. F. (1977). *Biochemistry of photosynthesis*, 2nd ed. Wiley, London.

Hall, D. O. (1972). *Nature, New Biol.* **235**, 125–6.

Halliwell, B. and Butt, V. S. (1972). *Biochem. J.* **129**, 1157–8.

Hampp, R. and Ziegler, H. (1980). *Planta* **147**, 485–494.

Hartman, G. C., Black, N., Sinclair, R., and Hinton, R. H. (1974). in *Methodological developments in biochemistry* (ed. E. Reid) pp. 93–110. Vol. 4, Longmans, London.

Heber, U. (1957). *Ber. Dt. Bot. Ges.* **70**, 371–8.

Hiller, R. G. and Raison, J. K. (1980). *Biochim. biophys. Acta* **599**, 63–72.

Huber, S. C. (1979). *Biochim. biophys. Acta* **545**, 131–40.

— and Edwards, G. E. (1975). *Physiol. Plant.* **35**, 203–9.

Jackson, C., Dench, J., Moore, A. L., Halliwell, B., Foyer, C. H., and Hall, D. O. (1978). *Eur. J. Biochem.* **91**, 339–44.

— — Hall D. O. and Moore, A. L. (1979). *Pl. Physiol., Lancaster* **64**, 150–3.

Jensen, R. G. and Bassham, J. A. (1966) *Proc. natn. Acad. Sci., U.S.A.* **56**, 1095–101.

Joyard, J. and Douce, R. (1976). *Physiol. Veg.* **14**, 31–48.

— — (1979). *FEBS Lett.* **102**, 147–50.

Larkum, A. W. D. and Wyn Jones, R. G. (1979). *Planta* **145**, 393–4.

Larsson, C. and Albertsson, P. (1974). *Biochim. biophys. Acta* **357**, 412–19.

Latzko, E. and Gibbs, M. (1968). *Z. Pflanzenphysiol.* **59**, 184–94.

Lea, P. J. and Thurman, D. A. (1972). *J. exp. Bot.* 23, 440–9.

Leech, R. M. and Murphy, D. J. (1976) in *The intact chloroplast, Topics in photosynthesis,* Vol. 1 (ed. J. Barber) pp. 365–401. Elsevier, Amsterdam.

Lilley, R. Mc. C., Fitzgerald, M. P., Rienits, K. G., and Walker, D. A. (1975). *New Phytol.* **75**, 1–10.

Mayer, A. M. and Harel, E. (1979). *Phytochemistry.* **18**, 193–215.

Miflin, B. J. and Beevers, H. (1974). *Pl. Physiol., Lancaster* **53**, 870–4.

Mills, W. R. and Joy, K. W. (1980) *Planta* **148**, 75–83.

Morgenthaler, J. J., Price, C. A., Robinson, J. M. and Gibbs, M. (1974). *Pl. Physiol., Lancaster* **54**, 532–4.

Murakami, S. and Strotman, H. (1978). *Archs Biochem. Biophys.* **185**, 30–8.

Nakatani, H. Y. and Barber, J. (1977). *Biochim. Biophys. Acta* **461**, 510–12.

Nelson, N., Nelson, H., and Racker, E. (1972). *Photochem. Photobiol.* **16**, 481–9.

Neuburger, M., Joyard, J., and Douce, R. (1977). *Pl. Physiol., Lancaster* **59**, 1178–81.

Pertoft, H., Laurent, T. C., Laas, T., and Kagedal, L. (1978). *Analyt. Biochem.* **88**, 271–82.

Poincelot, R. P. (1973). *Archs Biochem. Biophys.* **159**, 134–42.
— (1976). *Pl. Physiol., Lancaster* **58**, 595–8.
Quail, P. H. (1979). *A. Rev. Pl. Physiol.* **30**, 425–84.
Quinn, P. J. and Williams, W. P. (1978). *Prog. Biophys. molec. Biol.* **34**, 109–73.
Radmer, R. and Ollinger, O. (1980). *FEBS Lett.* **110**, 57–61.
Rathnam, C. K. M. and Edwards, G. E. (1976). *Plant Cell Physiol.* **17**, 177–86.
Robinson, S. P. and Walker, D. A. (1979). *Archs Biochem. Biophys.* **196**, 319–23.
Rocha, V. and Ting, I. P. (1970). *Archs Biochem. Biophys.* **140**, 398–407.
Schonfeld, M., Rahat, M., and Neumann, J. (1973). *Pl. Physiol., Lancaster* **52**, 283–7.
Shephard, D. C. and Bidwell, R. G. S. (1973). *Protoplasma* **76**, 289–307.
Singer, S. J. and Nicolson, G. L. (1972). *Science*, N.Y. **175**, 720–31.
Stocking, C. R. (1959). *Pl. Physiol., Lancaster* **34**, 56–8.
— (1971). *Meth. Enzymol.* **23A**, 221–8.
Takabe, T., Nishimura, M., and Akazawa, T. (1979). *Agr. biol. Chem.* **43**, 2137–42.
Thalacker, R. and Behrens, M. (1959). *Z. Naturforsch.* **14**, 443–8.
Thornber, J. P. and Barber, J. (1979). In *Photosynthesis in relation to model systems. Topics in photosynthesis*, Vol. 3 (ed. J. Barber) pp. 27–70. Elsevier, Amsterdam.
Trench, R. K. (1979). *A. Rev. Pl. Physiol.* **30**, 485–531.
— and Ohlhorst, S. (1976). *New Phytol.* **76**, 99–109.
— Boyle, J. E. and Smith, D. C. (1973a). *Proc. R. Soc.* **B184**, 51–61.
— — — (1973b). *Proc. R. Soc.* **B184**, 63–81.
Walker, D. A. (1964). *Biochem J.* **92**, 22–3C.
— (1971). *Meth. Enzymol.* **23A**, 211–20.
— (1980). *Meth. Enzymol.* **69C**, 94–104.
Wang, A. Y. and Packer, L. (1973). *Biochim. biophys. Acta* **305**, 488–92.
Whatley, F. R. and Arnon, D. J. (1963). *Meth. Enzymol.* **6**, 308–13.
Williams, J. P., Simpson, E. E., and Chapman, D. J. (1979). *Pl. Physiol., Lancaster* **63**, 669–73.
Wintermans, J. F. G. M., Van Besouw, A., and Bogemann, G. (1981). *Biochim. biophys. Acta* **663**, 99–107.

2 THE LIGHT REACTIONS OF PHOTOSYNTHESIS

2.1 The electron transport chain

The light reactions of photosynthesis, which take place in the thylakoid membranes of the chloroplast, generate the ATP and NADPH needed to drive carbon dioxide fixation in the stroma. As discussed in Chapter 1, absorption of light by chlorophyll and other pigment molecules produces a separation of positive and negative charges in special regions of the thylakoid membrane, namely reaction centres P700 and P680 at the heart of photosystems I and photosystems II respectively (eqn (2.1))

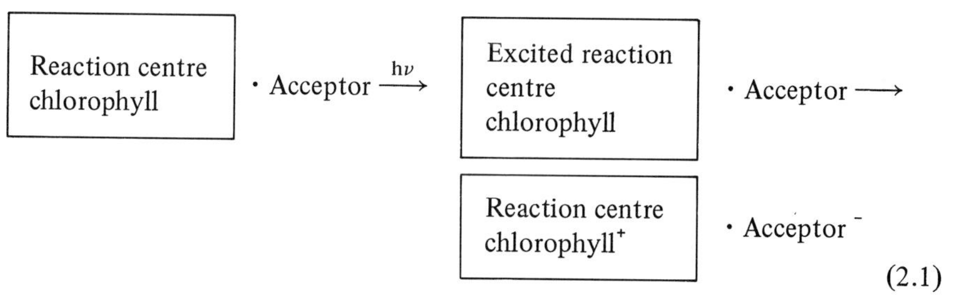

(2.1)

In 1960, Hill and Bendall suggested that the two photosystems act in series and the scheme which they proposed, often known as the Z-scheme, is still generally accepted today. An updated version of this scheme is shown in Fig. 2.1, from which it becomes clear why the term 'Z scheme' is used.

The electrons ejected from the P680 reaction centre of PSII are eventually accepted by a molecule known as 'Q', although they probably first pass to a molecule of pheophytin a (chlorophyll a without its Mg^{2+} ion) (Klimov, Dolan, and Ke 1980). The term Q was originally introduced to represent the 'quencher' of chlorophyll fluorescence, i.e. the accepting of excited electrons by Q prevents re-emission of photons from the excited chlorophyll molecules. The chemical identity of Q is not known, but it may be a special form of the plastoquinone molecule (Fig. 2.2). From Q electrons are passed onto a secondary electron acceptor ('B' or 'R') from which they are transferred onto a pool of plastoquinone molecules (Velthuys 1980). There seem to be about five molecules of plastoquinone per electron transport chain, corresponding to a capacity of ten electrons since this molecule is a two-electron acceptor (Fig. 2.2). A plastoquinone pool may serve to link several electron transport chains together. Transfer of electrons from Q to the plastoquinone pool can be blocked by the herbicide DCMU (Fig. 2.3) as shown in Fig. 2.1. Not all chloroplast plastoquinone is involved in electron transport, since large amounts occur in the plastoglobuli (Chapter 1).

From plastoquinone electrons flow on through cytochrome f and plastocyanin

Table 2.1. Representative distribution of substances in spinach thylakoid membranes

Compound	Amount (molecules/100 molecules of chlorophyll a)
Chlorophylls	
Chlorophyll a	(100)
Chlorophyll b	44
Carotenoids	
– β-carotene	9
Lutein	14
Violaxanthin	4
Neoxanthin	4
Quinones	
Plastoquinone A	10
Plastoquinone B	5
Plastoquinone C	3
α – tocopherol	6
α – tocopherylquinone	3
Vitamin K_1	3

Values are expressed per 100 molecules of chlorophyll a and are rounded to the nearest whole number. Data from Gregory (1977) and from Metzler (1977).

(Fig. 2.1). Plastocyanin is a protein containing one atom of copper per mole; it is blue in the oxidized state. Cytochrome f, named from the Latin *frons* for foliage, is a typical cytochrome molecule in which electrons are accepted and lost by changes in the oxidation state of the iron at the centre of the haem ring from Fe^{3+} to Fe^{2+} and back again. Both plastocyanin and cytochrome f are found in the thylakoid membranes at the same concentration as the reaction centre chlorophylls. Plastocyanin is loosely attached to the inner face of the thylakoid membrane, or possibly partially free in the intra-thylakoid space, and treatments that rupture the thylakoids cause loss of much plastocyanin from the membranes. It is possible that plastocyanin, like plastoquinone, serves to link together several electron transport chains through the intra-thylakoid space (Velthuys 1980).

Absorption of light by the P700 reaction centre chlorophyll of PSI also causes charge separation (eqn (2.1)). Electrons are passed on to an acceptor (Fig. 2.1) and are replaced by the feeding in of electrons from plastocyanin. The primary acceptor from PSI has not been conclusively identified, but there is evidence that it may be a membrane-bound protein which contains iron, not in the form of a haem ring, and sulphur. From the primary acceptor, electrons pass to other non-haem-iron–sulphur proteins bound to the membranes and finally on to a soluble protein of this type, known as ferredoxin. The soluble ferredoxin from higher plants contains two atoms of iron and two of

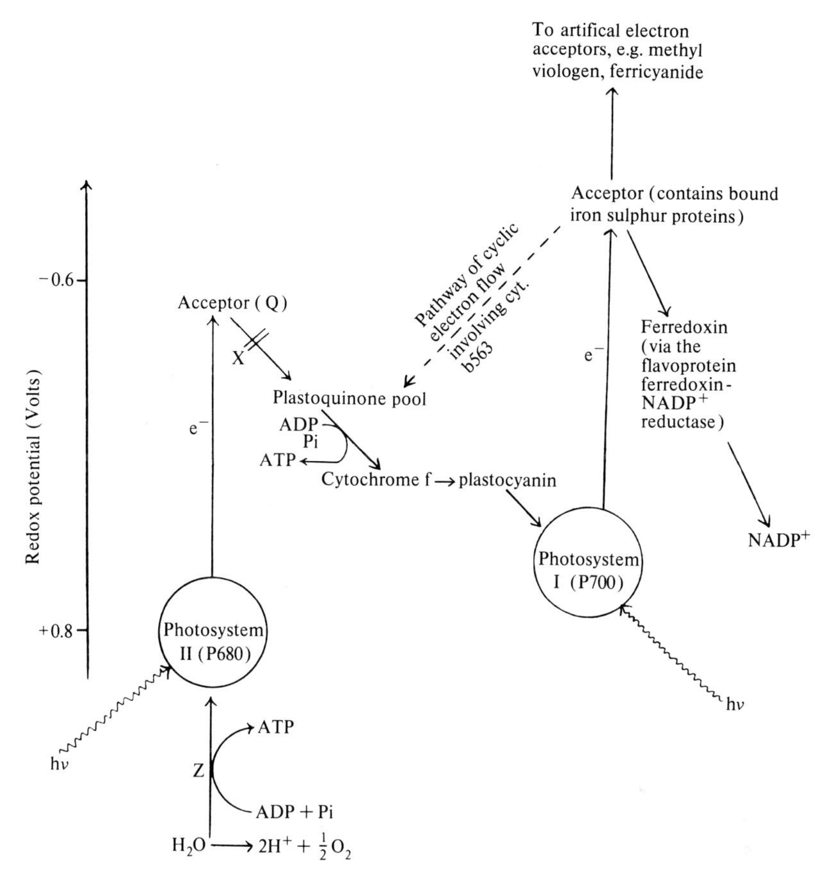

Fig. 2.1. An updated version of the Hill–Bendall ('Z') scheme of photosynthetic electron transport. The sequence of carriers is presented as an energy diagram using redox potentials. The lower the redox potential of a substance, the better is its electron-donating capacity. Hence a component with a negative redox potential is theoretically capable of donating electrons to another component with a less negative, zero, or positive redox potential. The redox potentials of the components are obtained by titrating them with reagents of known redox potential. X represents the site of action of the commonly used photosynthetic inhibitor DCMU (dichlorophenyldimethylurea). Based on data from Trebst (1974), Crofts and Wood (1978) and Blankenship and Parson (1978). Q is a special form of plastoquinone (PQ), possibly an iron (II)–PQ complex.

sulphur per molecule, has a molecular weight of about 10 500, a redox potential of −420 mV, and can transfer one electron per molecule. The iron-sulphide group at the active site is held by links between the iron and cysteine-SH groups on the protein, and it is decomposed into iron (II) salts and hydrogen sulphide under mildly acidic conditions (Hall and Rao 1977).

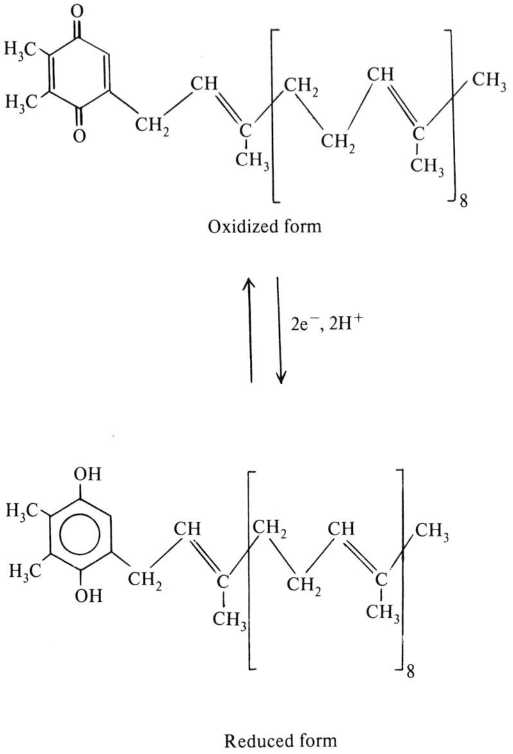

Oxidized form

$2e^-, 2H^+$

Reduced form

Fig. 2.2. Structure of plastoquinone. The structure shown is that of plastoquinone A, the predominant form in spinach chloroplasts (Table 2.1). Spinach chloroplasts contain at least six other plastoquinones, including plastoquinones C, which are hydroxylated in the side-chain and plastoquinones B in which the hydroxyl groups are acetylated. Many other modifications exist, including variations in the number of the isoprene side chains (Metzler 1977).

Reduced ferredoxin donates its electrons to several different systems in chloroplasts, as summarized in Table 2.2, but the principal route *in vivo* in the chloroplasts of higher plants is on to $NADP^+$ to form the NADPH needed to operate the Calvin cycle of carbon dioxide fixation in the stroma. Reduction of $NADP^+$ by reduced ferredoxin is catalysed by a flavoprotein enzyme, ferredoxin-NADP oxidoreductase, which contains one molecule of FAD in each protein molecule. Unlike the soluble ferredoxin, this enzyme is bound to the thylakoids, although it can be removed from them by repeated washing in hypotonic media. Hence even damaged thylakoids will usually reduce $NADP^+$ in the light if ferredoxin is provided.

We have now traced the flow of electrons from PSII on to $NADP^+$. Electrons ejected from the PSII reaction centre (P680) must be replaced by the splitting

DCMU
Dichlorophenyldimethylurea

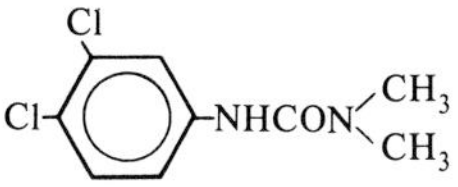

DMBIB
Dibromothymoquinone

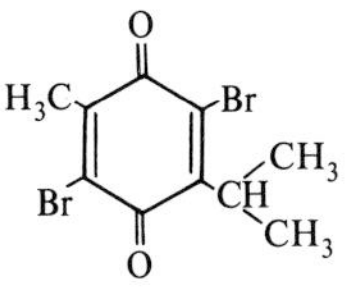

FCCP
Carbonylcyanide
p-trifluoromethoxyphenylhydrazone

DNP
2,4-Dinitrophenol

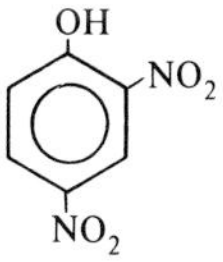

Antimycin A

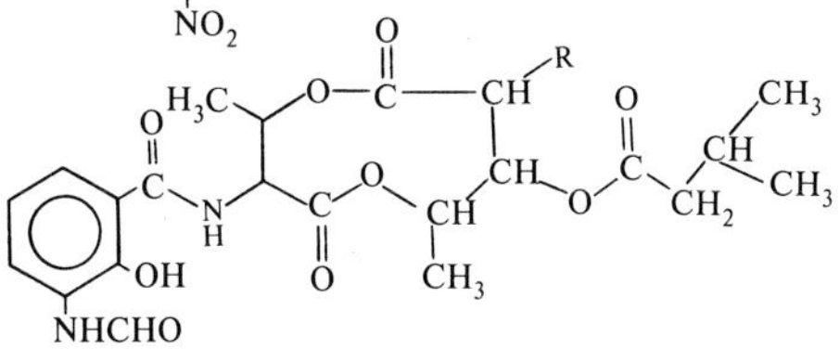

DCCD
Dicyclohexylcarbodiimide

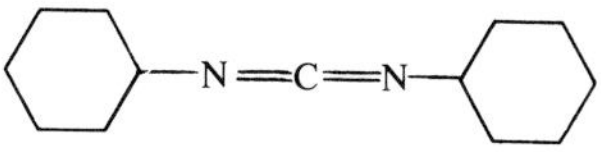

Fig. 2.3. Compounds that interfere with the light reactions of photosynthesis. DCMU inhibits close to PSII. DMBIB is an antagonist of plastoquinone. FCCP and DNP are uncouplers. Antimycin A interferes with cyclic photophosphorylation. DCCD interferes with transport of protons across the thylakoid membrane through the $CF_1 - F_0$ complex.

of water molecules. This is the part of the chloroplast electron transport system about which we know least. Formation of one molecule of oxygen generates four electrons (eqn (2.2)).

$$2H_2O \rightarrow 4e^- + 4H^+ + O_2 \qquad (2.2)$$

and at least four light quanta are necessary to drive this process. When isolated chloroplasts kept in the dark are suddenly subjected to a flash of bright light short enough to bring about only a single excitation of the PSII reaction centres, no oxygen is evolved. Indeed, not until three separate short flashes have been

Table 2.2. Physiological reactions involving ferredoxins in illuminated chloroplasts

Reaction	Other essential components	Discussed further in chapter
$NADP^+$ photoreduction and non-cyclic photophosphorylation	Ferredoxin-$NADP^+$ oxidoreductase	2 (this chapter)
Cyclic photophosphorylation	Chloroplast electron transport chain	2
Sulphite reduction $SO_3^{2-} \rightarrow S^{2-}$	Sulphite reductase	10
Nitrite reduction $NO_2^- \rightarrow NH_3$	Nitrite reductase	10
Reduction of chloroplast thioredoxins	Ferredoxin–thioredoxin reductase	3, 4
Desaturation of fatty acids	Oxygen, desaturase enzyme	9
Reduction of oxygen to give superoxide, O_2^- (pseudocyclic photophosphorylation)	None (non-enzymic reaction)	2 and 8

given is significant oxygen evolved. Another four flashes are required before oxygen uptake reaches a maximum again (Fig. 2.4). A possible explanation of these results was put forward by Kok, Forbush, and McGloin (1970) and is usually referred to as the 'S state model' or sometimes as the 'Kok-clock-hypothesis'. It suggests that the PSII reaction centre must accumulate four positive charges on its oxidizing side before a molecule of oxygen can be evolved, and therefore five possible states are open to it. These are designated S_0, S_1^+, S_2^{2+}, S_3^{3+}, and S_4^{4+}. The last four are formed by the successive loss of single electrons. Once S_4^{4+} is reached reaction can occur and the electrons are

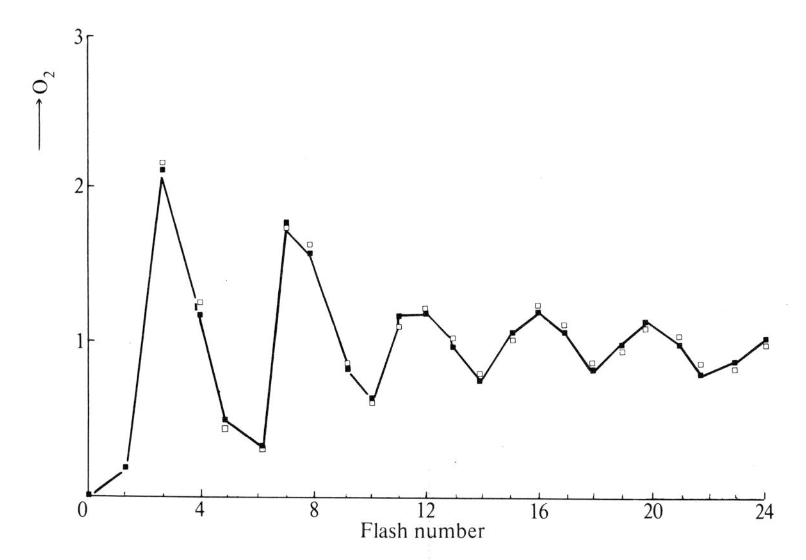

Fig. 2.4. Oxygen flash-yield sequence observed with isolated chloroplasts. $NADP^+$ was used as final electron acceptor. Before the experiment the chloroplasts were given a long, dark pre-treatment period. Data from Harriman and Barber (1979).

replaced (eqn (2.3)). An S^{-1} state, reduced below the level of S_0, can also be generated by treatment of the chloroplasts with 0.3 per cent H_2O_2 in the dark (Velthuys and Kok 1978).

$$S_4^{4+} + 2H_2O \rightarrow O_2 + 4H^+ + S_0 \qquad (2.3)$$

Equation (2.3) is probably an oversimplification in that it shows four protons being released during reduction of S_4^{4+} to S_0, whereas there is evidence for release of some of the protons in the other transitions of the S states (Crofts and Wood, 1978; Velthuys 1980).

In order to explain the observation that only three flashes are necessary to produce oxygen after a prolonged dark period (Fig. 2.4) it must be assumed that the predominant state in the dark is S_1^+. We do not know the chemical identity of these charged S states, or how many electron-transfer steps there are between the entry point of water and the species, known as 'Z' (Fig. 2.1), that actually donates electrons to the P680 reaction centre. It is possible, however, that the charge accumulation mechanism involves manganese, a transition metal that exists in several different oxidation states. Manganese is required for PSII activity (Cheniae 1980). Some chemical treatments of thylakoids, e.g. with hydroxylamine, release about two-thirds of the bound manganese and this is accompanied by a decrease in oxygen evolution. Addition of certain electron donors, such as tetra-phenyl boron, restores electron flow through PSII. Hence the extractable manganese, although required for oxygen release, is not required for the function of the P680 reaction centre (Spector and Winget 1980). Chloride ions are also necessary for the functioning of PSII with water as electron donor, but not when artificial electron donors are used (Harriman and Barber 1979). Photosystem II activity in chloroplast membranes additionally requires the presence of traces of HCO_3^- or CO_2 (Stemler 1980; Sarojini and Govindjee 1981).

Overall then, we have a flow of electrons with water as the ultimate electron donor and $NADP^+$ as the major final recipient of electrons. We can represent this by the equation

$$H_2O + NADP^+ \xrightarrow{h\nu} \tfrac{1}{2}O_2 + NADPH + H^+ \quad (2.4)$$

and the overall process is usually referred to as *non-cyclic electron flow*. Figure 2.1. also shows a cyclic electron pathway which does not involve PSII, although there must be a little PSII activity to feed in the electrons which cycle. In this cyclic pathway electrons from the acceptor complex of PSI pass through a special cytochrome (b_{563}) (although the role of this cytochrome has been disputed by Cox (1979)) plastoquinone, cytochrome f, and plastocyanin and eventually re-join the P700 reaction centre (Arnon and Chain 1977*a,b*, 1979; Chain 1979). NADP-ferredoxin reductase is not involved (Bohme 1977). Cyclic electron flow in isolated, illuminated thylakoid preparations can be made to occur by adding artifical redox carriers (e.g. FMN, phenazine methosulphate) and also by adding concentrations of ferredoxin similar to those present in the stroma *in vivo*. Demonstration of cyclic electron flow *in vitro* requires experimental conditions that reduce electron output from PSII, e.g. the use of far-red illumination, which is poorly utilized by PSII (Chapter 1) or by the addition of sufficient dichlorophenyldimethylurea (DCMU) to cause a partial inhibition of electron flow. The resulting increase in the reduction state of the PSI acceptor complex then presumably causes the cyclic pathway to become operative. Too much DCMU will prevent cyclic electron flow because a complete inhibition of electron output from PSII will not provide any electrons to cycle. It has been suggested that a similar redox poising could be achieved *in vivo* by an accumulation of NADPH causing a 'back-up' of electrons and reducing Q, the electron acceptor of PSII (Arnon and Chain 1979; Mills, Mitchell, and Barber 1979). Ferredoxin-dependent cyclic electron flow is inhibited by the plastoquinone antagonist DMBIB (Fig. 2.3) and by antimycin A (e.g. Robinson and Yocum 1980). There is still considerable debate as to whether cyclic electron flow occurs in chloroplasts *in vivo*, but the effects of antimycin A on illuminated type A chloroplasts, discussed in the following section, provide suggestive evidence that it might.

Chloroplasts also contain a b-type cytochrome known as cytochrome b_{559}. In freshly-prepared chloroplasts, b_{559} is mostly present in a state with a high (about 360–400 mV) redox potential (written as $b_{559_{HP}}$), but ageing or disruption of membrane structure can decrease this by as much as 300 mV to give the 'low potential' form $b\text{-}_{559_{LP}}$. The role of cytochrome b_{559} is unclear as yet, but there is some evidence that it might be involved in a cyclic electron-flow pathway around PSII (Cramer and Whitmarsh 1977; Butler, 1978; Heber, Kirk, and Boardman 1979).

2.2 Photophosphorylation

The photosynthetic light reactions also generate the ATP needed to drive the Calvin cycle, as was first observed for chloroplasts by Arnon, Allen, and Whatley (1954). The flow of electrons 'down' the electron transport chain (i.e. to more positive redox potentials) from Q to P700 provides energy to drive photophosphorylation. Energy is also provided by various events on the oxidizing side of photosystem II (Fig. 2.1). ATP synthesis associated with the flow of electrons from water to $NADP^+$, i.e. noncyclic electron flow, can be referred to as *non-cyclic photophosphorylation*. *Cyclic photophosphorylation* can also occur under conditions favouring cyclic electron flow, but little energy will then come from PSII to drive ATP synthesis. Non-cyclic photophosphorylation is accompanied by stoichiometric oxygen release from the chloroplasts (eqn (2.5)), whereas cyclic photophosphorylation is not.

$$H_2O + NADP^+ + ADP + \text{phosphate} \xrightarrow{h\nu} \tfrac{1}{2}O_2 + NADPH + ATP + H^+ \qquad (2.5)$$

A third type of phosphorylation, *pseudocyclic photophosphorylation*, can be made to occur in isolated chloroplasts by adding certain artificial electron acceptors, one good example being methyl viologen (MV). This compound is reduced by the electron-transport chain (Fig. 2.1), but the reduced form (MVH) rapidly reacts with oxygen. This autoxidation proceeds with the transfer of single electrons to oxygen molecules to give the superoxide radical, $O_2^{\cdot}$. Superoxide becomes converted non-enzymically at a significant rate into hydrogen peroxide (H_2O_2) and oxygen. Conversion of O_2^- to H_2O_2 and O_2 is greatly accelerated by the superoxide dismutase enzyme found both in the stroma and attached to the thylakoids (Chapter 8). The net effect of these reactions (eqns (2.6–8)) is shown in eqn 2.9

$$2MV + H_2O \xrightarrow{h\nu} \tfrac{1}{2}O_2 + 2MVH \quad \text{(reduction of MV by electron-transport chain)} \qquad (2.6)$$

$$2MVH + 2O_2 \rightarrow 2MV + 2O_2^- + 2H^+ \quad \text{(reaction of MVH with } O_2 \text{ to give } O_2^-\text{)} \qquad (2.7)$$

$$2O_2^- + 2H^+ \rightarrow H_2O_2 + O_2 \quad \text{(conversion of } O_2^- \text{ to } H_2O_2 \text{ and } O_2\text{, both spontaneously and catalysed by chloroplast superoxide dismutase)} \qquad (2.8)$$

$$\textit{Net}\ H_2O + \tfrac{1}{2}O_2 \rightarrow H_2O_2 \qquad (2.9)$$

In pseudocyclic photophosphorylation oxygen thus appears to act as a final electron acceptor, and ATP synthesis is accompanied by a light-induced oxygen *uptake*. In fact, as eqns (2.6–8) show, there is normal non-cyclic electron flow from water to the acceptor complex of PSI, but the methyl viologen reduced by this complex undergoes further reaction with oxygen. Methyl viologen is, of course, not present *in vivo*. However, isolated illuminated chloroplast thylakoids slowly take up oxygen in the absence of added electron acceptors, a reaction first observed by Mehler (1951) and hence often known as the 'Mehler reaction'. It appears to result from the reduction of O_2 to O_2^- by the bound electron acceptors associated with PSI (for further discussion see Chapter 8). Addition of ferredoxin to the thylakoids increases the oxygen uptake, since ferredoxin reduced by PSI can itself react with O_2 to generate O_2^- (eqn (2.10)).

$$\text{ferredoxin}_{(red)} + O_2 \rightarrow O_2^- + \text{ferredoxin}_{(ox)} \qquad (2.10)$$

Allen (1975) has proposed that reduced ferredoxin has a sufficiently low redox potential to further reduce O_2^- to H_2O_2 (eqn. (2.11)) but this has not yet been directly demonstrated

$$\text{ferredoxin}_{(red)} + O_2^- + 2H^+ \rightarrow H_2O_2 + \text{ferredoxin}_{(ox)} \qquad (2.11)$$

From eqn (2.10) it may be seen that ferredoxin is potentially capable of promoting pseudocyclic photophosphorylation *in vivo*. Reduced ferredoxin might therefore be regarded as a 'branch-point' in chloroplast electron-transport pathways (Table 2.2), being capable of passing electrons not only on to $NADP^+$ but also, at least *in vitro*, on to oxygen and into the cyclic electron-transport system (Fig. 2.1).

2.2.1 *How much ATP is made by non-cyclic photophosphorylation?*

Sufficient ATP must be made by photophosphorylation *in vivo* to operate the Calvin cycle. Fixation of one carbon dioxide molecule and its eventual incorporation into a hexose sugar requires three molecules of ATP and two of NADPH (Chapter 3). Hence, for perfect stoichiometry, the movement of four electrons from water on to $NADP^+$ to produce the two NADPH molecules required should generate enough energy to make three ATP molecules from ADP and phosphate, i.e. the $ATP/2e^-$ ratio should be 1.5 assuming only *non-cyclic photophosphorylation* occurs. The occurrence of either *cyclic* or *pseudocyclic* photophosphorylation would generate ATP without NADPH. It would thus be useful in assessing the likely physiological significance of the latter two processes to know how much ATP is made in *non-cyclic photophosphorylation* per two electrons transferred on to $NADP^+$. It must also be remembered, before placing too much stress on ATP/NADPH ratios, that there is considerable export of reducing power and ATP from chloroplasts to cytoplasm, not necessarily in a ratio similar to that required by the Calvin cycle (Chapter 6); that Calvin cycle intermediates are exported to the cytoplasm and also used for biosynthesis of starch within the chloroplast; that reduced ferredoxin is consumed in nitrite reduction by chloroplasts (Chapter 10); that both ATP and reduced ferredoxin are required for sulphate reduction (Chapter 10); that NADPH is consumed in the glutathione reductase reaction (Chapter 8) and that this list of processess consuming ATP and/or reducing power in the illuminated chloroplast is by no means complete! How much ATP and NADPH is required by the chloroplast *in vivo* to keep carbon dioxide fixation going obviously depends on what else it is doing at the same time. Should the $ATP/2e^-$ ratio only be 1.0, however, it seems unlikely that enough ATP could be made by non-cyclic photophosphorylation to fuel chloroplast metabolism. The flow of electrons 'down' the electron-transport chain of thylakoids is accelerated by the addition of ADP and phosphate, a phenomenon known as *photosynthetic control* that is essentially analogous to respiratory control in mitochondria. However, there is considerable electron transport in the absence of ADP. In other words, the photosynthetic control is 'loose' (West and Wiskich 1968). As discussed later in this chapter, some of the energy produced by this 'non-phosphorylating' electron transport may well actually be used to make ATP when ADP and phosphate are added.

Many workers have attempted to measure the number of ATP molecules synthesized per two electrons transported to $NADP^+$, or to artificial electron acceptors such as methyl viologen: values of $P/2e^-$ so obtained vary from 1.0 (Del Campo, Ramirez, and Arnon 1968; Ort and Izawa 1974) to almost 2.0 (Reeves and Hall 1973; Rosa 1979). Hall (1976) has summarized the work of various groups. He argues that measured $ATP/2e^-$ ratios depend very much on procedures used to isolate the chloroplasts and on the assay conditions, that the true ratio is 2.0 and that lower ratios reflect artefacts of assay or membrane damage, since many reports of ratios close to 1.0 were obtained with washed thylakoid preparations. Unfortunately, one cannot study the $P/2e^-$ ratio directly in type A spinach chloroplasts because the envelope is almost impermeable to adenine nucleotides (Chapter 6). Reeves and Hall (1973) prepared type A spinach chloroplasts and subjected them to osmotic shock to give type C preparations. They found $P/2e^-$ ratios of 1.5, 1.7, and 1.7 using ferricyanide, methyl viologen, or $NADP^+$ as electron acceptors respectively. If one subtracts from the electron transport rates that they measured the 'basal' rate of electron transport seen in the absence of ADP, the corrected $P/2e^-$ ratios are 2.0–2.4, and therefore Reeves and Hall (1973) argued that the true ADP/O ratio was 2.0, but a little extra ATP (no more than 10 per cent of the total) might have been made (possibly by cyclic photophosphorylation) in their preparations. Unfortunately, this correction for a 'basal' rate of electron flow is probably not a valid procedure since the energy generated by it can probably be used in ATP synthesis, as explained in the next section. Robinson and Wiskich (1976) isolated type A spinach chloroplast fractions and gradually decreased the integrity of the envelopes by lowering the osmolarity of the surrounding medium. The $P/2e^-$ ratios obtained using ferricyanide or methyl viologen as electron acceptors were, on average, 1.7 and 1.9 respectively and were *not* decreased in the osmotically shocked chloroplasts even when the remaining stromal components were removed by washing. Using isolated pea chloroplasts they also found that the measured $P/2e^-$ ratio was a maximum only if the inorganic phosphate concentration in the reaction mixture was 10 mM or more: lower concentrations gave markedly decreased $P/2e^-$ ratios. However, stromal phosphate concentrations are likely to be about this level *in vivo* (Chapters 3 and 4). In contrast, Arnon and Chain (1977*a*) using methods similar to those of Reeves and Hall (1973) found the $P/2e^-$ ratio in osmotically disrupted chloroplasts from young spinach leaves (14-days old) to be 1.0 with ferricyanide as terminal electron acceptor, but 1.4 with $NADP^+$ as electron acceptor, whereas in broken chloroplasts from mature (41-day old) leaves the $P/2e^-$ ratio with $NADP^+$ was 1.0. The higher ratio in young leaves using $NADP^+$ was decreased to 1.0 by adding antimycin A (an inhibitor of cyclic electron flow) to the chloroplast preparation, and the authors suggested that the extra ATP arose by cyclic photophosphorylation. The effects of antimycin A are discussed in detail in the next section.

How can one explain such contrasting results? Clearly leaf age, which is rarely specified in published papers, may be an important factor. It must also be realized that the use of methyl viologen as an electron acceptor can cause problems in studies of broken chloroplast preparations retaining variable amounts of stromal components, such as those used by Reeves and Hall (1973) and Robinson and Wiskich (1976). The rate of electron transport in the presence of this compound is usually assessed as a rate of oxygen uptake, using the stoichiometry of eqns (2.6–9). This assumes that all the O_2^- generated by reaction (2.7) is converted into oxygen and hydrogen peroxide (eqn (2.8)). However, the chloroplast stroma contains ascorbic acid in variable concentrations, together with glutathione. Both these compounds react with O_2^- (see Chapter 8). Hence if the broken chloroplast preparations contain ascorbic acid because of stromal contamination or because it was present in the isolation medium, some O_2^- will react with this molecule (eqn 2.12)) and will not be available to produce oxygen by reaction (2.8).

$$\text{ascorbate} + 2O_2^- \rightarrow H_2O_2 + \text{dehydroascorbate} \qquad (2.12)$$

Thus the measured oxygen uptake will be an overestimate of electron flow. Indeed, when extra ascorbate is added to disrupted chloroplasts illuminated in the presence of methyl viologen, oxygen uptake further increases (Epel and Neumann 1973; Elstner and Kramer 1973; Allen and Hall 1973). Perhaps the ratios obtained using methyl viologen are slightly less than 2.0 in the above experiments because of the ascorbate and glutathione effects. There are problems in using ferricyanide as well, since this molecule can be reduced by both PSI and PSII (Trebst 1974). Reduction of ferricyanide by PSII depends on its accessibility to the acceptor sites of PSII, which are largely 'covered up' in freshly-prepared chloroplast membranes, so that most reduction is achieved through PSI (Fig. 2.1). However, in damaged chloroplasts, some reduction by PSII can occur, giving an apparently low $P/2e^-$ ratio because the electron-transport chain is bypassed.

In an attempt to avoid the problems associated with 'partially intact' (partially degraded!) chloroplasts a different approach to determination of the $P/2e^-$ ratio was pioneered by Heber and his group (e.g. Heber and Kirk 1975; Krause and Heber 1976). Type A spinach chloroplasts were supplied with various substrates that require different proportions of ATP and NADPH for their reduction (Table 2.3). The fact that illuminated chloroplasts can carry out reduction of these various substances for long periods *in vitro* must mean that they can regulate their 'effective' ATP/NADPH, as shown in the last column of Table 2.3. Quantitative estimates of maximal $ATP/2e^-$ ratios were made by measuring the quantum requirement of oxygen evolution in these different chloroplast reactions at low light intensities. The number of light quanta required to evolve one molecule of oxygen during oxaloacetate reduction by chloroplasts was found to be 8.0. Two molecules of oxaloacetate were reduced by two molecules of NADPH, utilizing four electrons obtained by the splitting of two water molecules and so releasing one oxygen molecule. Absorption of two light quanta, one by each photosystem, is required to move each of the four electrons through the electron-transport chain. The quantum requirement for phosphoglycerate reduction per molecule of oxygen evolved was also found to be 8.0 One molecule of oxygen evolved indicates reduction of 2 mol of phosphoglycerate to phosphoglyceraldehyde at the expense of 2 NADPH and 2 ATP mols.

Table 2.3. Photoreduction of substrates by isolated type A spinach chloroplasts

Compound supplied	Observed reduction rate (μmol h^{-1} mg chlorophyll^{-1})	Pathway of metabolism	moles ATP used per mole of substance reduced	moles NADPH used per mole of substance reduced	Theoretical ATP/NADPH ratio needed for continued substrate reduction
CO_2	80–150	Calvin cycle (Chapter 3)	3	2	1.5
Oxaloacetate	30–130	Reduction by malate dehydrogenase plus NADPH (Chapter 3)	0	1	0
Phosphoglycerate	120–450	Reduction to phosphoglyceraldehyde by glyceraldehyde-3-phosphate dehydrogenase (Chapter 3)	1	1	1.0
Glycerate	15–40	Conversion to phosphoglycerate by glycerate kinase plus ATP (Chapter 7), then as for phosphoglycerate	2	1	2.0

Numerical data from Krause and Heber (1976)

Therefore the $P/2e^-$ ratio in this case must have been at least 1.0. When glycerate is reduced a further two ATP molecules are required per mol of oxygen evolved (Table 2.3). If the $ATP/2e^-$ ratio *in vivo* were 2.0, both of these ATP molecules could be provided during transport of four electrons from water to $NADP^+$, i.e. the quantum requirement should still be 8.0. However, the quantum requirement was found to be much higher (12–13), implying that further light quanta are needed in extra electron-transfer reactions to supply the extra ATP, presumably in cyclic and/or pseudocyclic photophosphorylation. It follows that the ATP/NADPH ratio *in vivo* during glycerate reduction must be much lower than 2.0. A similar argument can be applied to carbon dioxide fixation by chloroplasts, for which the quantum requirement per oxygen molecule evolved is 10–12. Krause and Heber (1976) also point out that when carbon dioxide is supplied to illuminated chloroplasts under limiting light intensities, the internal ATP level decreases, whereas that of NADPH remains constant. It is difficult to see why this should happen if the $P/2e^-$ ratio were greater than 1.5.

The results of Heber's group clearly indicate that chloroplasts can vary their 'effective' ratio *in vivo* (Table 2.3). This can be explained easily if the $P/2e^-$ ratio of non-cyclic photophosphorylation is 1.0, and extra ATP is made when required by other processes, presumably cyclic or pseudocyclic photophosphorylation. It would also explain why many experiments with isolated thylakoids have shown a $P/2e^-$ ratio of 1.0. Heber argues that if one accepts the view of Hall (1976) that the $P/2e^-$ ratio *in vivo* is 2.0, it is difficult to see how the experimental observations in Table 2.3 can be accommodated. For example, he argues, how could phosphoglycerate be rapidly reduced by chloroplasts, since accumulation of the unconsumed ATP should decrease NADPH generation by photosynthetic control? One must not forget the existence of other metabolic activities within chloroplasts, however. Indeed, Robinson and Wiskich (1976) found an unidentified 'ATP-consuming reaction' in the stroma of type A pea chloroplasts. If such processes dispose of considerable light-generated ATP *in vivo*, then even with a $P/2e^-$ ratio of 2.0 for non-cyclic photophosphorylation there might not be sufficient ATP left over to reduce glycerate or even to fix carbon dioxide without other sources.

If the results of Reeves and Hall (1973) and Robinson and Wiskich (1976) are taken at their face value, the measured $P/2e^-$ ratios are less than 2.0 but greater than 1.0. There seems to be no reason for not supposing that the $P/2e^-$ ratio *in vivo* is actually approximately 1.5 and can be modulated either by consumption of excess ATP in other synthetic processes or by dissipating a variable part of the proton gradient that is used to power ATP synthesis (Section 2.2.3) by allowing a back-flow of protons across the thylakoid membrane to an extent depending on the demands made on the electron-transport chain. This point will be discussed further in Section 2.2.3.

2.2.2 *Do cyclic and pseudocyclic photophosphorylation occur* in vivo?

Since we do not actually know the P/2e$^-$ ratio of non-cyclic electron transport in chloroplasts, we cannot rely on theoretical arguments as to whether cyclic or pseudocyclic photophosphorylation must occur *in vivo* in order to meet the ATP requirements of the Calvin cycle. We must therefore turn to direct experimental evidence. Pseudocyclic photophosphorylation involves reduction of an electron acceptor, probably ferredoxin, which then reacts with oxygen to yield the superoxide radical, which is converted into hydrogen peroxide. The fact that chloroplasts contain high activities of a superoxide dismutase enzyme (Chapter 8) suggests that O_2^- is indeed formed *in vivo*. Egneus, Heber, Matthiesen, and Kirk (1975) showed that type A spinach chloroplasts exhibit a light-dependent uptake of ^{18}O-labelled oxygen during carbon dioxide fixation, which they attributed to pseudocyclic photophosphorylation. Hydrogen peroxide was produced in their chloroplast preparations in amounts sufficient to inhibit carbon dioxide fixation: this could be overcome by addition of catalase to the reaction mixture. Several other workers have shown that catalase is necessary to obtain maximum rates of carbon dioxide fixation by illuminated type A chloroplasts (e.g. Kaiser 1976; Hind, Mills, and Slovacek 1978; Sicher and Jensen 1979). Chloroplasts contain no catalase activity in the stroma although they do have other mechanisms, of variable efficiency, for coping with hydrogen peroxide (Chapter 8). However, type A chloroplast fractions often contain substantial catalase activity because of cytoplasmic contamination (Chapter 1), which could easily prevent detection of any hydrogen peroxide generated by pseudocyclic photophosphorylation in the preparations used by some workers (e.g. Rosa 1979). Egneus *et al.* (1975) found that chloroplasts reducing phosphoglycerate, which require an effective P/2e$^-$ ratio of 1.0 (Table 2.3), did not accumulate hydrogen peroxide or show ^{18}O-oxygen uptake. This therefore implies that pseudocyclic photophosphorylation supplies some of the ATP needed to drive carbon dioxide fixation, which further means that the '*in vivo*' P/2e$^-$ ratio of non-cyclic electron transport must have been less than 1.5 in their preparations. Jennings and Forti (1975) and Marsho, Behrens, and Radmer (1979) have demonstrated considerable oxygen uptake during the 'induction period' when chloroplasts are first illuminated, in which electron transport is operating at normal rates but carbon dioxide fixation is low (Chapters 3 and 4). They suggest that oxygen serves as terminal electron acceptor during this phase. Radmer and Kok (1976) have demonstrated a similar phenomenon in several algae, and the blue-greeen alga *Anacystis nidulans* has been shown to release hydrogen peroxide into the surrounding medium upon illumination (Patterson and Myers 1973).

In view of the above evidence, of the fact that reduced ferredoxin and possibly other components of the electron-transport chain reduce O_2 to O_2^-, and that illuminated chloroplasts contain a high internal concentration of oxygen, it seems almost certain that some pseudocyclic electron flow will occur *in vivo* (this point is further discussed in Chapter 8). It is difficult to see any role for chloroplast superoxide dismutase if O_2^- is not generated *in vivo*. The *rate* of this process is more difficult to assess, however. Measurement of the uptake of $^{18}O_2$ by chloroplasts or cells during photosynthesis will measure not only pseudocyclic photophosphorylation but also oxygen uptake due to glycollate synthesis (Chapter 7) and to the metabolism of phenolic compounds (Chapter 9). Glycollate synthesis can be corrected for (as was done in the experiments of Egneus *et al.* (1975)) but the contribution of phenol oxidase reactions to oxygen uptake is at present completely unknown. Measurement of the rate of hydrogen peroxide production by chloroplasts, in the presence of azide to inhibit endogenous catalase, would be an alternative index of rates of pseudocyclic electron flow were it not that some hydrogen peroxide is probably metabolized in the stroma by an ascorbate/glutathione cycle before it escapes from the chloroplast (Chapter 8). The efficiency of this cycle in removing hydrogen peroxide appears to vary between different chloroplast preparations.

To continue the tale of our ignorance, the physiological significance of cyclic photophosphorylation will be examined. As discussed in Section 2.1, cyclic electron flow can be made to occur in chloroplasts *in vitro* by experimental conditions that decrease electron output from PSII, causing an increase in the reduction state of ferredoxin. Ferredoxin-dependent cyclic electron flow is inhibited by antimycin A in a concentration range that does not appear to inhibit non-cyclic electron flow or photophosphorylation. This compound might therefore be used to probe the role of cyclic photophosphorylation in chloroplast metabolism, as was done by Arnon and Chain (1977 *a, b*) in their measurements of the $P/2e^-$ ratio discussed previously. When antimycin A is added to type A spinach chloroplasts fixing carbon dioxide at high rates under aerobic conditions, carbon dioxide fixation is partially inhibited (Hind, Nakatani, and Izawa 1978; Mills, Slovacek, and Hind 1978; Heber, Egneus, Hanck, Jensen, and Koster 1978). This could be taken to mean that some cyclic photophosphorylation is necessary to meet the ATP requirements of the Calvin cycle, from which it follows that the '*in vivo*' $P/2e^-$ ratio of non-cyclic electron flow is less than 1.5. In all of the above experiments pseudocyclic electron flow was also probably occurring, as evidenced by $^{18}O_2$ uptake (Heber *et al* 1978) or accumulation of hydrogen peroxide in the reaction mixture (Hind *et al.* 1978; Mills *et al.* 1978). One might predict that inhibition of cyclic electron flow by antimycin A should increase the concentration of reduced ferredoxin in the chloroplast and so stimulate pseudocyclic electron flow, although this is difficult to assess because rates of pseudocyclic electron flow cannot be measured accurately (see above). Accepting this argument, it follows that in these chloroplast preparations, pseudocyclic photophosphorylation could not by itself meet that part of the 'ATP demand' of carbon dioxide fixation that was not supplied by non-cyclic photophosphorylation.

The effects of antimycin A are probably more complex than merely an inhibition of cyclic electron flow. When it was added to type A spinach chloroplasts fixing carbon dioxide at low rates, either because catalase had been omitted from the reaction mixture and hydrogen peroxide allowed to accumulate to an extent that inhibits the Calvin cycle (see Chapter 8), or because a high phosphate concentration was present in the medium, which decreases the concentration of Calvin cycle intermediates in the stroma (Chapter 6), antimycin A actually stimulated the low rates of carbon dioxide fixation (e.g. Mills *et al.* 1978), although the stimulated rate was not greater than the 'inhibited' rates obtained when chloroplasts fixing carbon dioxide at high rates were treated with antimycin A. It is difficult to see why this should happen, although it has been suggested that inhibition of cyclic electron flow could increase the pool of reduced ferredoxin and so increase the activity of some Calvin cycle enzymes (Mills *et al.* 1978), (Chapters 3 and 4). In the absence of added catalase or in the presence of inorganic phosphate, the activity of enzymes such as fructose diphosphatase may indeed limit the rate of carbon dioxide fixation by chloroplasts (Chapters 3 and 4).

The delicate 'redox poising' needed to demonstrate cyclic photophosphorylation *in vitro* might not occur *in vivo*, or it might only be possible if pseudocyclic electron flow operates simultaneously to 'bleed off' electrons to oxygen and prevent over-reduction of the electron transport chain (Heber *et al.* 1978; Ziem-Hanck and Heber 1980).

Although it is believed that pseudocyclic electron flow occurs *in vivo* in the chloroplasts of higher plants, it is not certain, according to the evidence presented so far, that cyclic photophosphorylation occurs as well. However, cyclic photophosphorylation may be an important source of ATP in some algae (Simonis and Urbach 1973; Raven 1976) and in certain orchids, which have low activities of PSII (Schmid, Jankowitz, and Menke 1976; Menke and Schmid, 1976).

2.2.3 The mechanism by which electron flow generates ATP

Whatever the $P/2e^-$ ratio, the mechanism by which the movement of electrons in the electron-transport chain is coupled to ATP synthesis in chloroplasts is also a major question. A similar coupling of electron flow and ATP synthesis is seen during the process of *oxidative phosphorylation* carried out by plant and animal mitochondria and by some aerobic bacteria, and it may be that the coupling mechanism is similar in all three systems. Essentially, there have been two major approaches to understanding the coupling mechanism, the *chemical approach* and the *chemi-osmotic approach*, which are summarized in Fig. 2.5. The key difference between them is that in the chemi-osmotic theory the 'energy intermediate' of coupling is a difference in the number of protons on each side of a membrane, setting up both a membrane potential and a pH difference which can be used to drive ATP synthesis. In the chemical theory

(1) *Mitchells' Chemi-osmotic theory*

A. The components of the electron-transport chain are arranged asymmetrically in the coupling membrane (e.g. the thylakoid membrane or mitochondrial inner membrane) so that electron movement causes uptake of H^+ on one side of the membrane and release of H^+ on the other side. The coupling membrane is essentially impermeable to protons.

B. This vectorial movement of H^+ sets up an electrochemical gradient, Δp, sometimes called a *protonmotive force* where

$$\Delta p = \Delta\psi - \frac{2.303RT}{F} \cdot \Delta\mathrm{pH} \left[\frac{2.303RT}{F} \equiv \frac{59\ \mathrm{mV}}{\text{at } 25°\mathrm{C}} \right]$$

The ΔpH term reflects the difference in concentration of protons on either side of the membrane. The membrane potential term, $\Delta\psi$, is due to unilateral movement of H^+. If movement of H^+ is accompanied by electroneutral movement of other ions, then $\Delta\psi = 0$ and Δp is governed only by ΔpH.

C. Unless dissipated, Δp builds up to a level which stops further movement of H^+, as H^+ cannot flow easily through the coupling membrane. Uncouplers such as FCCP or dinitrophenol stimulate electron transport and inhibit ATP synthesis by catalysing the removal of Δp: they are usually weak acids that can cross the membrane in the unionized form, so transporting H^+.

D. The coupling membrane contains sites where movement of H^+ down the electrochemical gradient can drive ATP synthesis, i.e. a proton-driven asymmetrically-arranged ATP synthetase.

(2) *Chemical hypothesis*

Electron transport (et) generates a reactive chemical intermediate (symbolized $\sim X$) which has sufficient reactivity to drive ATP synthesis (through one or more additional reactions) and other energy-linked processes, including ion movements (e.g. H^+).

$$\text{et} \longrightarrow \sim X \rightleftharpoons \begin{cases} \text{Ion pumping (e.g. a pump of } H^+) \\ \text{ATP synthesis} \end{cases}$$

Fig. 2.5. Theories to explain coupling of electron flow to ATP synthesis. (Based on the review by Greville 1969.) The chemi-osmotic theory was pioneered by Peter Mitchell.

the energy intermediate is a high energy compound (symbolized $\sim X$), the decomposition of which liberates sufficient energy to drive ATP synthesis and the movements of ions, including those of protons. Since no candidate so far proposed for $\sim X$ has stood up to rigorous analysis, and since the evidence for the postulates listed in Fig. 2.5 is quite strong for chloroplasts, as reviewed below, most workers now interpret the mechanism of energy coupling in chemi-osmotic terms. It must be noted that the maintenance of a protonmotive force requires a closed intact membrane system so as to maintain differences in the number of protons between the aqueous phases on either side. If ATP synthesis could ever be demonstrated in open thylakoid membrane fragments the chemi-osmotic theory, in this form, would be disproved.

Let us now consider how well the postulates of the chemi-osmotic theory as listed in Fig. 2.5 apply to chloroplasts.

2.2.3.1 Postulate One: Asymmetric arrangement of the electron-transport chain to achieve proton gradients during electron transport

The exact arrangement of the electron-transport chain components in the thylakoid membrane is not known well enough to provide a rigorous test of this postulate, but there is considerable evidence for asymmetric location of at least some of them. Such information has come from the use of radioactive compounds that label membrane proteins, from studies of the accessibility of components to artificial electron donors or acceptors, and by the use of specific antibodies. For example, ferredoxin-NADP reductase is exposed at the outer side of the thylakoids. The oxygen-evolving apparatus associated with PSII is bulky and spans the thylakoid membrane, but the sites of proton release are on the inner thylakoid face. Hence the splitting of water acidifies the aqueous space inside the thylakoids, releasing $4H^+$ per oxygen evolved. The primary electron acceptors of both photosystems are sited near the outer surface of the thylakoid membrane, whereas the primary donors are on the inner side (Crofts and Wood 1978). Figure 2.6 shows a mechanism, consistent with our current knowledge, by which electron transport in chloroplast thylakoids could generate a protonmotive force. It is based on the fact that P680, P700, cytochrome f, plastocyanin, and ferredoxin are electron acceptors, whereas plastoquinone and $NADP^+$ accept protons at the same time as accepting electrons (Fig. 2.2 shows this for plastoquinone). The plastoquinone shuttle proposed in Fig. 2.6 seems feasible, since this non-polar low-molecular-weight compound would be expected to be relatively mobile within the hydrophobic membrane

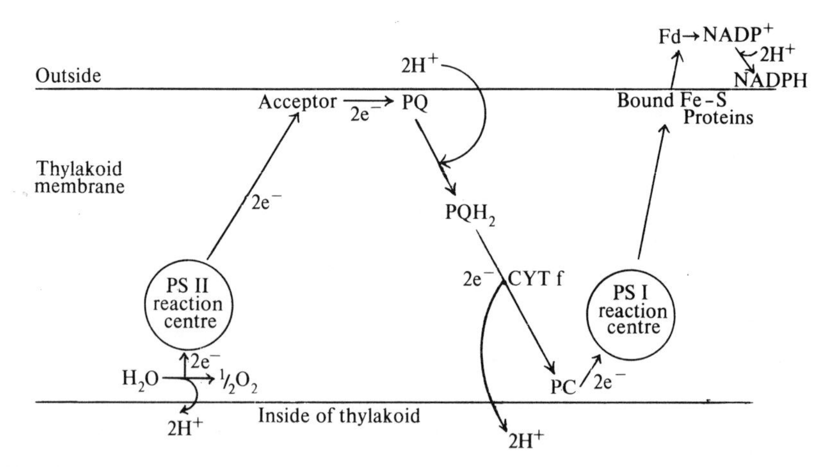

Fig. 2.6 A mechanism by which electron transport in chloroplasts might generate a protonmotive force. (Based on data and interpretation by Trebst (1974)). The existence of a shuttle of plastoquinone (PQ) is proposed in which PQ accepts H^+ on one side of the membrane and loses them on the other (PC – plastocyanin: Fd – ferredoxin).

interior (Tiemann, Renger, Graber, and Witt 1979). According to this scheme, movement of two electrons from water on to $NADP^+$ causes four protons to be transported from the outside to the inside of the thylakoid.

2.2.3.2 Postulate Two: As a result of the asymmetric arrangement of the electron-transport chain, illumination of chloroplasts sets up a protonmotive force

Although the exact arrangement of the electron-transport-chain components may not be known, it has been well established that illumination of an unbuffered or weakly-buffered suspension of thylakoids causes an increase in the pH of the suspension medium, which is prevented by inhibitors of electron transport. Figure 2.7 shows a typical result. This uptake of protons is accompanied by acidification of the internal spaces of the thylakoids (Avron and Neumann 1978; Avron 1978) as can be shown by the use of indicator dyes that enter the thylakoid membrane. In many cases, it is not clear exactly where these dyes become located, but it has been shown that neutral red binds to the inside of the thylakoid membrane and reports pH changes in the internal space (Junge, Auslander, McGeer, and Runge 1979). Alternative estimates of the proton gradient have come from measurements of the distribution of weak acids and bases, which penetrate the thylakoid membrane only as an uncharged form. For example, methylamine enters thylakoids only in the unprotonated form, CH_3NH_2, and so at equilibrium the concentration of this species will be the same on both sides of the membrane. Measurement of the amine concentration on either side of the membrane, e.g. by the use of ^{14}C-labelled compounds, enables calculation of the ratio $[H^+]_{IN}/[H^+]_{OUT}$ across the thylakoid membrane. Fluorescent amines such as 9-aminoacridine, whose fluorescence is quenched when they enter the thylakoids, have also been used (Heldt 1980). Such studies suggest that the pH difference between the inside and the outside of illuminated thylakoids can be over three pH units, although there is some controversy as to how many protons actually move per electron transported through the non-cyclic electron-transport chain. Numbers range from two protons per electron (see Fig. 2.6) up to four, although a value of two is favoured (Hall 1976; Reeves and Hall 1978). The size of the proton gradient appears to be decreased when ADP and phosphate are provided to the thylakoids to allow ATP synthesis, although in such experiments one must bear in mind the buffering effects of both ADP and inorganic phosphate (Pick, Rottenberg, and Avron 1973). Addition of ADP and phosphate increases the rate of electron transport by relieving photosynthetic control and should therefore increase actual proton movements provided that the terminal electron acceptor is not present in limiting amounts. Indeed, the phenomenon of photosynthetic control may be accounted for by a rise in the protonmotive force to a level sufficient to decrease electron transport, perhaps because a low value of the intra-thylakoid pH is unfavourable for this process (Kobayashi, Inoue, Shibata,

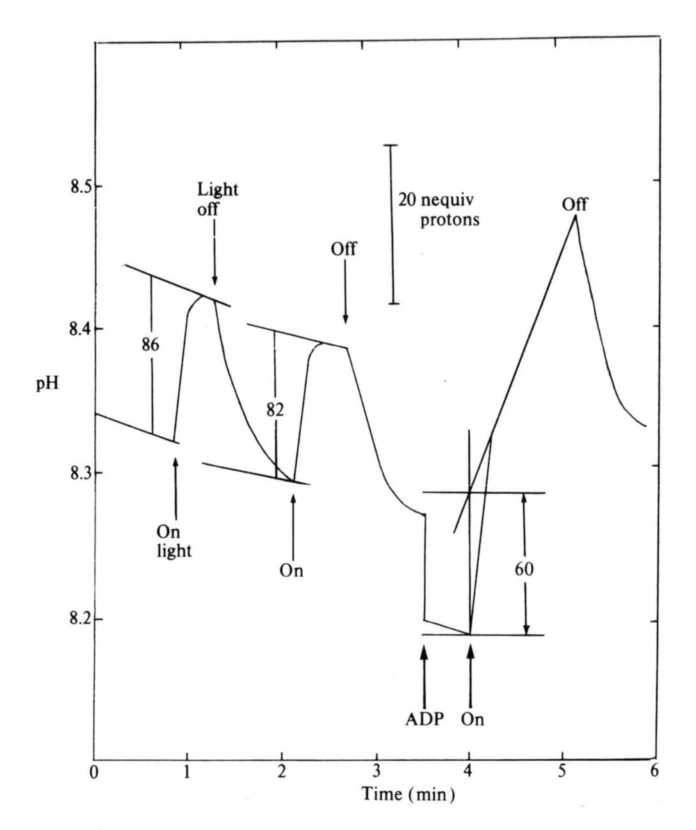

Fig. 2.7. pH changes induced in the medium by illumination of spinach thylakoids. From Telfer and Evans (1972). The reaction mixture contained 200 μg chlorophyll in 3 ml together with 150 μmol KCl, 6 μmol $MgCl_2$, 3 μmol sodium azide, 1.5 μmol potassium phosphate, and 0.375 nmol of methyl viologen as electron acceptor. 0.08 μmol ADP (pH 8.3) was added where indicated. Construction lines show the extent of the pH change due to proton gradient formation, taking into account the general drift downwards of pH seen both in the light and in the dark.

and Heber 1979). Addition of ADP and phosphate, by allowing protons to leave the thylakoids, speeds up the electron transport.

Association of proton movements with electron transport is further demonstrated by the observation of proton-dependent reverse electron transport (Avron 1978). If chloroplast suspensions are treated with a penetrant buffer at pH 5.3, so as to acidify the inside of the thylakoids, and then quickly placed in a solution at pH 9.6, transient reduction of the PSII acceptor 'Q' is observed and even fluorescence from the PSII reaction centre, which can be inhibited by DCMU. The abnormally large proton gradient imposed on the membranes in these experiments seems to be able to drive electrons back through the electron-transport chain and generate an excited state of PSII, which then decays with light emission.

Uncoupling agents such as dinitrophenol or ammonium chloride do not affect the electron-transport chain of chloroplasts but decrease or prevent the ATP synthesis associated with it. Such compounds stimulate the decay of the light-induced proton gradient in chloroplasts, i.e. they act to increase the permeability of the thylakoid membrane to protons and prevent establishment of a proton gradient that can be used to drive ATP synthesis. For example, addition of ammonium chloride to the outside of illuminated thylakoids will produce some ammonia in the suspension medium by the reaction

$$NH_4^+ \rightleftarrows NH_3 + H^+ \qquad (2.12)$$

The small, covalent, lipid-soluble NH_3 molecule will penetrate the thylakoid membrane rapidly, whereas the charged NH_4^+ ion will penetrate much more slowly. In the interior of the thylakoids the high proton concentration will push the equilibrium shown to the left of (2.12). Hence the ammonium salt acts in absorbing protons on the inside of the thylakoid membrane and releasing them on the outside, so decreasing the proton gradient.

2.2.3.3 Postulate Three: The coupling membrane contains an ATP synthesizing system which utilizes the energy released by movement of protons back down the electrochemical gradient

Washing chloroplast thylakoids in solutions of low ionic strength in the presence of EDTA removes a protein complex, known as coupling factor 1 or 'CF_1'. The complex is approximately spherical, about 9 nm in diameter, and appears to be identical with the 9 nm spheres seen on the outer surface of unstacked thylakoid membranes under the electron microscope (Chapter 1). If this complex is gently heated, subjected to mild proteolysis with trypsin or incubated with thiol compounds such as dithiothreitol, it gains the ability to hydrolyse ATP to ADP and phosphate in the presence of divalent metal ions as cofactors. EDTA-washed thylakoids cannot carry out photophosphorylation, but this ability can be restored to them by addition of CF_1 in the presence of Mg^{2+} ions. Antibodies raised against CF_1 inhibit photophosphorylation in intact thylakoids. The antibiotic Dio-9 inhibits the ATPase activity of CF_1, and also photophosphorylation and coupled electron transport (i.e. that part of electron transport that is induced by addition of ADP and phosphate) in thylakoids. It therefore seems very likely that CF_1 is involved in photophosphorylation *in vivo* (Nelson 1976; Nelson, Eytan, and Julian 1978). Analysis of CF_1 by polyacrylamide gel electrophoresis in the presence of sodium dodecyl sulphate shows it to contain five different polypeptide chains, which are designated as the α, β, γ, δ, and ϵ chains in order of decreasing molecular weight. There may be 2α,

2β, 1γ, 1δ, and 2ϵ chains in the complex, although this is not certain (Nelson 1976). The active site for the ATPase activity seems to be associated with the β-subunits and contains an essential arginine residue (Andreo and Vallejos 1977), whereas the purified ϵ subunits inhibit the ATPase activity of the complete CF_1 complex. It seems likely that the treatments required to activate the ATPase activity of CF_1 cause a conformational change which moves the ϵ subunit away from the active site. The α, β, γ, and ϵ subunits seem to be made on chloroplast ribosomes (Chapter 3), whereas the δ subunit is imported from the cytoplasm, probably as a larger precursor form (Nelson, Nelson, and Schatz 1980) in spinach.

The ATPase activity of CF_1 can also be induced *in situ*. Isolated thylakoids show little ATPase activity in the dark, but it can be induced by trypsin treatment or by illumination in the presence of thiol compounds. Such induced ATP hydrolysis is accompanied by movement of protons into the thylakoids, and can drive a reverse flow of electrons along the electron-transport chain under certain conditions (Avron 1978). In type A chloroplasts, however, ATPase activity can be induced by light alone, although activity is lost again in the dark (Mills and Hind 1979; Inoue, Kobayashi, Shibata, and Heber 1978). Presumably this 'dark inactivation' is a mechanism by which hydrolysis of the ATP generated in chloroplasts in the dark by metabolism of stored carbohydrates (Chapter 4) is avoided. Intact chloroplasts contain various thiol compounds, including the dithiol protein thioredoxin (Chapter 3) which might act *in vivo* similarly to the dithiothreitol needed to promote light-activation of the ATPase *in vitro* (McKinney, Buchanan, and Wolosiuk 1979; Mills, Mitchell, and Schurmann 1980). Illumination must produce conformational changes in the CF_1 complex that facilitate activation. Indeed, exposure of previously buried thiol and amino groups in CF_1 occurs on illumination of thylakoids (Jagendorf 1975; Oliver and Jagendorf 1976; Andreo and Vallejos 1976). Small amounts of ATP and ADP are known to be tightly attached to the CF_1 complex *in situ*. The exchange of these bound nucleotides with nucleotides in the surrounding medium is very slow in the dark, but much faster in the light, again suggesting conformational changes (Harris and Slater 1975; Rosing, Smith, Kayalar, and Boyer 1976).

If the CF_1 complex is active as an ATPase in illuminated chloroplasts *in vivo*, then, according to Mitchell's chemi-osmotic theory, a flow of protons through it from the interior of the thylakoid to the stroma driven by the protonmotive force set up by electron transport, must force CF_1 to act as an ATP *synthesizing* system. The chemical mechanism by which movement of protons through an ATPase might force it to synthesize ATP is hotly debated (e.g. see Mitchell 1974; Boyer 1975; Williams 1979), yet the evidence that ATP can be made in this way is convincing. It is possible to acidify the internal space of thylakoids by incubating them in the dark in the presence of a penetrant acid. If such thylakoids are transferred to a solution at high pH, so that an artificial proton gradient is set up, significant amounts of ATP can be synthesized from ADP and phosphate. The synthesis of ATP in this 'acid-bath' experiment is prevented by inhibitors of, or antibodies directed against, CF_1 (Jagendorf 1975). Spinach thylakoid membranes pre-incubated with ferricyanide to allow it to enter the intra-thylakoid space can make ATP in the dark from ADP and phosphate, provided that they are supplied with an external reducing agent plus a lipid-soluble component that can mediate a flow of electrons across the membrane. ATP synthesis only occurs, however, if the lipophilic electron carrier is of the type that releases protons into the intra-thylakoid space as it donates electrons to $Fe(CN)_6^{3-}$. Carriers that transfer only electrons are ineffective (Selman and Ort 1977). There is as yet, however, no agreement on the number of protons which have to be moved through CF_1 in order to make one ATP molecule.

Although both these experiments are fully consistent with the chemi-osmotic theory, the chemical hypothesis can explain them by proposing that $\sim X$ drives a proton pump as well as ATP synthesis. If the link between $\sim X$ and the proton gradient were reversible (Fig. 2.5) then any artificial ΔpH set up across the thylakoid membrane could generate some $\sim X$ and hence could make ATP. Such possibilities are very difficult to rule out completely, but are unconvincing for two reasons, apart from the inability of scientists to isolate $\sim X$. Firstly, CF_1 is associated with the outer surfaces of thylakoid membranes only in unstacked membrane regions, whereas many electron-transport chains are located in the various membranes of the granal stacks. This spatial separation is no problem if the energy intermediate is a difference in the number of protons on each side of the thylakoid membrane, but it seems difficult to believe that a highly labile, reactive $\sim X$ produced by electron transport should have to move such distances through the membrane. A second reason for rejecting the chemical hypothesis comes from studies on a very different system. If the salt-tolerant bacterium *Halobacterium halobium* is grown at low oxygen tensions, it develops purple patches on its cell membrane, which contain only one protein, bacteriorhodopsin. Illumination of the cells generates a protonmotive force across the cell membrane, apparently as the bacteriorhodopsin cycles between different conformational states (Stoeckenius 1976). It is possible to generate closed membrane vesicles (liposomes) by sonication of phospholipids under appropriate conditions. They consist of an inner aqueous phase bounded by a spherical lipid bilayer. If bacteriorhodopsin is added to the mixture from which liposomes are prepared, one can obtain an incorporation of this molecule into the lipid bilayer of the vesicles. Illumination of such vesicles sets up a protonmotive force across the bilayer. If a CF_1–ATPase complex is also incorporated into the bilayer, ATP is made from ADP and phosphate on illumination (Winget, Kanner, and Racker 1977). ATP synthesis is inhibited by uncouplers and by inhibitors of the CF_1–ATPase complex, so it must be driven by the protonmotive force generated by bacteriorhodopsin. No other chloroplast components were present in these experiments, which leaves no room for a $\sim X$.

In the thylakoid membrane the CF_1–ATPase complex is associated with at least three other membrane proteins, two of which are synthesized on chloroplast ribosomes, whilst one is imported from the cytoplasm (Nelson *et al.* 1978, 1980; Pick and Racker 1979). One of these membrane proteins is a hydrophobic lipoprotein, sometimes known as F_0, which is thought to represent a 'proton channel' across the membrane, since its incorporation into liposomes causes the bilayer membrane to become permeable to protons. This can be prevented by adding *N,N*^{1}-dicyclohexylcarbodiimide, a compound which becomes covalently bound to the protein. Under conditions in which it has little effect on the ATPase activity of purified CF_1 complex, dicyclohexylcarbodiimide (DCCD) inhibits the ATPase activity of thylakoids (Nelson 1976). Membranes depleted of CF_1 by washing with EDTA are unable to build up a proton gradient upon illumination, but addition of DCCD restores this ability. Protons pumped by the electron-transport chain presumably leak back through F_0 until the hole is blocked by DCCD or 'capped' by CF_1. The function of the other membrane proteins associated with CF_1 is at present unknown.

2.2.3.4 Postulate Four: The thylakoid membrane is essentially impermeable to protons

In illuminated thylakoids large numbers of protons must be translocated into the intra-thylakoid space to achieve ΔpH values greater than three. It may be calculated that movement of only two protons across the thylakoid membrane would set up a membrane potential of approximately 32 mV (Barber 1978). Experimental imposition of a membrane potential, which can contribute to the protonmotive force, does increase ATP synthesis by thylakoids. For example, Schuldinger, Rottenberg, and Avron (1972, 1973) exposed thylakoid membranes to high external concentrations of KCl or NaCl in the presence of the antibiotics valinomycin or monactin. These antibiotics act as 'ionophores', i.e. they can bring about the transport of certain ions across the thylakoid membranes. Valinomycin is specific for K^+ whereas monactin will transport Na^+. In their presence, K^+ or Na^+ will be driven across the thylakoid membranes because of their high external concentrations, which will set up a positive membrane potential inside the thylakoids. ATP synthesis under conditions where the pH difference across the membrane is small was greatly enhanced by the imposed membrane potential. Witt, Schlodder, and Graber (1976) have shown that small amounts of ATP can be made by chloroplasts exposed to an external electric field, i.e. only a membrane potential.

However, other evidence indicates that a membrane potential makes little contribution to the protonmotive force in chloroplasts synthesizing ATP in the light at a steady rate (Avron 1978). Part of this evidence comes from the observation that the ionophore nigericin is a powerful inhibitor of photophosphorylation. This antibiotic catalyses a strict one-for-one exchange of K^+ and H^+ ions across biological membranes. Although it will destroy a pH gradient by replacing the excess of protons on one side of the thylakoid membrane with K^+ ions, it cannot abolish a membrane potential and its inhibitory effect thereby indicates the lack of importance of such a potential. If the membrane potential is small, it follows that the large numbers of protons entering the thylakoids in the light must be electrically balanced by a simultaneous influx of anions, an efflux of cations from the thylakoids, or both. The nature of these ion movements will be considered in Section 2.3.

In isolated illuminated thylakoids under phosphorylating or non-phosphorylating conditions, the proton gradient quickly reaches a maximum steady-state value (Fig. 2.7), even though oxygen evolution and reduction of electron acceptors continues. In this situation, the number of protons being absorbed from the medium by the electron-transport chain must be equalled by the number passing back across the thylakoid membrane from inside to outside. Such movement could conceivably occur by a passive leakage of protons across the lipid bilayer of the membrane, by a leakage through the CF_1-F_0 complex not coupled to ATP synthesis or, under phosphorylating conditions, by proton movement through CF_1-F_0 during ATP synthesis. One or other, or both, of the first two processes should account for the rate of electron transport seen in the absence of ADP and phosphate. Analysis of proton movements in illuminated thylakoids has shown that passive leakage is a slow process, but that protons can leak out through the CF_1-F_0 complex to an extent which increases with light intensity (Schonfeld and Neumann 1977; Ho, Liu, Saunders, and Wang 1979). Treatment of the thylakoids with DCCD or with adenylyl imidodiphosphate, an inactive analogue of ADP which binds to CF_1, decreased this leakage. Spinach chloroplasts treated with glutaraldehyde, a reagent which cross-links protein molecules, no longer show photophosphorylation but still translocate protons in the light. Thylakoids treated with this reagent in the dark have only a low proton permeability, whereas thylakoids treated in the light have a proton permeability that depends on the illumination used during the treatment. These results strongly suggest that the 'uncoupled leakage' of protons through CF_1-F_0 is increased by light-induced conformational changes and probably decreased when ADP and phosphate are added, which further means that the rate of 'uncoupled' electron transport (i.e. in the absence of phosphorylation) does not continue unchanged when phosphorylation occurs. This conclusion affects some assessments of $P/2e^-$ ratios (Section 2.2.1). Consistent with it, the rate of photosynthetic electron-transport measured in the absence of ADP and phosphate can be stimulated by addition of certain reagents that react with -SH groups (e.g. *N*-ethylmaleimide or low concentrations of Hg^{2+} ions). Stimulation is reversed by inhibitors of CF_1 (e.g. phlorizin) or by addition of ADP, ATP, or inorganic phosphate (Underwood and Gould 1980). Modification of some sulphydryl residues on CF_1 presumably increases the proton conductivity of the complex, which can be reversed by binding adenine nucleotides or phosphate on to it.

2.2.3.5 Does a membrane potential ever contribute to the protonmotive force?

The electric potential across the thylakoid membrane in chloroplasts carrying out photophosphorylation at a steady rate is virtually zero (Avron 1978). However, when thylakoids are first illuminated ATP synthesis begins before a signficant pH gradient has built up (e.g. Avron 1978; Davenport and McCarty 1980; Vinkler, Avron, and Boyer 1980). Addition of valinomycin plus K^+ ions, which will abolish a membrane potential because K^+ ions can move across the membrane to neutralize it, inhibits this initial ATP synthesis. Although nigericin is a powerful inhibitor of steady-state photophosphorylation, it does allow ATP synthesis to continue for the first 1–2 s of illumination. It may be concluded that during this period a membrane potential makes a significant contribution to the protonmotive force. The electric-field strength associated with the membrane potential would be expected to induce a phenomenon known as *electrochromism*, i.e. to distort the electronic absorption bands of the photosynthetic pigments. Such a shift, with a maximum of 518 nm, has been detected in illuminated thylakoids and is largely due to carotenoids (e.g. Witt 1975; Junge 1977; Barber 1978; Schlodder and Witt 1980). Illumination probably causes the rapid establishment of a membrane potential because of the primary photochemical reactions occurring within PSI and PSII. An electron is passed to an acceptor located on one side of the membrane and 'chl.$^+$' is neutralized by electrons from a donor on the other side (Fig. 2.6), i.e. there is a negative charge on one side of the membrane and a positive charge on the other. A proton gradient is then established as the ejected electrons are accepted, e.g. by plastoquinone, and water splits to replace the electron lost from the donor of PSII. Studies of the light energy ('luminescence') re-emitted by chloroplasts after the light has been turned off (Vredenberg 1976) also suggest that brief illumination sets up a membrane potential of 75–105 mV, positive inside, across the thylakoid membrane, roughly in agreement with Witt's results (Vredenberg 1976). Potentials comparable to these have also been measured in experiments designed to determine chloroplast membrane potentials directly using micro-electrodes, but in such experiments it is often unclear exactly where the electrode tip is located within the chloroplast. The effect of membrane damage caused by electrode penetration on any potential present would probably be significant in an organelle as small as the chloroplast (Vredenberg 1976).

A certain minimum protonmotive force, to which the membrane potential contributes, is required to drive the conformational changes in CF_1 necessary for its activation at the onset of illumination (Junge 1977; Harris and Crofts 1978). Indeed, an externally-imposed electric potential can bring about this change (Graber, Schlodder, and Witt 1978).

2.2.3.6 Alternatives to Mitchell's chemi-osmotic theory

The chemi-osmotic theory in its current form accounts for most of the observed properties of photophosphorylation in chloroplasts (e.g. Avron 1978; Vinkler *et al.* 1980; Davenport and McCarty 1980). Some problems remain, however. Giersch, Heber, Kobayashi, Inoue, Shibata, and Heldt (1980) found that the ATP/ADP ratios in type A chloroplasts were much lower than would be expected from the protonmotive force that can be built up by thylakoids. Decreasing the size of the proton gradient in chloroplasts by adding small amounts of uncouplers such as NH_4Cl did not lower the ATP/ADP ratio, i.e. the 'energy charge' of the chloroplasts did not appear to be in equilibrium with the proton gradient (Kobayashi *et al.* 1979).

A modified version of the chemi-osmotic theory, originally due to Williams (e.g. Williams 1979) might help to account for such results. He proposes that free energy differences of H^+ *within the membrane* drive ATP formation and that osmotic components, such as ΔpH, are not required and may act as 'energy stores' outside the main path of synthesis. Hence the 'bulk' proton gradient across the membrane would not be expected to be in equilibrium with the energy charge. It might be expected, however, that the mobility of protons would be restricted in the hydrophobic membrane interior, and it is difficult to see how 'localized proton pools' generated by electron transport chains in the grana could migrate rapidly to the CF_1 complexes on the stromal thylakoids. It is true to say, however, that the thylakoid membrane cannot be regarded merely as a static, inert barrier to protons. Since the intra-thylakoid volume is quite small and its pH does not seem to decrease below 4.0 upon illumination, the large numbers of protons taken up ($\geqslant 300$ nmol H^+ mg chlorophyll^{-1}) must become largely bound to the inside of the membrane (Barber 1980). At all pH values above 4.3 the thylakoid membrane carries a net negative charge, apparently due to the side-chain carboxyl groups of aspartate and glutamate residues on some membrane proteins (Nakatani, Barber, and Forrester 1978). As translocated protons bind to these groups, counter-ions will be displaced (see below). Indeed, the thylakoid membrane is known to undergo conformational changes on illumination (Chapter 1), some of which occur in the CF_1–ATPase complex, as discussed previously.

Boyer and Slater have proposed that the protonmotive force causes ATP synthesis *because* it drives conformational changes in the CF_1 complex. It has been suggested that these changes cause the release of pre-formed, tightly-bound ATP from CF_1 and the tight binding of previously loosely-bound ADP and phosphate. The 'energized' CF_1 is then capable of forming ATP from this ADP and phosphate and the ATP is displaced in the next cycle. Small amounts of tightly bound ATP and ADP are known to be attached to CF_1: their exchange with nucleotides in the surrounding medium is very slow in the dark, but much faster in the light (Boyer 1974; Harris and Slater 1975; Rosing, Smith, Kayalar, and Boyer 1976). However, conversion of tightly-bound ADP to tightly-bound ATP is too slow to be a step on the major pathway of chloroplast ATP synthesis (Rosing *et al.* 1976; Shavit 1980) and the exact role of 'bound nucleotides' has yet to be clearly defined.

2.3 Ion movements in illuminated chloroplasts

Since the thylakoid membranes bear a net negative charge, they will attract cations to both the exterior and interior membrane surfaces. When protons enter the thylakoids and bind to $-COO^-$ groups, some of the internal ions will be displaced and can presumably traverse the thylakoid membrane in exchange for the protons, so minimizing the membrane potential. These ions have two possible fates: they could bind to the exterior of the thylakoids if binding sites are available (and they may be available because of conformational changes in the membrane on illumination (Schapendonk, Hemrika-Wagner, Theuvenet, Wong Fong Sang, Vredenberg and Kraayenhof 1980)), because the negative groups are not completely saturated with cations or because the rise in pH of the stroma as protons are removed from it will increase the degree of ionisation of the external carboxyl groupings (Gross and Hess 1974)) or they can enter the stroma, in which they could remain 'free' or bind to stromal proteins.

Hind *et al.* (1974) studied ion movements in illuminated spinach thylakoids at pH 6.6 and found that they depended on the ions present in the surrounding medium. Depending on the exact conditions used, the import of protons could be balanced by a simultaneous influx of Cl^- and effluxes of K^+, Mg^{2+}, or Ca^{2+}. Gimmler, Schafer, and Heber (1975) estimated the total ion concentrations in type A spinach chloroplasts to be $[Na^+] = 23$ mM, $[K^+] = 20$–30 mM, $[Mg^{2+}] = 27$ mM (Ca^{2+} was not measured). From the results of Hind *et al.* (1974) this would mean that *in vivo* proton uptake would be partially balanced by release of Mg^{2+} from the thylakoids. Barber (1980) carried out a more extensive analysis of the ion concentrations inside type A pea chloroplasts: finding $[Na^+] = 9$ mM, $[K^+] = 107$ mM, $[Mg^{2+}] = 72$ mM, $[Ca^{2+}] = 32$ mM, and $[Cl^-] = 9$ mM, which led him to similar conclusions. Such measurements are possible because the chloroplast envelope is almost impermeable to cations. One must be very careful, however, in assessing data on the total ion concentration of chloroplasts, since many of these ions will be bound *in vivo*. Stromal enzymes, which are present at high concentrations, may well bind considerable amounts of Mg^{2+} ion. For example, ribulose diphosphate carboxylase is present in the stroma at millimolar concentrations (Chapter 3) and has eight binding sites for this ion on each molecule. The exterior of the thylakoids preferentially binds divalent metal ions, such as Mg^{2+} and Ca^{2+}, although Ca^{2+} seems to be more strongly bound than is Mg^{2+} (Hind *et al.* 1974). The export of Mg^{2+} is insufficient to balance the proton uptake however (Hind *et al.* 1974) and little attention has been paid to other ion movements. Mg^{2+}/H^+ exchange across the thylakoid membrane is necessary to regulate the 'spillover' of excitation energy between the two photosystems, which equalizes their quantal intakes (Barber 1976, 1980: see Chapter 1).

Does the Mg^{2+} exported enter the stroma or bind to the outside of the thylakoid? This point is crucial to an understanding of the regulation of the Calvin cycle, since it has been proposed that a light-induced increase in stromal Mg^{2+} ion concentration activates several Calvin cycle enzymes (Chapters 3 and 4). Portis and Heldt (1976) measured both the loss of Mg^{2+} from illuminated spinach thylakoids at pH 8.0, and the rise in the stromal Mg^{2+} concentration on illumination of type A chloroplasts. From an analysis of the results of both types of experiment, they concluded that illumination causes an increase in the Mg^{2+} concentration in the stroma of 1-3 mM. Krause (1977) obtained similar results in a study of the absorbance changes of a Mg^{2+}-specific indicator dye, eriochrome blue SE, within spinach chloroplasts. In contrast, Ben-Hayyim (1978), using lettuce chloroplasts, concluded that Mg^{2+} exported from the thylakoids remained bound to the outside of them. These contradictions remain to be resolved.

Ca^{2+} is also present at high concentrations within chloroplasts but it is more tightly bound than is Mg^{2+}, which perhaps makes it less liable to act as a major counter-ion to protons, and it is not involved in regulation of energy 'spillover' (Barber 1980). It may have other important roles, however. For example, incubation of spinach chloroplasts with the specific calcium chelator EGTA inhibits electron transport through both photosystems (Barr, Troxel, and Crane 1980). Added Ca^{2+} can restore activity under certain conditions, but Mg^{2+} cannot. Calcium ions are powerful inhibitors of chloroplast protein synthesis and so the amount free in the stroma (if any) must be carefully controlled (Jagendorf 1981).

Although the chloroplast envelope in the dark has a low permeability to cations, it has been shown that protons move across it in the light. They leave the chloroplasts, charge being partially balanced by import of K^+ ions from the surrounding medium (Gimmler *et al.* 1975; Barber 1980). However the extent of this proton release is only a few per cent of the amount of proton uptake into the thylakoids, and its significance has yet to be established. The rate of K^+/H^+ exchange appears to be stimulated by addition of Mg^{2+} ions to the reaction medium, and the resulting acidification of the stroma may inhibit carbon dioxide fixation (Huber and Maury 1980).

References

Allen, J. F. (1975). *Biochem. Biophys. Res. Commun.* **66**, 36–43.
— and Hall, D. O. (1973). *Biochem. Biophys. Res. Commun.* **52**, 856–62.
Andreo, C. S. and Vallejos, R. H. (1976). *Biochim. Biophys. Acta* **423**, 590–601.

Andreo, C. S. and Vallejos, R. H. (1977). *FEBS Lett* **78**, 207–10.
Arnon, D. I. and Chain, R. K. (1977*a*). *Plant Cell Physiol.* Special Issue on photosynthetic organelles, pp. 129–47.
— — (1977*b*). *FEBS Lett.* **82**, 297–302.
— — (1979). *FEBS Lett.* **102**, 133–8.
— Allen, M. B., and Whatley, F. R. (1954). *Nature, Lond.* **174**, 394–6.
Avron, M. (1978). *FEBS Lett.* **96**, 225–32.
— and Neumann, J. (1968). *A. Rev. Pl. Physiol.* **19**, 137–66.
Barber, J. (1976). In *The intact chloroplast, Topics in photosynthesis,* Vol. 1 (ed. J. Barber.) pp. 89–134. Elsevier, Amsterdam.
— (1978). *Rep. Prog. Phys.* **41**, 1157–1199.
— (1980). in *Plant membrane transport, current conceptual issues* (eds. R. M. Spanswick, W. J. Lucas and J. Dainty) pp. 83–96. Elsevier, Amsterdam.
Barr, R., Troxel, K. S., and Crane, F. L. (1980). *Biochem. Biophys. Res. Commun.* **92**, 206–12.
Ben-Hayyim, G. (1978). *Eur. J. Biochem.* **83**, 99–104.
Blankenship, R. E. and Parson, W. W. (1978) *A. Rev. Biochem.* **47**, 635–53.
Bohme, H. (1977). *Eur. J. Biochem.* **72**, 283–9.
Boyer, P. D. (1974). *Fedn Proc. Fedn Am. Socs exp. Biol.* **34**,1711–17.
— (1975). *FEBS Lett.* **50**, 91–4.
Butler, W. L. (1978). *FEBS Lett.* **95**, 19–25.
Chain, R. K. (1979). *FEBS Lett.* **105**, 365–9.
Cheniae, G. M. (1980). *Meth. Enzymol.* **69C**, 349–63.
Cox, R. P. (1979). *Biochem. J.* **184**, 39–44.
Cramer, W. A. and Whitmarsh, J. (1977). *A. Rev. Pl. Physiol.* **28**, 133–172.
Crofts, A. R. and Wood, P. M. (1978). *Curr. Top. Bioenerg.* **7**, 175–245.
Davenport, J. W. and McCarty, R. E. (1980). *Biochim. biophys. Acta* **589**, 353–7.
Del Campo, F. F., Ramirez, J. M., and Arnon, D. I. (1968). *J. biol. Chem.* **243**, 2805–9.
Egneus, H., Heber, U., Matthiesen, U., and Kirk, M. (1975). *Biochim. biophys. Acta* **408**, 252–68.
Elstner, E. F. and Kramer, R. (1973). *Biochim. biophys. Acta* **314**, 340–53.
Epel, B. L. and Neumann, J. (1973). *Biochim. biophys. Acta* **325**, 520–529.
Giersch, C., Heber, U., Kobayashi, Y., Inoue, Y., Shibata, K., and Heldt, H. W. (1980). *Biochim. biophys. Acta* **590**, 59–73.
Gimler, H., Schafer, G., and Heber, U. (1975). In *Proceedings of the third international congress on photosynthesis* (ed. M. Avron) pp. 1381–92. Elsevier, Amsterdam.
Graber, P., Schlodder, E., and Witt, H. T. (1978). In *Photosynthesis '77* (ed. D. O. Hall, J. Coombs, and T. W. Goodwin) pp. 197–210. Biochemical Society, London.
Gregory, R. P. F. (1977). *Biochemistry of photosynthesis*, 2nd edn. John Wiley, London.
Greville, G. D. (1969). *Curr. Top. Bioenerg.* **3**, 1–78.
Gross, E. L. and Hess, S. C. (1974). *Biochim. biophys. Acta* **339**, 334–46.
Hall, D. O. (1976). In *The intact chloroplast, Topics in photosynthesis,* Vol. 1 (ed. J. Barber) pp. 135–170. Elsevier, Amsterdam.
— Rao, K. K. (1977) in *Encyclopaedia of plant physiology*. New Series, Vol. 5 (eds. A. Trebst and M. Avron) pp. 206–216. Springer, Berlin.
Harriman, A. and Barber, J. (1979). In *Photosynthesis in relation to model*

systems. Topics in photosynthesis, Vol. 3 (ed. J. Barber) pp. 243-380. Elsevier. Amsterdam.
Harris, D. A. and Slater, E. C. (1975). *Biochim. biophys. Acta* **387**, 335–48.
— Crofts, A. R. (1978). *Biochim. biophys. Acta* **502**, 87-102.
Heber, U. and Kirk, M. R. (1975). *Biochim. biophys. Acta* **376**, 136-50.
— Egneus, H., Hanck, U., Jensen, M., and Koster, S. (1978). *Planta* **143**, 41-9.
— Kirk, M. R., and Boardman, N. K. (1979). *Biochim. biophys. Acta* **546**, 292-306.
Heldt, H. W. (1980). *Meth. Enzymol.* **69C**, 604-13.
Hill, R. and Bendall, F. (1960). *Nature, Lond.* **186**, 136-7.
Hind, G., Nakatani, H. Y., and Izawa, S. (1974). *Proc. natn. Acad. Sci., U.S.A.* **71**, 1484-8.
— Mills, J. D., and Slovacek, R. E. (1978). In *Photosynthesis '77* (eds. D. O. Hall, J. Coombs, and T. W. Goodwin) pp. 591-600. Biochemical Society, London.
Ho, Y. K., Liu, C. J., Saunders, D. R., and Wang, J. H. (1979). *Biochim. biophys. Acta* **547**, 149-60.
Huber, S. C. and Maury, W. (1980). *Pl. Physiol., Lancaster* **65**, 350-4.
Inoue, Y., Kobayashi, Y., Shibata, K., and Heber, U. (1978). *Biochim. biophys. Acta* **504**, 142-52.
Jagendorf, A. (1975). In *Bioenergetics of photosynthesis* (ed. Govindjee) pp. 414-492. Academic Press, New York.
— (1981). in *Proceedings of the Fifth International Congress on Photosynthesis*, in press.
Jennings, R. C. and Forti, G. (1975). In *Proceedings of the third international congress on photosynthesis* (ed. M. Avron) pp. 735-743. Elsevier, Amsterdam.
Junge, W. (1977). *A. Rev. Pl. Physiol.* **28**, 503-36.
— Auslander, W., McGeer, A. J., and Runge, T. (1979). *Biochim. biophys. Acta* **546**, 121-41.
Kaiser, W. (1976). *Biochim. biophys. Acta* **440**, 476-82.
Klimov, V. V., Dolan, E., and Ke, B. (1980). *FEBS Lett.* **112**, 97-100.
Kobayashi, Y., Inoue, Y., Shibata, K., and Heber, U. (1979). *Planta* **146**, 481-6.
Kok, B., Forbush, B., and McGloin, M. (1970). *Photochem. Photobiol.* **11**, 457-75.
Krause, G. H. (1977). *Biochim. biophys. Acta* **460**, 500-10.
— Heber, U. (1976). In *The intact chloroplast, Topics in photosynthesis*, Vol. 1 (ed. J. Barber) pp. 171-214. Elsevier, Amsterdam.
Marsho, T. V., Behrens, P. W., and Radmer, R. J. (1979). *Pl. Physiol., Lancaster* **64**, 656-9.
McKinney, D. W., Buchanan, B. B., and Wolosiuk, R. A. (1979). *Biochem. Biophys. Res. Commun.* **86**, 1178-84.
Mehler, A. H. (1951). *Archs Biochem. Biophys.* **33**, 65-77.
Menke, W. and Schmid, G. H. (1976). *Pl. Physiol., Lancaster* **57**, 716-19.
Metzler, D. E. (1977). *Biochemistry. The chemical reactions of living cells.* Academic Press, New York.
Mills, J. D. and Hind, G. (1979). *Biochim. biophys. Acta* **547**, 455-62.
— Slovacek, R. E. and Hind, G. (1978). *Biochim. biophys. Acta* **504**, 298-309.
— Mitchell, P. D., and Barber, J. (1979). *Photobiochem. Photobiophys.* **1**, 3-9
— — Schurmann, P. (1980). *FEBS Lett* **112**, 173-7.
Mitchell, P. (1974). *FEBS Lett.* **43**, 189-94.
Nakatani, H. Y., Barber, J., and Forrester, T. A. (1978). *Biochim. biophys. Acta* **504**, 215-25.

Nelson, N. (1976). *Biochim. biophys. Acta* **456**, 314–38.
— Eytan, E., and Julian, C. (1978). In *Photosynthesis '77* (eds. D. O. Hall, J. Coombs, and T. W. Goodwin) pp. 559–70. Biochemical Society, London.
— Nelson, H. and Schatz, G. (1980). *Proc. natn. Acad. Sci. U.S.A.* **77**, 1361–4.
Oliver, D. and Jagendorf, A. (1976). *J. biol. Chem.* **251**, 7168–75.
Ort, D. R. and Izawa, S. (1974). *Pl. Physiol., Lancaster* **53**, 370–6.
Patterson, C. O. P. and Myers, J. (1973). *Pl. Physiol., Lancaster* **51**, 104–9.
Pick, U. and Racker, E. (1979). *J. biol. Chem.* **254**, 2793–9.
— Rottenberg, H., and Avron, M. (1973). *FEBS Lett.* **32**, 91–4.
Portis, A. R. and Heldt, H. W. (1976). *Biochim. biophys. Acta* **449**, 434–46.
Radmer, R. J. and Kok, B. (1976). *Pl. Physiol., Lancaster* **58**, 336–40.
Raven, J. A. (1976). In *The intact chloroplast, Topics in photosynthesis,* Vol. 1 (ed. J. Barber) pp. 403–43. Elsevier, Amsterdam.
Reeves, S. G. and Hall, D. O. (1973). *Biochim. biophys. Acta* **314**, 66–78.
— — (1978). *Biochim. biophys. Acta* **463**, 275–97.
Robinson, H. H. and Yocum, C. F. (1980). *Biochim. biophys. Acta* **590**, 97–106.
Robinson, S. P. and Wiskich, J. T. (1976). *Biochim. biophys. Acta* **440**, 131–46.
Rosa, L. (1979). *Biochem. Biophys. Res. Commun.* **88**, 154–63.
Rosen, G., Gresser, M., Vinkler, C., and Boyer, P. D. (1979). *J. biol. Chem.* **254**, 10654–61.
Rosing, J., Smith, D. J., Kayalar, C., and Boyer, P. D. (1976). *Biochem. Biophys. Res. Commun.* **72**, 1–8.
Sarojini, G. and Govindjee (1981). *Biochim. biophys. Acta* **634**, 340–3.
Schapendonk, A. H. C. M., Hemrika-Wagner, A. M., Theuvenet, A. P. R., Wong Fong Sang, H. W., Vredenberg, W. J., and Kraayenhof, R. (1980). *Biochemistry* **19**, 1922–7.
Schlodder, E. and Witt, H. T. (1980). *FEBS Lett.* **112**, 105–13.
Schmid, G. H., Jankowicz, M., and Menke, W. (1976). *J. Microsc. Biol. Cell.* **26**, 25–8.
Schonfeld, M. and Neumann, J. (1977). *FEBS Lett.* **73**, 51–4.
Schuldinger, S., Rottenberg, H., and Avron, M. (1972). *FEBS Lett.* **28**, 173–6.
— — — (1973). *Eur. J. Biochem.* **39**, 455–562.
Selman, B. R. and Ort, D. R. (1977). *Biochim. biophys. Acta* **460**, 101–12.
Shavit, N. (1980). *A. Rev. Biochem.* **49**, 111–38.
Sicher, R. C. and Jensen, R. G. (1979). *Pl. Physiol., Lancaster* **64**, 880–3.
Simonis, W. and Urbach, W. (1973). *A. Rev. Pl. Physiol.* **24**, 89–114.
Spector, M. and Winget, G. D. (1980). *Proc. natn. Acad. Sci. U.S.A.* **77**, 957–9.
Stemler, A. (1980). *Pl. Physiol., Lancaster* **65**, 1160–5.
Stoeckenius, W. (1976). in *Structural basis of membrane function* (eds. Y. Hatefi and L. Djavadi-Ohaniance) pp. 39–44. Academic Press, New York.
Telfer, A. and Evans, M. C. W. (1972). *Biochim. biophys. Acta* **256**, 625–37.
Tiemann, R., Renger, G., Graber, P., and Witt, H. T. (1979) *Biochim. biophys. Acta* **546**, 498–519.
Trebst, A. (1974). *A. Rev. Pl. Physiol.* **25**, 423–58.
— (1980). *Meth. Enzymol.* **69C**, 675–715.
Underwood, C. and Gould, J. M. (1980). *Biochim. biophys. Acta* **589**, 287–98.
Velthuys, B. R. (1980). *A. Rev. Pl. Physiol.* **31**, 545–67.
— and Kok, B. (1978). in *Photosynthesis '77* (eds. D. O. Hall, J. Coombs, and T. W. Goodwin) pp. 397–405. Biochemical Society, London.
Vinkler, C., Avron, M., and Boyer, P. D. (1980). *J. biol. Chem.* **255**, 2263–6.
Vredenberg, W. J. (1976). In *The intact chloroplast, Topics in photosynthesis,* Vol. 1. (ed. J. Barber) pp. 53–88. Elsevier, Amsterdam.

West, K. R. and Wiskich, J. T. (1968). *Biochem. J.* **109**, 527–32.
Williams, R. J. P. (1979). *FEBS Lett.* **102**, 126–32.
Winget, G. D., Kanner, N., and Racker, E. (1977). *Biochim. biophys. Acta* **460**, 490–9.
Witt, H. T. (1975). In *Bioenergetics of photosynthesis,* (ed. Govindjee) pp. 498–554. Academic Press, New York.
— Schlodder, E., and Graber, P. (1976). *FEBS Lett.* **69**, 272–6.
Ziem-Hanck, U. and Heber, V. (1980). *Biochim. biophys. Acta* **591**, 266–74.

3 ENZYMES OF CARBON DIOXIDE FIXATION. THE CALVIN CYCLE

3.1 Introduction

The ATP and NADPH generated by the light reactions of photosynthesis are used by chloroplasts to drive fixation of carbon dioxide into sugars by a cycle of reactions first postulated by Melvin Calvin and hence often known as the 'Calvin cycle'. This cycle, which takes place in the stroma, is the only process of *net* photosynthetic carbon dioxide fixation in green plant tissues: in C_4 plants (see Chapter 5) the carbon dioxide is first fixed by other reactions, but is then released again within the leaf to allow the Calvin cycle to operate.

Figures 3.1 and 3.2 show the reactions of the Calvin cycle in the form

1. *Carboxylation*

   6-Ribulose 1,5-diphosphate + $6CO_2$ + $6H_2O$ → 12 3-phosphoglycerate (enzyme: *ribulose diphosphate carboxylase*)

2. *Reduction of phosphoglycerate*

   12 3-Phosphoglycerate + 12ATP → 12 1,3-diphosphoglycerate + 12ADP (*phosphoglycerate kinase*)

   12 1,3-diphosphoglycerate + 12 NADPH + $12H^+$ → 12 glyceraldehyde 3-phosphate (*NADP-glyceraldehyde 3-phosphate dehydrogenase*)

3. *Regeneration of* CO_2 *acceptor*

   5 Glyceraldehyde 3-phosphate ⇄ 5 dihydroxyacetone phosphate (*triose phosphate isomerase*)

   3 Glyceraldehyde 3-phosphate + 3 dihydroxyacetone phosphate ⇄ 3 fructose 1,6-diphosphate (*aldolase*)

   3 Fructose 1,6-diphosphate + $3H_2O$ → 3 fructose 6-phosphate + 3Pi (*fructose diphosphatase*)

   2 Fructose 6-phosphate + 2 glyceraldehyde 3-phosphate ⇄ 2 xylulose 5-phosphate + 2 erythrose 4-phosphate (*transketolase*)

   2 Erythrose 4-phosphate + 2 dihydroxyacetone phosphate ⇄ 2 sedoheptulose 1,7-diphosphate (*aldolase*)

   2 Sedoheptulose 1,7-diphosphate + $2H_2O$ → 2 sedoheptulose 7-phosphate + 2Pi (*sedoheptulose diphosphatase*)

   2 Sedoheptulose 7-phosphate + 2 glyceraldehyde 3-phosphate ⇄ 2 ribose 5-phosphate + 2 xylulose 5-phosphate (*transketolase*)

   2 Ribose 5-phosphate ⇄ 2 ribulose 5-phosphate (*ribose phosphate isomerase*)
   4 xylulose 5-phosphate ⇄ 4 ribulose 5-phosphate (*ribulose phosphate epimerase*)
   6 ribulose 5-phosphate + 6ATP → 6 ribulose 1,5 diphosphate + 6ADP (*phosphoribulokinase*)

   Net: $6CO_2$ + 18 ATP + 12NADPH + $12H^+$ + $11H_2O$ → fructose 6-phosphate + 18 ADP + $12NADP^+$ + 17Pi.

Fig. 3.1. Reactions of the Calvin cycle: the enzymes involved.

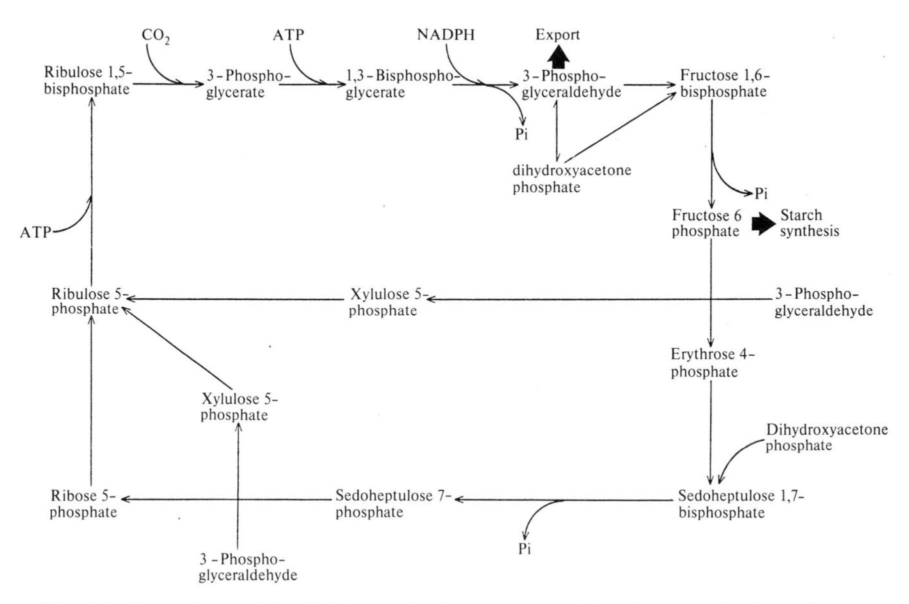

Fig. 3.2. Reactions of the Calvin cycle. Interaction with other metabolic pathways.

conventionally presented in textbooks. Carbon dioxide is fixed by the action of the enzyme ribulose diphosphate carboxylase, which produces two molecules of phosphoglycerate that are then reduced to phosphoglyceraldehyde in reactions using ATP and NADPH. ATP is also required by phosphoribulokinase, so that two molecules of NADPH and three of ATP are needed to fix one carbon dioxide molecule into a hexose sugar, a stoichiometry which has implications when considering the amount of ATP synthesized during photophosphorylation (see Chapter 2).

The carboxylation of three molecules of ribulose diphosphate produces six molecules of phosphoglycerate, containing 18 carbon atoms overall. Only 15 of these need to be used to regenerate the ribulose diphosphate. The other three, corresponding to one phosphoglycerate molecule, can also be fed into the later reactions of the cycle, so resulting in an increase in the total carbon pool. However, C_3 compounds are also involved in regulating the synthesis of starch within the chloroplast (see Chapter 4), and they can be exported from the chloroplast through the porter systems in the envelope (see Chapter 6). Export occurs in exchange for the inorganic phosphate that will be needed to maintain the rate of photophosphorylation if organic phosphate compounds are being exported. Hence the phosphoglycerate/dihydroxyacetone phosphate/phosphoglyceraldehyde region of the Calvin cycle represents an important metabolic control-point between export, hexose sugar formation, and starch synthesis, and one would expect careful regulation of the enzymes involved (Fig. 3.2).

Before discussing the overall regulation of the Calvin cycle (in the next chapter) the properties of the individual enzymes will be reviewed.

3.2 Ribulose diphosphate carboxylase

Ribulose diphosphate carboxylase, discovered in 1947 by Wildman and Bonner and initially named 'fraction I protein', catalyses the first reaction of the Calvin cycle. This one enzyme can comprise up to 50 per cent of the soluble protein content of green leaves, so it is easily obtained in pure crystalline form. The enzyme from higher plants has a molecular weight of about 550 000 and consists of eight large subunits (molecular weight 51 000–58 000) and eight small subunits (molecular weight 12 000–18 000) (Gray, Hooper, and Perham 1980). The exact arrangement of these subunits is unknown, but Baker, Eisenberg, and Eiserling (1977) have proposed a model, consistent with current microscopic and X-ray diffraction evidence, which suggests that the enzyme has a bi-layered structure, each layer containing four large and four small subunits. The amino acid sequence of the small subunit of the spinach enzyme has been reported (Martin 1979).

Since isolated large subunits retain about 20 per cent of the carboxylase activity of the whole enzyme, whereas isolated small subunits show no activity, it is believed that the active sites of the enzyme are located on the large subunits. In agreement with this, antibodies directed against the large subunits inhibit the carboxylase activity of the whole enzyme, whereas antibodies against the small subunits do not. Active-site-directed irreversible inhibitors of the carboxylase modify amino acid residues on the large subunits (Akazawa 1979). Mg^{2+} is required for the action of the enzyme and carbon dioxide, not HCO_3^-, is the true substrate. However, chloroplasts contain a high activity of carbonic anhydrase so that there will be a rapid equilibration between HCO_3^- and dissolved carbon dioxide (Poincelot 1972; Jacobson, Fong, and Heath 1975). Carbonic anhydrase contains zinc at its active site, and lack of active enzyme may account for some of the symptoms of zinc deficiency in plants. Purified ribulose diphosphate carboxylase contains variable amounts of Cu^{2+} (values of 0.11 up to 1.0 g-atom per mole of enzyme have been reported), but this metal is not involved in the catalytic activity of the enzyme and may become spuriously bound to it during the isolation procedure. Ryan and Tolbert (1975*a*) reported that the purified carboxylase showed superoxide dismutase activity (see Chapter 8) on polyacrylamide gels, but this may well have been related to the presence of bound copper, since Cu^{2+} ions and their complexes react rapidly with the superoxide radical (Johal, Bourque, Smith, Won Suh, and Eisenberg 1980).

3.2.1 Synthesis and assembly of the carboxylase in higher plants

Chloroplasts contain all the components necessary to carry out protein synthesis. These comprise 10–30 multiple copies of a double-helical circular DNA molecule, a DNA polymerase for its replication, a DNA-dependent RNA polymerase for transcription, ribosomes to translate the mRNA so produced, and tRNA molecules together with the aminoacyl-tRNA synthetase enzymes needed for their activation. The ribosomes of the chloroplast are very different from those of the leaf cell cytoplasm, both in their sedimentation coefficient (70S as opposed to 80S) and in their sensitivity to antibiotics (chloroplast protein synthesis is inhibited by D-*threo* chloramphenicol, spectinomycin, lincomycin, and erythromycin but not by cycloheximide, whereas protein synthesis by cytoplasmic (80S) ribosomes is inhibited by cycloheximide only). In the above respects, chloroplast ribosomes are more similar to bacterial ribosomes than to eukaryotic (80S) ribosomes. However, the structures of the RNA and proteins found in chloroplast ribosomes are very different from those of bacterial ribosomes. Initiation of protein synthesis in chloroplasts requires a formyl-methionine tRNA molecule, again a feature characteristic of bacterial protein synthesis (Ellis, Highfield, and Silverthorne 1978). Chloroplast ribosomes can be found attached to the outside of the thylakoids as well as free in the stroma. The genes for chloroplast rRNA (5S, 4.5S, 16S, and 23S species) and tRNA are located in the chloroplast DNA. The rRNA genes are transcribed as a single precursor that is later cleaved. By contrast, the chloroplast RNA polymerase is encoded in the nuclear DNA (Burkard *et al.* 1980).

The products of protein synthesis by isolated chloroplasts consist of two major components and about ninety minor components. One major component is the large subunit of ribulose diphosphate carboxylase (Link and Bogorad 1980) and the second comprises components of the coupling factor involved in photophosphorylation (Chapter 2). Othere products of chloroplast protein synthesis may include cytochrome f (Doherty and Gray 1979) and cytochrome b_{559} (Zielinski and Price 1980).

In contrast, the small subunits of ribulose diphosphate carboxylase are synthesized in the cytoplasm of the leaf cell, as are the other Calvin cycle enzymes. The small subunit appears to be synthesized as a higher-molecular-weight precursor. It is taken up into the chloroplast, a process which requires energy (Grossman, Bartlett, and Chua 1980), and the excess amino acids are trimmed off, perhaps by a protease located in the stroma (Highfield and Ellis 1978; Smith and Ellis 1979). This processing is not prevented by inhibitors of chloroplast protein synthesis.

3.2.2 Oxygenase activity of the enzyme

In 1971, Ogren and Bowes observed that oxygen inhibits carbon dioxide fixation by ribulose diphosphate carboxylase. Later work showed that this occurs because oxygen itself is a substrate for the enzyme, ribulose diphosphate being cleaved to phosphoglycollate and phosphoglycerate (Fig. 3.3). The oxygen-dependent reaction is often referred to as the 'oxygenase' activity of the enzyme, and the enzyme itself is now frequently described as ribulose diphosphate carboxylase/

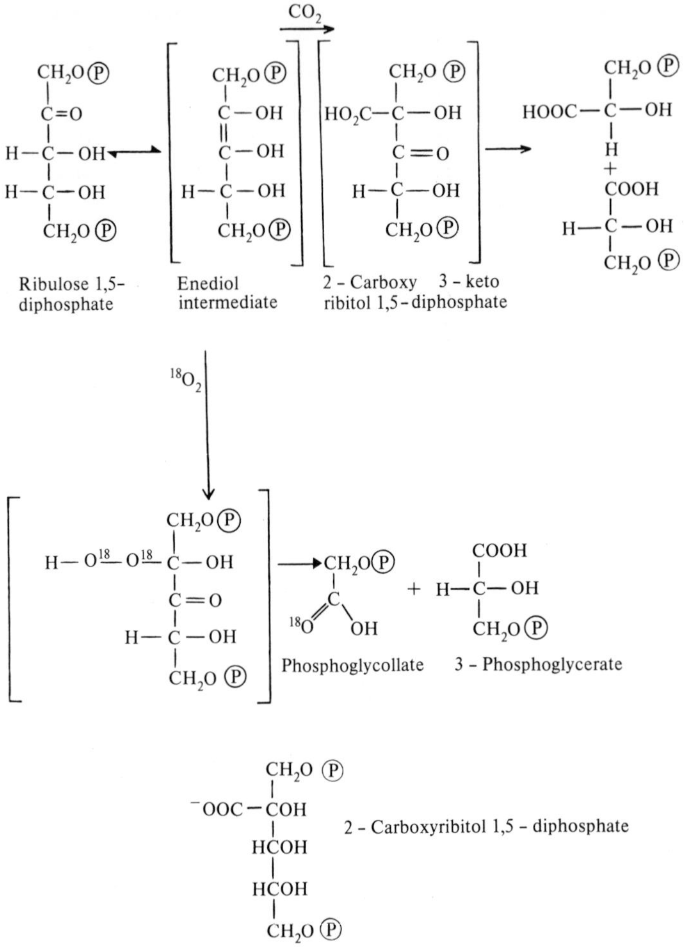

CO_2

Ribulose 1,5-diphosphate

Enediol intermediate

2 - Carboxy 3 - keto ribitol 1,5-diphosphate

$^{18}O_2$

Phosphoglycollate

3 - Phosphoglycerate

2 - Carboxyribitol 1,5 - diphosphate

Fig. 3.3. Proposed mechanism for carboxylase and oxygenase activities of ribulose diphosphate carboxylase.

oxygenase. It is now completely accepted, despite a few recent claims to the contrary, that carboxylase and oxygenase activities are an integral part of the same enzyme. The major points of evidence supporting this assertion are listed below (Andrews and Lorimer 1978).

(a) Oxygenase activity has been observed in all ribulose diphosphate carboxylases studied, including those from anaerobic bacteria such as *Chromatium* and *Rhodospirillum*, which should rarely be exposed to oxygen in their normal environments. Hence oxygenase activity seems to be an essential feature of the enzyme.

(b) Both oxygenase and carboxylase activities are located on the large subunit of the enzyme and they are both inhibited by the compound 2-carboxyribitol 1,5-diphosphate at a ratio of 1 mol of inhibitor per large subunit (Ryan and Tolbert 1975*a*, *b*). Carbon dioxide and oxygen compete with each other in a classically competitive manner.

(c) Provided that care is taken to use proper assay techniques (see below), carboxylase and oxygenase activities co-purify and cannot be separated by any chromatographic procedures.

3.2.3 *Assay of the enzyme*

When ribulose diphosphate carboxylase was first isolated and studied, its Michaelis constant (K_m) for carbon dioxide was found to be much higher than that of isolated chloroplasts or of whole leaves for carbon dioxide fixation (equivalent to a 1.8 per cent carbon dioxide concentration in the gas phase in equilibrium with the reaction mixture as opposed to 0.03 per cent – see Walker 1973) and the V_{max} was too low to account for observed rates of carbon dioxide fixation. The oxygenase activity was low and had the highly-alkaline pH-optimum of 9–9.5. However, if type A chloroplasts were isolated, disrupted, and assayed at once for the carboxylase/oxygenase, it was found that the enzyme has an affinity for carbon dioxide much closer to that of the intact chloroplast, higher oxygenase activity, and pH optima of 8.0–8.8 for both carboxylase and oxygenase activities (Bahr and Jensen 1974*a*, *b*; Andrews, Badger, and Lorimer 1975; Walker 1976). The affinity of the enzyme for oxygen is always much smaller than that for carbon dioxide, e.g. in the experiments of Andrews *et al.* (1975) their carboxylase preparation had a K_m for carbon dioxide of about 20 μM, which is close to the carbon dioxide concentration of 10 μM present in water in equilibrium with normal air (0.03 per cent carbon dioxide), but the K_m for oxygen was 0.22 mM.

The existence of a 'high-affinity' or 'low K_m' form of carboxylase in chloroplast extracts is not some special feature of the organization of the enzyme within the chloroplast, but instead is probably related to the environment to which the enzyme has been exposed *in vivo*. In the absence of carbon dioxide and Mg^{2+} the isolated carboxylase/oxygenase has no activity. When the enzyme is mixed with ribulose diphosphate and the assay then started by adding Mg^{2+} and carbon dioxide there is a marked lag period before carboxylation begins. In contrast, preincubation of the enzyme with Mg^{2+} and carbon dioxide gives a much faster initial rate when ribulose diphosphate is subsequently added to start the reaction. Such activation effects were first noted in 1963 by Pon, Rabin, and Calvin but it is only recently that all workers have begun to take sufficient notice of them. Activation is also required for correct assay of the oxygenase activity.

Activation, like the carboxylation reaction itself, requires carbon dioxide rather than HCO_3^- and may take several minutes for completion, depending on the concentration of Mg^{2+} and carbon dioxide. It can be reversed if the Mg^{2+}

and carbon dioxide are removed by gel filtration. The inactive carboxylase/oxygenase appears to react slowly with carbon dioxide to give a complex, which then reacts rapidly with Mg^{2+} to give the active species, i.e.

$$\text{E (inactive)} + CO_2 \overset{\text{slow}}{\rightleftharpoons} \text{E} \cdot CO_2 \text{ (inactive)} \tag{3.1}$$

$$\text{E} \cdot CO_2 + Mg^{2+} \overset{\text{fast}}{\rightleftharpoons} \text{E} \cdot CO_2 \cdot Mg^{2+} \text{ (active)} \tag{3.2}$$

The activator carbon dioxide molecule might be bound at a site different from the active site of the enzyme (Miziorko 1979; Lorimer 1979) although this is not certain. The rate of activation is increased by raising the pH of the incubation medium and decreased by the presence of ribulose diphosphate. Indeed, it has been proposed that the Mg^{2+} requirement of the enzyme is explained solely by its role in activation, i.e. Mg^{2+} is not needed for the actual catalytic process (Laing and Christeller 1976; Purohit and McFadden 1979), although there is considerable debate on this point (Miziorko and Sealy 1980).

Since carbon dioxide is an activator as well as a substrate of the enzyme, the greatest care must be taken during kinetic studies to distinguish the kinetics of activation from those of catalysis. Jensen and Bahr (1977) recommend that the enzyme should be fully activated by pre-incubation for 5–10 min at saturating Mg^{2+} (20 mM $MgCl_2$) and carbon dioxide (10 mM $NaHCO_3$) at pH 8.0–8.6. The assay should be started by adding ribulose diphosphate and allowed to run for no longer than 2 min, in case the enzyme should begin to deactivate in the assay mixture. Addition of carbonic anhydrase to the reaction mixture to ensure rapid conversion of HCO_3^- to CO_2 is also recommended (Bird, Cornelius, and Keys 1980; Okabe, Lindlar, Tsuzuki, and Miyachi 1980). The necessity for activation creates a particular problem in assays of the oxygenase activity of the enzyme. Preincubation with Mg^{2+} and carbon dioxide is necessary to obtain maximum activity, yet carbon dioxide is a powerful competitive inhibitor of the oxygenase. Hence the preincubation must be carried out using a very concentrated enzyme solution, a small volume of which is then transferred into the assay mixture to start the reaction, so that the amount of HCO_3^- carried over with it is minimized. Thiol compounds such as dithiothreitol or glutathione have sometimes been included in oxygenase assays, but it must be borne in mind that these compounds autoxidize (Chapter 8) and the resulting oxygen uptake can lead to confusion. The fully activated carboxylase, like the enzyme of illuminated chloroplasts, has a K_m for carbon dioxide of 10–20 μM, a V_{max} sufficient to support observed rates of carbon dioxide fixation and similar pH-profiles for both carboxylase and oxygenase activities.

What actually happens to the enzyme during the activation process is not yet clear. Irreversible inactivation of carboxylase/oxygenase by the affinity label *N*-bromoacetylethanolamine phosphate is correlated with modification of one lysine residue per large subunit in the presence of Mg^{2+}, or with modification of two different cysteine residues in the absence of Mg^{2+}. One of these cysteine residues is only three residues distant from the lysine in the primary structure of the protein. These results are consistent with a conformational change induced by Mg^{2+}. Work with other affinity labels has also shown that lysine, histidine, and arginine residues are present at or near the active site of the enzyme (Schloss, Norton, Stringer, and Hartman 1978*a*; Schloss, Stringer, and Hartman 1978*a, b*; Stringer and Hartman 1978; Saluja and McFadden 1980). The CO_2 bound during activation forms a carbamate with the side-chain amino group of lysine (Lorimer and Miziorko 1980), i.e. lys—NH—COO^-.

The small subunits of the carboxylase/oxygenase enzyme may play some role in activation by carbon dioxide and Mg^{2+} ions. Activation of the carboxylation reaction is decreased if the spinach enzyme is treated with antiserum directed against the small subunits and the extent of activation in isolated large subunits is smaller than that in the enzyme as a whole (Kobayashi, Takabe, Nishimura, and Akazawa 1979). Carboxylase/oxygenase from the photosynthetic bacterium *Rhodospirillum rubrum* has no small subunits; it consists merely of a dimer of large subunits. Activation by carbon dioxide and Mg^{2+} ions can be demonstrated, but the K_m of the activated enzyme for carbon dioxide is higher than that of the spinach enzyme (Whitman, Martin, and Tabita 1979; Kobayashi *et al.* 1979). The bacterium *Rhodopseudomonas sphaeroides* contains two types of carboxylase; form I resembles the higher plant enzyme, whereas form II contains large subunits only. Form II is activated more slowly by Mg^{2+} and carbon dioxide than is form I (Gibson and Tabita 1979). Interestingly, the activation of the form II enzyme or of the *R. rubrum* enzyme is not inhibited by ribulose diphosphate, whereas activation of the form I enzyme is, resembling the enzyme from higher plants.

3.2.4 Mechanism of the carboxylase and oxygenase reactions

Figure 3.3 shows a mechanism for the carboxylation reaction which was first proposed by Calvin in 1954, but is still generally accepted. The proposed intermediate 2-carboxy-3-ketoribitol 1,5-diphosphate has been synthesized and is cleaved by the carboxylase. A synthetic compound of similar structure, 2-carboxyribitol 1,5-diphosphate (Fig. 3.3), is a powerful competitive inhibitor of the carboxylase, which provides strong evidence for the proposed mechanism (Siegel and Lane 1973; Pierce, Tolbert, and Barker 1980*a*). Isotopic studies have confirmed that the carbon-oxygen bonds at C2 and C3 of the substrate are retained during the carboxylation process (Lorimer 1978; Pierce *et al*, 1980*b*). Affinity-labelling techniques have shown that the ribulose diphosphate binding sites of the enzyme contain lysine and arginine residues, which has led to suggestions that such residues might form Schiff bases with the >C=O group of the ribulose diphosphate during catalysis. This is not compatible with carbon-oxygen bond retention, however.

3.2.5 Effect of metabolites on carboxylase/oxygenase activity

There have been many studies of the effects of Calvin cycle intermediates and other metabolites on the carboxylase/oxygenase enzyme. In such studies, it is essential to distinguish effects on activation from effects on the carboxylase/oxygenase reaction itself. For example, if 6-phosphogluconate is present during the activation process, it greatly increases the rate of activation, but if present during the assay it inhibits the activated enzyme competitively with ribulose diphosphate (Chu and Bassham 1973). Fructose 6-phosphate was once thought to stimulate carboxylase activity, but it has no effect on the fully-activated enzyme. Several other effectors have been reported to stimulate or inhibit carboxylase/oxygenase activity, but the physiological significance of many of these effects is debateable. Often the effects are very small at pH 8, the pH of the stroma in the illuminated chloroplast (Chollet and Anderson 1976), and Jensen and Bahr (1977) have pointed out that the concentration of carboxylase/oxygenase in the chloroplast stroma may be equal to or greater than that of most of the metabolites suggested to regulate it. An enzyme concentration of 2 mM in the stroma has been calculated (Wildman 1979). A recently proposed candidate for a regulatory metabolite is sedoheptulose diphosphate, which is a powerful competitive inhibitor of activated carboxylase/oxygenase (Saluja and McFadden 1978). The physiological significance of this remains to be evaluated.

No Calvin cycle intermediate has yet been rigorously proven to have a differential effect on the carboxylase and oxygenase activities of the enzyme, although replacement of Mg^{2+} by other divalent metal ions, such as Mn^{2+} or Co^{2+}, has been reported to produce different ratios of carboxylase/oxygenase activity (Christeller and Laing 1979; Robison, Martin, and Tabita 1979). Claims that glycidic acid or hydroxylamine inhibit oxygenase but not carboxylase activity have not been substantiated (P. V. Sane, personal communication).

Nevertheless, the possibility of differential regulation is suggested by the discovery of mutant strains of tobacco and of the alga *Chlamydomonas* that contain enzymes which appear to have an altered ratio of carboxylase to oxygenase activity (Kung and Marsho 1976; Nelson and Surzycki 1976). Unfortunately it is not clear from either report whether correct activation procedures were followed before assay of the enzyme activities. More recently Garrett (1978), using the proper procedures, discovered that the carboxylase/oxygenase enzyme in tetraploid strains of ryegrass has a lower K_m for carbon dioxide than the enzyme from diploid strains, although the V_{max} values are identical. He claims that this might be related to the more rapid growth rate of the tetraploid strains. Both the diploid and the tetraploid enzymes have the same molecular weight, antigenic properties, and oxygenase activity (Rathnam and Chollet 1980). These results have been the subject of fierce debate, however, and cannot yet be regarded as firmly established.

3.2.6 Light activation of the enzyme

In 1968, Jensen and Bassham observed that carbon dioxide fixation by illuminated cells of the green alga *Chlorella* ceases almost immediately when the light is turned off, even though significant amounts of carbon dioxide and ribulose diphosphate are still present within the chloroplast. They suggested that ribulose diphosphate carboxylase/oxygenase activity is somehow 'switched off' in the dark. A reduction in the actual amount of enzyme activity detected during assay of isolated spinach chloroplasts placed in the dark, and an increase on re-illumination have been demonstrated by Bahr and Jensen (1978) and by Heldt *et al.* (1978), although Robinson, McNeil, and Walker (1979) could demonstrate no such light-dependent increase in the enzyme activity in wheat protoplasts.

The 'switching off' of the carboxylase activity in the dark has also often been attributed to the regulatory effect of metabolites such as 6-phosphogluconate, NADPH, or fructose diphosphate, but we have already seen that these proposals are not very convincing. However, if the carboxylase/oxygenase *in vivo* in the dark is in an incompletely activated state, the rise in pH of the stroma on illumination from 7.0 to 8.0 (see Chapter 2) and the efflux of Mg^{2+} from the thylakoids, which might increase the stromal Mg^{2+} concentration by 1-3 mM, could increase the degree of activation and so produce more active enzyme. The extent to which this happens might well depend on the previous history of the material studied, e.g. length of dark exposure or, if isolated chloroplasts are used, the exact isolation procedure followed. One can perhaps understand, therefore, the differences in the observed extents of the 'light activation'.

3.3 Phosphoglycerate kinase

Phosphoglycerate kinase catalyses the phosphorylation of 3-phosphoglycerate to 1,3-diphosphoglycerate (Fig. 3.1). Unlike most kinases, it catalyses a freely reversible reaction, so the rate of conversion of phosphoglycerate to diphosphoglycerate is very sensitive to the ATP/ADP ratio within the chloroplasts. This may provide a means for regulation of the enzyme *in vivo* (Robinson and Walker 1979). The enzyme, as expected for a kinase, requires Mg^{2+} for its action. Its activity has not been found to be directly influenced by light in chloroplasts from any plant species, although changes in light intensity can, of course, influence the enzyme indirectly via changes in the ATP/ADP ratio.

3.4 Glyceraldehyde 3-phosphate dehydrogenase

There are three different glyceraldehyde 3-phosphate dehydrogenase enzymes in leaf tissues, viz. a cytoplasmic NAD-specific enzyme involved in glycolysis, a cytoplasmic NADPH-linked enzyme which catalyses an irreversible oxidation of glyceraldehyde 3-phosphate into phosphoglycerate (Kelly and Gibbs 1973) and is involved in export of reducing power from the chloroplast (Chapter 6), and a chloroplast enzyme, which functions in the Calvin cycle (Fig. 3.1). It uses NADPH generated by the photosynthetic light reactions to convert 1,3-diphosphoglycerate into 3-phosphoglyceraldehyde. Although the isolated enzyme is capable of using NADH as an electron donor, this does not seem to be of physiological significance (Smillie and Fuller 1960; Preiss and Kosuge 1970).

The purified chloroplast enzyme exists in various stages of aggregation (Cerff 1978; 1979*a*, *b*; Ferri, Comerio, Iadarola, Zapponi, and Speranza 1978; Pupillo and Faggiani 1979). Most studies have been carried out on the spinach enzyme although the enzymes from other plants are essentially similar in structure. The spinach enzyme exists in both oligomeric (molecular weight approximately 600 000) and monomeric (molecular weight 100 000–145 000) forms, although higher states of aggregation have also been observed (Pupillo and Guiliani-Piccari 1975; Wolosiuk and Buchanan 1976). The 600 000 molecular weight oligomer has mainly NAD(H)-specificity *in vitro*, but the monomers show a slight preference for NADP(H). Dissociation of the enzyme, which leads to an increase in the NADP(H)-linked activity, is promoted by incubating the enzyme with $NADP^+$, NADPH, inorganic phosphate or ATP. The synthetic dithiol compound dithiothreitol enhances the dissociating effect of NADP(H). In contrast, incubation of the enzyme with NAD^+ or with glyceraldehyde 3-phosphate promotes association of the monomers and so favours the NAD(H)-linked reaction. It has been reported that the isolated enzyme contains bound NAD^+ and $NADP^+$ (Ferri *et al.* 1978).

Although the enzyme in the chloroplast catalyses a reduction of 1,3-diphosphoglycerate by NADPH, many kinetic studies on the purified enzyme have been carried out using assays in the reverse direction, measuring the oxidation of glyceraldehyde 3-phosphate. This compound has a highly-reactive carbonyl group which can readily form complexes with thiols, such as the dithiothreitol often used in activation experiments, and with amines, such as the Tris buffer usually used in the enzyme assay (Kelly, Latzko, and Gibbs 1976*a*; McArmstrong and Trentham 1976). The kinetic complexities introduced by such reactions may well be responsible for the great variation in the reported affinities of the various forms of the enzyme for NAD(H) or NADP(H). In general, however, the conclusions presented above have also been confirmed by studies in which the enzyme was assayed in the physiological direction.

The state of aggregation of glyceraldehyde 3-phosphate dehydrogenase within the chloroplast is not at all clear. However, the NADPH-dependent reduction of diphosphoglycerate by leaf homogenates or by isolated chloroplasts is usually increased on incubation of the preparation with ATP, NADPH, and dithiothreitol, which does suggest that at least some oligomeric form is present *in vivo* (Muller, Ziegler, and Ziegler 1969; Bradbeer 1976; Schwarz, Maretzka, and Schonheim 1976).

Illumination of isolated chloroplasts or of whole leaves also increases the amount of NADPH-dependent glyceradehyde 3-phosphate dehydrogenase activity detected on subsequent assay and activity decreases again in the dark. However, the rate of this 'light activation' is usually slow (often taking 10–15 min for completion), it rarely exceeds a two- or threefold increase in activity and sometimes it cannot be demonstrated (Leegood and Walker 1980). Light activation of the enzyme in isolated chloroplasts, when it occurs, is prevented by CMU or by DCMU and so must require electron transport (Champigny and Bismuth 1976). Anderson and Avron (1976) found that light activation of the enzyme in isolated pea (*Pisum sativum*) chloroplasts could be demonstrated even if the envelopes were disrupted. A membrane-fraction centrifuged down from the disrupted chloroplasts could also mediate light-activation of the enzyme in a stromal extract. Incubation of the membrane fraction with arsenite prevents light activation of the enzyme in such experiments, but only if the treatment is carried out in the light, which suggests that arsenite may react with adjacent thiol groups on the membrane that appear only on illumination. Consistent with this, Andreo and Vallejos (1976) have reported the appearance of new thiol groups upon illumination of chloroplast membranes. Activation of the NADPH-linked glyceraldehyde 3-phosphate dehydrogenase activity of chloroplasts by dithiothreitol *in vitro* was suggested to be mimicking the action of light-generated membrane-bound dithiol groups *in vivo*. Figure 3.4 shows a mechanism originally suggested by Anderson's group to account for light activation. It is assumed throughout that the thiol groups of the 'membrane fraction' are associated with the thylakoid membranes. Although this seems likely, it has not been proved. Other soluble thiol-containing molecules ('protein modulases') might be involved in mediating enzyme activation by the membrane fraction (L. E. Anderson, personal communication), i.e. the light generated thylakoid dithiol groups first reduce the protein modulase, which then reacts with the enzyme, rather than a direct action as shown in Fig. 3.4.

Buchanan's group have studied isolated spinach chloroplasts and have produced a different mechanism for light activation of chloroplast glyceraldehyde 3-phosphate dehydrogenase, which involves the protein thioredoxin. Thioredoxins are small thermostable proteins containing a single disulphide bridge that have been shown to take part in several redox reactions in animal, plant, and bacterial systems. During such reactions, the disulphide bridge undergoes a reversible reduction to two cysteine residues. At least two different thioredoxins (types f and m) are present in spinach chloroplasts (Wolosiuk, Crawford, Yee, and Buchanan 1979), which also contain a ferredoxin-dependent thioredoxin reductase enzyme (De La Torre, Lara, Wolosiuk, and Buchanan 1979) that catalyses the reaction shown in eqn (3.3).

$$\underset{\text{S}\!-\!\!-\!\text{S}}{\text{Thioredoxin}} + \underset{\text{(reduced)}}{\text{Ferredoxin}} \longrightarrow \underset{\text{SH}\quad\text{SH}}{\text{Thioredoxin}} + \underset{\text{(oxidized)}}{\text{Ferredoxin}} \tag{3.3}$$

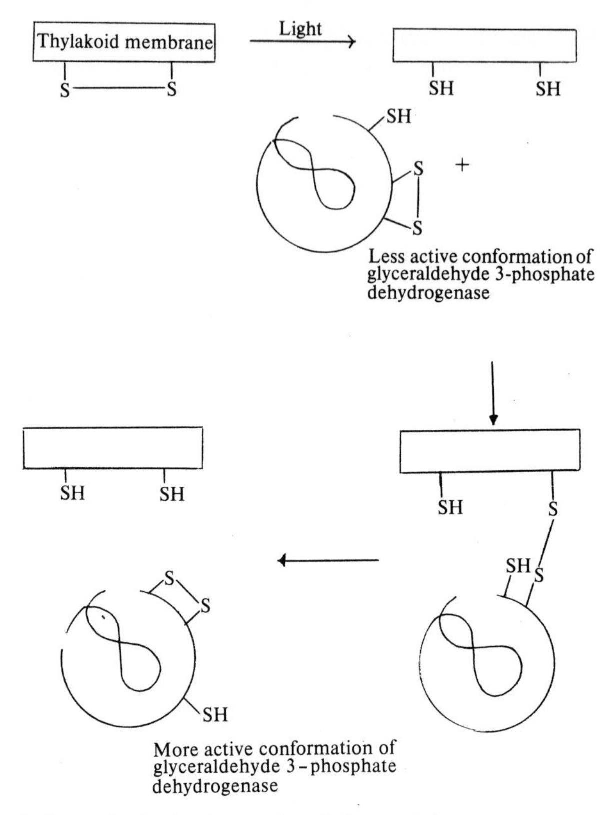

Fig. 3.4. A hypothetical scheme for light-modulation of chloroplast glyceraldehyde 3-phosphate dehydrogenase activity. Modified from Anderson, Nehrlich, and Champigny (1978).

Two forms of thioredoxin m (m_b and m_c) may exist (Schurmann (1981). Reduced thioredoxin f is believed to activate the glyceraldehyde 3-phosphate dehydrogenase by a disulphide exchange process that perhaps resembles the mechanism shown in Fig. 3.4. Hence in Buchanan's scheme illumination of the chloroplast generates reduced ferredoxin, which reduces thioredoxin f and so activates glyceraldehyde 3-phosphate dehydrogenase. The action of dithiothreitol *in vitro* may be due to its ability to reduce the oxidized forms of thioredoxins directly (Wolosiuk and Buchanan 1978*a*; Holmgren 1979).

Thioredoxin f in spinach chloroplasts seems to be a soluble stromal protein. Its relationship to the work of Anderson using pea chloroplasts is not as yet clear. Perhaps thioredoxin and its reductase are tightly membrane-bound in pea chloroplasts. 'Protein modulase' is apparently not identical with thioredoxin.

How the 'thiol-activated' glyceraldehyde 3-phosphate dehydrogenase loses its extra activity again in the dark has not been established. Wolosiuk and

Buchanan (1977) suggested that oxidized glutathione (GSSG) is involved in the deactivation process, but this is unlikely to be the case since little oxidized glutathione is present in chloroplasts either in the light or in the dark (Chapter 8). Chloroplast glutathione remains almost entirely in the reduced form (GSH) under both light and dark conditions, and GSH does not affect glyceraldehyde 3-phosphate dehydrogenase (Halliwell and Foyer 1978). Perhaps oxidized thioredoxin decreases the activity of the enzyme in the dark by a reversal of the disulphide exchange mechanism (Ashton, Brennan, and Anderson 1980).

Light activation of glyceraldehyde 3-phosphate dehydrogenase in isolated barley chloroplasts seems to be inhibited by the presence of inorganic phosphate (Huber 1978). Whether or not this is a general phenomenon, and the reason for it, remain to be discovered. Huber's results may be contrasted with the observation that inorganic phosphate *activates* the NADPH-dependent activity of the purified enzyme by promoting the dissociation reaction (see above).

3.5 Fructose diphosphatase and sedoheptulose diphosphatase

3.5.1 Fructose diphosphatase

The reactions catalysed by aldolase, transketolase, triose phosphate isomerase, and the other isomerases of the Calvin cycle seem to be freely reversible and not subject to regulation by other than mass action effects. The same aldolase enzyme is involved in the formation of both fructose and sedoheptulose diphosphates (Fig. 3.1). The next likely candidate for a control point would be fructose diphosphatase. Spinach leaves contain both a cytoplasmic and a chloroplast fructose diphosphatase, both of which require Mg^{2+} for activity (Latzko, Zimmermann, and Fuller 1974). Various non-specific phosphatases capable of hydrolysing fructose 1,6-diphosphate are also present. However, both the cytoplasmic fructose diphosphatase and the non-specific phosphatases are active at pH 7.5 in the presence of 10 mM Mg^{2+}, whereas the chloroplast enzyme is not, having an optimum of pH 8–9 (unless it has been preincubated with dithiothreitol – see below). The cytoplasmic fructose diphosphatase, like the animal enzymes, is completely inhibited by 2 mM AMP, whereas the chloroplast enzyme is not subject to adenylate control (Zimmermann, Kelly, and Latzko 1978). Hence activities can be clearly distinguished. It is obvious that work on the fructose diphosphatase activity of isolated chloroplast fractions should be carried out on preparations washed to decrease cytoplamic contamination, and the possibility of such contamination must be checked by test assays at pH 7.5.

Purified spinach-chloroplast fructose diphosphatase has a molecular weight of about 160 000. At pH 8.8, in the absence of Mg^{2+}, it dissociates into subunits of molecular weight 80 000 (Zimmermann, Kelly, and Latzko 1976), although this seems unlikely to occur *in vivo* because the pH in illuminated chloroplasts *in vivo* is not so alkaline (see Chapter 2). The high molecular weight form is probably a tetramer of subunits with identical molecular weight and the smaller form is a dimer. Prolonged incubation of the high molecular weight form of the enzyme with dithiothreitol slowly changes its pH profile, causing activity to appear at pH 7.5 and eventually to equal the activity at higher pH values. Incubation with dithiothreitol increases the number of thiol (—SH) groups on the surface of the enzyme from four to twelve (Zimmermann *et al.* 1976). Some earlier workers (e.g. El-Badry 1974; Preiss, Biggs, and Greenberg 1967) included dithiothreitol in the homogenizing media used to prepare spinach-chloroplast fructose diphosphatase, and so the properties of the purified enzyme that they described were those of the thiol-treated form, since the treated enzyme is stable in that form even on prolonged storage at 2 °C (Zimmermann *et al.* 1976).

Spinach chloroplast fructose diphosphatase purified in the absence of thiol compounds has a specific activity of 100–200 μmol substrate hydrolysed min^{-1} mg $protein^{-1}$ at pH 8. Its K_m for Mg^{2+} is 9 mM and that for fructose diphosphate is 0.8 mM at this pH. The enzyme is strongly inhibited by inorganic phosphate, 10 mM phosphate producing over 50 per cent inhibition. Estimates of the phosphate concentration in the chloroplast stroma vary from 10 to 138 mM (Hall 1976) although more recent evidence suggests that values of 5–10 mM in the light are more representative (Lilley, Chon, Mosbach, and Heldt 1977; Heldt, Chon, and Lorimer 1978; Kaiser and Bassham 1979). The stromal concentration of fructose diphosphate in illuminated chloroplasts is in the range of 0.1–0.4 mM (Heldt *et al.* 1978; Heldt, Portis, Lilley, Mosbach, and Chon 1980), insufficient to saturate the enzyme. These observations suggest that this form of fructose diphosphatase would be virtually completely inactive *in vivo*. Incubation of the enzyme with dithiothreitol does not change its V_{max}, but it decreases the K_m for fructose diphosphate to 33 μM and the K_m for Mg^{2+} to 2 mM and it substantially alleviates inhibition by inorganic phosphate. Hence this 'thiol-treated' form of the enzyme would be more suited to a role *in vivo* (Charles and Halliwell 1980). Schurmann and Wolosiuk (1978) found that fructose diphosphatase, assayed at 1 mM Mg^{2+} and 6 mM fructose diphosphate, shows an increase in activity after incubation with reduced thioredoxin. At higher Mg^{2+} concentrations the activation is much less marked, which suggests that thioredoxin and/or membrane bound dithiol groups *in vivo* may affect the enzyme in the same way as dithiothreitol *in vitro*. The action of thioredoxin on the isolated enzyme is a slow process, taking 15–20 min for completion.

The fructose diphosphatase activity of isolated chloroplasts, like that of glyceraldehyde 3-phosphate dehydrogenase, has been reported to increase on illumination and decreases again in the dark. These changes are usually slow, however, taking minutes for completion. The extent of the reported light activation varies widely, from a two- to threefold rise in activity to over sixfold. For example, it takes 10 min of illumination to achieve a 50 per cent increase in the fructose diphosphatase activity of pea chloroplasts (Anderson, Chin, and

Gupta 1979) and 5 min illumination to produce a sixfold activation of this enzyme in type A wheat chloroplasts (Leegood and Walker 1980). Light-activation of the enzyme in chloroplasts is decreased by the compound disalicylidene-propanediamine, which interferes with the generation of reduced ferredoxin (Anderson *et al.* 1979). This does suggest a role for thioredoxin and ferredoxin-dependent thioredoxin reductase in the light activation process, although the specificity of disalicylidenepropanediamine as an inhibitor has been questioned (Ireland and Goldwin 1979; Laasch, Kaiser, and Urbach 1979).

All the above activation experiments used assay conditions that were insufficient to saturate the enzyme either with substrate or with Mg^{2+}, so the increase in activity detected can probably be attributed to changes in the enzyme similar to those induced by incubating the purified fructose diphosphatase with dithiothreitol, viz. converting it into a 'low K_m' form that exhibits maximal velocity under physiological conditions. Indeed, when the enzyme is assayed in spinach chloroplasts at very high substrate and Mg^{2+} concentrations, activity is equally high under both light and dark conditions (Charles and Halliwell 1981), no 'activation' being demonstrable.

A chloroplast protein other than thioredoxin ('ferralterin') has been claimed to regulate spinach fructose diphosphatase (Lara, De la Torre, and Buchanan 1980) but its exact function *in vivo* has not yet been established. Ferralterin is an iron-sulphur protein.

3.5.2 Sedoheptulose diphosphatase

Sedoheptulose diphosphatase activity is also increased to variable extents, often only twofold, when whole leaves or intact chloroplasts are illuminated (Anderson 1974; Champigny and Bismuth 1976; Schurmann, Wolosiuk, Breazeale, and Buchanan 1976; Gupta and Anderson 1978). Light activation of the enzyme in pea chloroplasts is inhibited by the ferredoxin antagonist disalicylidene-propanediamine (Anderson and Avron 1976), which suggests that reduced ferredoxin might play a role in the activation process, presumably via the thioredoxin system. Light activation was also prevented by incubation of Anderson's membrane fraction from pea chloroplasts (Section 3.4) with arsenite, suggesting that membrane-bound dithiol groups are involved. Breazeale, Buchanan, and Wolosiuk (1978) have isolated a sedoheptulose diphosphatase from spinach chloroplasts, which requires incubation with reduced thioredoxin f before significant activity develops. Its optimum pH is 8-9, with little activity below pH 7.5. Mg^{2+} is an essential co-factor of the activated enzyme, but even high Mg^{2+} concentrations are unable to activate the enzyme in the absence of thioredoxin, in contrast with spinach chloroplast fructose diphosphatase activity (see Section 3.5.1). Detailed kinetic studies of the thioredoxin-treated form of the enzyme have not yet been performed.

Antibodies raised against spinach-chloroplast fructose diphosphatase do not inhibit the sedoheptulose diphosphatase, which suggests that the two are different enzymes. However, the sedoheptulose diphosphatase preparation of Breazeale *et al.* (1978) did have a small amount of phosphatase activity on several other sugar phosphates, including fructose diphosphate. Perhaps this was due to contamination by non-specific phosphatases.

Purified spinach chloroplast fructose diphosphatase does not hydrolyse sedoheptulose diphosphate provided that the enzyme remains in the oligomeric form, but the dissociated enzyme obtained by treatment at pH 8.8 (see Section 3.5.2) has been reported to show sedoheptulose diphosphatase activity in the presence of Mg^{2+} and reduced thioredoxin (Buchanan, Schurmann, and Wolosiuk 1976). The physiological significance of the pH 8.8 form is debateable, however, since the stroma does not become so alkaline *in vivo*.

3.6 Ribulose 5-phosphate kinase

The amount of phosphoribulokinase activity detected in extracts of leaves or in chloroplasts is increased by their illumination before assay, a phenomenon not seen with phosphoglycerate kinase. However, the activation rarely exceeds two- or threefold (Latzko and Gibbs 1969; Gupta and Anderson 1978; Leegood and Walker 1980). Dithiothreitol also activates phosphoribulokinase in chloroplasts or in leaf extracts.

Light activation of phosphoribulokinase in isolated pea chloroplasts is prevented by DCMU and by arsenite, so Anderson and Avron (1976) have suggested that membrane-bound dithiol groups regulate phosphoribulokinase as well as some of the other Calvin cycle enzymes. The spinach chloroplast enzyme can be slowly activated by incubation with reduced thioredoxin f (Wolosiuk and Buchanan 1978*b*).

A kinase enzyme would be expected to be subject to adenylate control. Indeed, the spinach enzyme is inhibited by ADP (Lavergne, Bismuth, and Champigny 1974; Walker 1976). In contrast, Anderson (1973) reported that the pea chloroplast enzyme was unaffected by ADP or by AMP and its K_m for ATP was so low that it should be saturated even at the ATP concentrations present in chloroplasts in the dark. She suggested that the enzyme was regulated by 6-phosphogluconate, which is formed in chloroplasts in the dark (see Section 3.6.3) and is an inhibitor of the pea chloroplast kinase.

Phosphoribulokinases require 5 mM Mg^{2+} for activity and have pH optima close to 8.0. Hence the light-induced changes in pH and Mg^{2+} concentration in the stroma (Chapter 4) may provide a simple means of regulating the activity *in vivo*.

3.7 Other chloroplast enzymes regulated by light and thiol compounds

Several chloroplast enzymes not involved in the Calvin cycle have been reported to be light-activated in chloroplast preparations. Since the mechanism of activation appears to be very similar to that used by Calvin cycle enzymes, it is convenient to discuss these other enzymes here.

3.7.1 NADP(H)-linked malate dehydrogenase

The mitochondria, peroxisomes, and cytosol of green leaf tissues contain isoenzymes of malate dehydrogenase that are specific for NAD(H). Their activities are unaffected by light or by thiol compounds. In contrast, chloroplasts contain an NADP(H)-specific enzyme which catalyses the reaction

$$\text{oxaloacetate} + \text{NADPH} + \text{H}^+ \rightleftarrows \text{L-malate} + \text{NADP}^+ \qquad (3.4)$$

Like all malate dehydrogenases, its equilibrium position greatly favours reduction of oxaloacetate. It is believed to be involved in the transport of reducing power out of the chloroplast into the cytoplasm (see Chapter 6).

The NADP(H)-linked malate dehydrogenase activity of leaves or of isolated chloroplasts is increased on illumination and decreases again in the dark (e.g. Scheibe and Beck 1979). Light-activation may be mimicked by incubating leaf extracts with a dithiol compound such as dithiothreitol or reduced lipoic acid. Light activation of the enzyme in pea chloroplasts was prevented by DCMU or by arsenite in the system of Anderson and Avron (1976). Buchanan's group found that spinach chloroplast malate dehydrogenase was activated by incubation with reduced thioredoxin, although the thioredoxins required (types m_b and m_c) were found to be different from that which mediates activation of Calvin cycle enzymes (type f) (Wolosiuk, Buchanan, and Crawford 1977; Wolosiuk *et al.* 1979; Jacquot, Vidal, Gadal, and Schurmann 1978; Schurmann 1981).

3.7.2 Phenylalanine-ammonia lyase

Phenylalanine-ammonia lyase is the first enzyme in the biosynthetic pathway leading to lignin and to flavonoids. Although some activity is usually found in isolated chloroplast fractions, the true subcellular location of the enzyme is not entirely clear from the data available at present (see Chapter 9). Illumination of leaves increases their lyase activity, through mechanisms that involve both increased enzyme synthesis and activation of pre-existing inactive enzyme (Camm and Towers 1973). Nishizawa, Wolosiuk, and Buchanan (1979) have shown that the activity of phenylalanine-ammonia lyase from isolated spinach chloroplast preparations is significantly increased upon incubation with reduced thioredoxin. The relevance of this observation to regulation of the enzyme *in vivo* is not yet known, but it would certainly be interesting to examine the possible regulation of the lignin biosynthetic pathway by the thioredoxin system.

3.7.3 Enzymes of the 'oxidative' pentose phosphate pathway

The Calvin cycle has often been called the 'reductive pentose phosphate pathway' because of the similarity of many of its enzymes to those of the 'oxidative pentose phosphate pathway' used by animal tissues for the provision of NADPH and pentose sugars. The first two stages in the 'oxidative' pathway are the conversion of glucose 6-phosphate into 6-phosphogluconate (eqn (3.5)), followed by oxidation of the latter to ribulose 5-phosphate (eqn (3.6)).

$$\text{glucose 6-phosphate} + NADP^{+} \underset{\text{dehydrogenase}}{\overset{\text{glucose 6-phosphate}}{\rightleftarrows}} NADPH + \text{6 phosphoglucono } \delta\text{-lactone} \tag{3.5}$$

(rearranges to 6-phosphogluconate)

$$\text{6-phosphogluconate} + NADP^{+} \underset{\text{dehydrogenase}}{\overset{\text{6-phosphogluconate}}{\rightarrow}} \text{D-ribulose 5-phosphate} + CO_2 + NADPH + H^{+} \tag{3.6}$$

Isoenzymes of glucose 6-phosphate and 6-phosphogluconate dehydrogenases are present in both chloroplasts and cytoplasm of leaf tissues (Schnarrenberger, Oeser, and Tolbert 1973) and chloroplasts should be capable of operating the oxidative pathway. However, 6-phosphogluconate normally cannot be detected in illuminated chloroplasts, leaves or algal cells but appears rapidly when the light is turned off, so the oxidative pathway must be prevented from operating in the light (Bassham 1973).

Two theories have been proposed to account for this. The first suggests that the glucose 6-phosphate dehydrogenase activity of the chloroplast is converted to an inactive form upon illumination. Although the measured activity of the enzyme in leaves or in isolated chloroplasts is indeed decreased upon illumination, the residual activity is usually at least 50 per cent of the activity in the dark. If vitamin K is added to illuminated algal cells or to spinach chloroplasts, 6-phosphogluconate appears at once (Bassham 1973). Since vitamin K does not directly affect the activity of glucose 6-phosphate dehydrogenase, this means that at least some of this enzyme must be present in a 'non-deactivated' form in the light, which is consistent with the observation above that 'light-deactivation' rarely exceeds 50 per cent. The decrease in the glucose 6-phosphate dehydrogenase activity on illumination of isolated pea chloroplasts has been suggested to involve the same membrane-bound dithiol system that mediates *activation* of enzymes such as fructose diphosphatase and glyceraldehyde 3-phosphate dehydrogenase (Anderson and Avron 1976). No doubt thioredoxin will also be implicated in due course.

The second theory involves inhibition of the enzyme by NADPH. Chloroplast glucose 6-phosphate dehydrogenase is powerfully inhibited by NADPH and by ribulose diphosphate. The $NADPH/NADP^{+}$ ratio in chloroplasts in the dark is about 0.3, but it rises to 2.0 or more on illumination, which would be sufficient to shut off the dehydrogenase completely (Wildner 1975; Lendzian and Bassham 1975). The effect of vitamin K mentioned above seems to be due to its ability to decrease chloroplast $NADPH/NADP^+$ ratios by interfering with the electron-transport chain. 6-Phosphogluconate dehydrogenase would also be inhibited by high $NADPH/NADP^+$ ratios. It is not light-inactivated, but it obviously cannot work if its substrate is not provided by the first dehydrogenase.

In darkened chloroplasts, the $NADPH/NADP^+$ ratio is lower, the glucose 6-phosphate dehydrogenase is more active and so the oxidative pentose phosphate pathway can operate, enabling the chloroplast to obtain some NADPH and pentose sugars. One reaction which presumably uses NADPH in the dark is that catalysed by the chloroplast glutathione reductase enzyme (eqn (3.7)).

$$\underset{\text{oxidized glutathione}}{\text{GSSG}} + \text{NADPH} + \text{H}^+ \rightarrow \underset{\text{reduced glutathione}}{2\text{GSH}} + \text{NADP}^+ \qquad (3.7)$$

The chloroplast stroma contains glutathione at mM concentrations (Foyer and Halliwell 1976), which remains almost entirely in the reduced form (GSH) under both light and dark conditions (Halliwell and Foyer 1978). GSH, a monothiol compound, cannot replace dithiothreitol or thioredoxin in the activation of any of the enzymes discussed in this chapter. Its precise role *in vivo* is unknown, although it may help to stabilise chloroplast enzymes and to remove hydrogen peroxide (see Chapter 8).

References

Akazawa, T. (1979) in *Photosynthesis II, Encyclopaedia of plant physiology* Vol. 6 (eds. M. Gibbs and E. Latzko) pp. 208–29. Springer, Berlin.

Anderson, L. E. (1973). *Biochim. biophys. Acta* **321**, 484–8.

— (1974). *Biochem. Biophys. Res. Commun.* **59**, 907–13.

— and Avron, M. (1976). *Pl. Physiol., Lancaster* **57**, 209–13.

— Chin, H. M., and Gupta, V. K. (1979). *Pl. Physiol., Lancaster* **64**, 491–4.

— Nehrlich, S. C., and Champigny, M. L. (1978). *Pl. Physiol., Lancaster* **61**, 601–5.

Andreo, C. S. and Vallejos, R. H. (1976). *Biochim. biophys. Acta* **423**, 590–601.

Andrews, T. J. and Lorimer, G. H. (1978). *FEBS Lett.* **90**, 1–9.

— Badger, M. R. and Lorimer, G. H. (1975). *Archs Biochem Biophys.* **171**, 93–103.

Ashton, A. R., Brennan, T., and Anderson, L. E. (1980). *Pl. Physiol., Lancaster* **66**, 605–8.

Bahr, J. T. and Jensen, R. G. (1974*a*). *Archs Biochem. Biophys.* **164**, 408–13.

— — (1974*b*). *Pl. Physiol., Lancaster* **53**, 39–44.

— — (1978). *Archs Biochem. Biophys.* **185**, 39–48.

Baker, T. S., Eisenberg, D., and Eiserling, F. (1977). *Science, N.Y.* **196**, 293–5.

Bassham, J. A. (1973) in *Rate Control of Biological Processes*, pp. 461–83 SEB Symposium No. XXVII, Cambridge University Press, England.

Bird, I. F., Cornelius, M. J., and Keys, A. J. (1980). *J. exp. Bot.* **31**, 365–9.

Bradbeer, J. W. (1976) in *Perspectives in Experimental Biology*, Vol. II (ed. N. Sunderland) pp. 131–4. Pergamon Press, Oxford.

Breazeale, V. D., Buchanan, B. B., and Wolosiuk, R. A. (1978). *Z. Naturforsch.* **33C**, 521–8.

Buchanan, B. B., Schurmann, R. and Wolosiuk, R. A. (1976). *Biochem. Biophys. Res. Commun.* **69**, 970–8.

Burkard, G., Canaday, J., Crouse, E., Guillemaut, P., Imbault, P., Keith, G., Keller, M., Mubumbila, M., Osorio, L., Sarantoglou, V., Steinmetz, A., and Weil, J. H. (1980) in *Genome organization and expression in plants* (ed. C. J. Leaver). Plenum Press, New York.

Calvin, M. (1954). *Fedn Proc. Fedn Am. Socs exp. Biol.* **13**, 697.

Camm, E. L. and Towers, G. H. N. (1973). *Phytochemistry.* **12**, 961–73.

Cerff, R. (1978). *Eur. J. Biochem.* **82**, 45–53.

— (1979*a*). *Eur. J. Biochem.* **94**, 243–7.

— (1979*b*). *J. biol. Chem.* **254**, 6094–8.

Champigny, M. L. and Bismuth, E. (1976). *Physiol. Plant.* **36**, 95–100.

Charles, S. A. and Halliwell, B. (1980). *Biochem J.* **185**, 689–93.

— — (1981) in *Proceedings of the Fifth International Congress on Photosynthesis*. In press.

Chollet, R. and Anderson, L. I. (1976). *Archs Biochem. Biophys.* **176**, 344–51.

Christeller, J. T. and Laing, W. A. (1979). *Biochem. J.* **183**, 747–50.

Chu, D. K. and Bassham, J. A. (1973). *Pl. Physiol., Lancaster* 52, 373–9.

De La Torre, A., Lara, C., Wolosiuk, R. A., and Buchanan, B. B. (1979). *FEBS Lett.* **107**, 141–5.

Doherty, A. and Gray, J. C. (1979). *Eur. J. Biochem.* **98**, 87–92.

El-Badry, A. M. (1974). *Biochim. biophys. Acta* **333**, 366–77.

Ellis, R. J., Highfield, P. E., and Silverthorne, J. (1979) in *Photosynthesis '77* (eds. D. O. Hall, J. Coombs, and T. W. Goodwin) pp. 497–506. Biochemical Society, London.

Ferri, G., Comerio, G., Iadarola, P., Zapponi, C. M., and Speranza, M. L. (1978). *Biochim. biophys. Acta* **522**, 19–31.

Foyer, C. H. and Halliwell, B. (1976). *Planta* **133**, 21–25.

Garrett, M. K. (1978). *Nature, Lond.* **274**, 913–5.

Gibson, J. L. and Tabita, F. R. (1979). *J. Bact.* **140**, 1023–7.

Gray, J. C. Hooper, E. A., and Perham, R. N. (1980). *FEBS Lett.* **114**, 237–9.

Grossman, A., Bartlett, S. and Chua, N. H. (1980). *Nature, Lond.* **285**, 625–8.

Gupta, V. K. and Anderson, L. E. (1978). *Pl. Physiol., Lancaster* **61**, 469–71.

Hall, D. O. (1976) in *The intact chloroplast, Topics in photosynthesis* Vol. 1 (ed. J. Barber) pp. 135–70. Elsevier, Amsterdam.

Halliwell, B. and Foyer, C. H. (1978). *Planta* **139**, 9–17.

Heldt, H. W., Chon, C. J., and Lorimer, G. H. (1978). *FEBS Lett.* **92**, 234–40.

— Portis, A. R., Lilley, R. McC, Mosbach, A., and Chon, C. J. (1980). *Analyt. Biochem.* **101**, 278–87.

Highfield, P. E. and Ellis, R. J. (1978). *Nature, Lond.* **271**, 420–4.

Holmgren, A. (1979). *J. biol. Chem.* **254**, 9627–32.

Huber, S. C. (1978). *FEBS Lett.* **92**, 12–16.

Ireland, C. R. and Goldwin, G. K. (1979). *Pl. Physiol., Lancaster* **63**, 1210–11.

Jacobson, B. S., Fong, F., and Heath, R. L. (1975). *Pl. Physiol., Lancaster* **55**, 468–74.

Jacquot, J. P., Vidal, J., Gadal, P., and Schurmann, P. (1979). *FEBS Lett.* **96**, 243–6.

Jensen, R. G. and Bahr, J. T. (1977). *A. Rev. Pl. Physiol.* **28**, 379–400.

— and Bassham, J. A. (1968). *Biochim. biophys. Acta* **153**, 227–34.

Johal, S., Bourque, D. P., Smith, W. W., Won Suh, S., and Eisenberg, D. (1980) *J. biol. Chem.* **255**, 8873–80.
Kaiser, W. M. and Bassham, J. A. (1979). *Pl. Physiol., Lancaster* **63**, 105–108.
Kelly, G. J. and Gibbs, M. (1973). *Pl. Physiol., Lancaster* **52**, 111–18.
— Latzko, E., and Gibbs, M. (1976*a*). *A. Rev. Pl. Physiol.* **27**, 181–205.
— Zimmerman, G., and Latzko, E. (1976*b*). *Biochem. Biophys. Res. Commun.* **70**, 193–9.
Kobayashi, H., Takabe, T., Nishimura, M., and Akazawa, T. (1979). *J. Biochem., Tokyo* **85**, 923–30.
Kung, S. D. and Marsho, T. V. (1976). *Nature, Lond.* **259**, 325–6.
Laasch, N., Kaiser, W., and Urbach, W. (1979). *Pl. Physiol., Lancaster* **63**, 605–8.
Laing, W. A. and Christeller, J. T. (1976). *Biochem. J.* **159**, 563–70.
Lara, C., De La Torre, A., and Buchanan, B. B. (1980). *Biochem. Biophys. Res. Commun.* **93**, 544–51.
Latzko, E. and Gibbs, M. (1969). *Prog. Photosynth. Res.* **3**, 1624–30.
— Zimmerman, G. and Fuller, U. (1974). *Hoppe-Seyler's Z. Physiol Chem.* **355**, 321–6.
Lavergne, D., Bismuth, E., and Champigny, M. L. (1974). *Plant Sci. Lett.* **3**, 391–7.
Leegood, R. C. and Walker, D. A. (1980). *Archs Biochem. Biophys.* **200**, 575–82.
Lendzian, K. and Bassham, J. A. (1975). *Biochim. biophys. Acta* **396**, 260–75.
Lilley, R. McC., Chon, C. J., Mosbach, A., and Heldt, H. W. (1977). *Biochim biophys. Acta* **460**, 259–72.
Link, G. and Bogorad, L. (1980). *Proc. natn. Acad. Sci., U.S.A.* **77**, 1832–6.
Lorimer, G. H. (1978). *Eur. J. Biochem.* **89**, 43–50.
— (1979). *J. biol. Chem.* **254**, 5599–601.
— and Miziorko, H. M. (1980). *Biochemistry* **19**, 5321–8.
Martin, P. G. (1979). *Aust. J. Pl. Physiol.* **6**, 401–8.
McArmstrong, J. and Trentham, D. R. (1976). *Biochem J.* **159**, 513–27.
Miziorko, H. M. (1979). *J. biol. Chem.* **254**, 270–2.
— and Sealy, R. C. (1980). *Biochemistry.* **19**, 1167–71.
Muller, B., Ziegler, I., and Ziegler, H. (1969). *Eur. J. Biochem.* **9**, 101–6.
Nelson, P. E. and Surzycki, S. J. (1976). *Eur. J. Biochem.* **61**, 475–80.
Nishizawa, A. N., Wolosiuk, R. A., and Buchanan, B. B. (1979). *Planta* **145**, 7–12.
Ogren, W. L. and Bowes, G. (1971). *Nature, Lond.* **230**, 159–60.
Okabe, K., Lindlar, A., Tsuzuki, M., and Miyachi, S. (1980). *FEBS Lett.* **114**, 142–4.
Pierce, J., Tolbert, N. E., and Barker, R. (1980*a*). *Biochemistry.* **19**, 934–42.
— — — (1980*b*). *J. biol. Chem.* **255**, 509–11.
Poincelot, R. P. (1972). *Biochim. biophys. Acta* **258**, 637–42.
Pon, N. G., Rabin, B. R., and Calvin, M. (1963). *Biochem. Z.* **338**, 7–19.
Preiss, J. and Kosuge, T. (1970). *A. Rev. Pl. Physiol.* **21**, 433–66.
— Biggs, M. L., and Greenberg, E. (1967). *J. biol. Chem.* **242**, 2292–4.
Pupillo, P. and Giuliani-Piccari, G. (1975). *Eur. J. Biochem.* **51**, 475–82.
— and Faggiani, R. (1979). *Archs Biochem. Biophys.* **194**, 581–92.
Purohit, K. and McFadden, B. A. (1979). *Archs Biochem. Biophys.* **194**, 101–6.
Rathnam, C. K. M. and Chollet, R. (1980). *Pl. Physiol., Lancaster* **65**, 489–94.
Robinson, S. P. and Walker, D. A. (1979). *Biochim. biophys. Acta* **545**, 528–36.
— McNeil, P. H. and Walker, D. A. (1979). *FEBS Lett.* **97**, 296–300.

Robison, P. D., Martin, M. N., and Tabita, F. R. (1979). *Biochemistry, Easton* **18**, 4453–8.
Ryan, F. J. and Tolbert, N. E. (1975*a*). *J. biol. Chem.* **250**, 4229–33.
Ryan, F. J. and Tolbert, N. E. (1975*b*). *J. biol. Chem.* **250**, 4234–8.
Saluja, A. K. and McFadden, B. A. (1978). *FEBS Lett.* **96**, 361–3.
— — (1980). *Biochem. biophys. Res. Commun.* **94**, 1091–7.
Scheibe, R. and Beck, E. (1979). *Pl. Physiol., Lancaster* **64**, 744–8.
Schloss, J. V., Norton, I. L., Stringer, C. D., and Hartman, F. C. (1978*a*). *Biochemistry, Easton* **17**, 5626–31.
— Stringer, C. D. and Hartman, F. C. (1978*b*). *J. biol. Chem.* **253**, 5707–11.
Schnarrenberger, C., Oeser, A. and Tolbert, N. E. (1973). *Archs Biochem. Biophys.* **154**, 438–48.
Schurmann, P. (1981) in *Proceedings of the Fifth International Congress on Photosynthesis.* In press.
— and Wolosiuk, R. A. (1978). *Biochim. biophys. Acta* **522**, 130–8.
— — Breazeale, V. D., and Buchanan, B. B. (1976). *Nature, Lond.* **263**, 257–8.
Schwarz, Z., Maretzka, D., and Schonheim, J. (1976). *Biochem. Physiol. Pflanzen.* **170**, 537–50.
Siegel, M. J. and Lane, M. D. (1973). *J. biol. Chem.* **248**, 5486–98.
Smillie, R. M. and Fuller, R. C. (1960). *Biochem. Biophys. Res. Commun.* **3**, 368–72.
Smith, S. M. and Ellis, R. J. (1979). *Nature, Lond.* **278**, 662–4.
Stringer, C. D. and Hartman, F. C. (1978). *Biochem. Biophys. Res. Commun.* **80**, 1043–48.
Walker, D. A. (1973). *New Phytol.* 72, 209–35.
— (1976) in *The intact chloroplast, Topics in photosynthesis* Vol. 1 (ed. J. Barber) pp. 235–278. Elsevier, Amsterdam.
Whitman, W. B., Martin, M. N., and Tabita, F. R. (1979). *J. biol. Chem.* **254**, 10184–9.
Wildman, S. G. (1979). *Archs Biochem. Biophys.* **196**, 598–610.
— and Bonner, J. (1947). *Archs Biochem. Biophys.* **14**, 381–413.
Wildner, G. F. (1975). *Z. Naturforsch.* **30C**, 756–60.
Wolosiuk, R. A. and Buchanan, B. B. (1976). *J. biol. Chem.* **251**, 6456–61.
— — (1977). *Nature, Lond.* **266**, 565–7.
— — (1978*a*). *Pl. Physiol., Lancaster* **61**, 669–71.
— — (1978*b*). *Archs Biochem. Biophys.* **189**, 97–101.
— — and Crawford, N. A. (1977). *FEBS Lett.* **81**, 253–8.
— Crawford, N. A., Yee, B. C., and Buchanan, B. B. (1979). *J. biol. Chem.* **254**, 1627–32.
Zielinski, R. E. and Price, C. A. (1980). *J. Cell. Biol.* **85**, 435–45.
Zimmermann, G., Kelly, G. J., and Latzko, E. (1976). *Eur. J. Biochem.* **70**, 361–7.
— — — (1978). *J. biol. Chem.* **253**, 5952–6.

4 REGULATION OF THE CALVIN CYCLE. SYNTHESIS OF STARCH AND SUCROSE

4.1 Introduction

In the preceding chapter, the information presently available on the properties of the enzymes of the Calvin cycle was reviewed. In the present chapter, an attempt will be made to evaluate the significance of such information to the means by which the cycle is regulated *in vivo*, in so far as this is possible from current knowledge.

Regulation of the Calvin cycle has been studied by many techniques. A favourite technique is to examine the rate of synthesis of cycle intermediates in whole leaves by isotopic methods (e.g. Bassham 1979). Isolated type A chloroplasts may also be used, but it is essential to ensure that the preparations are capable of high rates of carbon dioxide fixation, to avoid spurious results. For example, antimycin A stimulates the rate of carbon dioxide fixation only in preparations in which the original rate of fixation is sub-normal (Chapter 2). A recent development is the use of 'reconstituted' chloroplast systems, in which carbon dioxide fixation can be demonstrated using washed thylakoids (type D chloroplasts) mixed with substrates and purified enzymes or stromal extracts (Stokes and Walker 1971; Bassham, Levine, and Forger 1974; Walker 1976). The requirements of such systems for carbon dioxide fixation *in vitro* should give an insight into regulatory mechanisms *in vivo*.

When whole leaves, protoplasts, or isolated type A chloroplasts are illuminated after a prolonged dark period, the rate of carbon dioxide fixation may take several minutes to reach its maximum value, even though electron transport and photophosphorylation begin immediately. Figure 4.1 shows some typical results. The length of this 'induction period' can be shortened by supplying Calvin cycle intermediates to protoplasts or chloroplasts (provided that they can penetrate – see Chapter 6), or lengthened by adding inorganic phosphate, which promotes export of C_3 compounds from the chloroplast by an exchange porter (Chapter 6). In contrast, if chloroplasts or leaves are illuminated for, say, 30 min and the light turned off briefly, the rate of carbon dioxide fixation when the light is turned on again reaches its maximum value almost at once. It seems that after a prolonged dark period the concentrations of Calvin cycle metabolites in the chloroplasts *in vivo* are below those needed to support maximal rates of carbon dioxide fixation, and that during the induction period any carbon dioxide fixed is being used to raise the concentration of cycle intermediates (Walker 1976; Leegood and Walker 1980, 1981) rather than for other processes (Chapter 3). Obviously, the mechanisms which regulate carbon dioxide fixation during the induction period may be very different

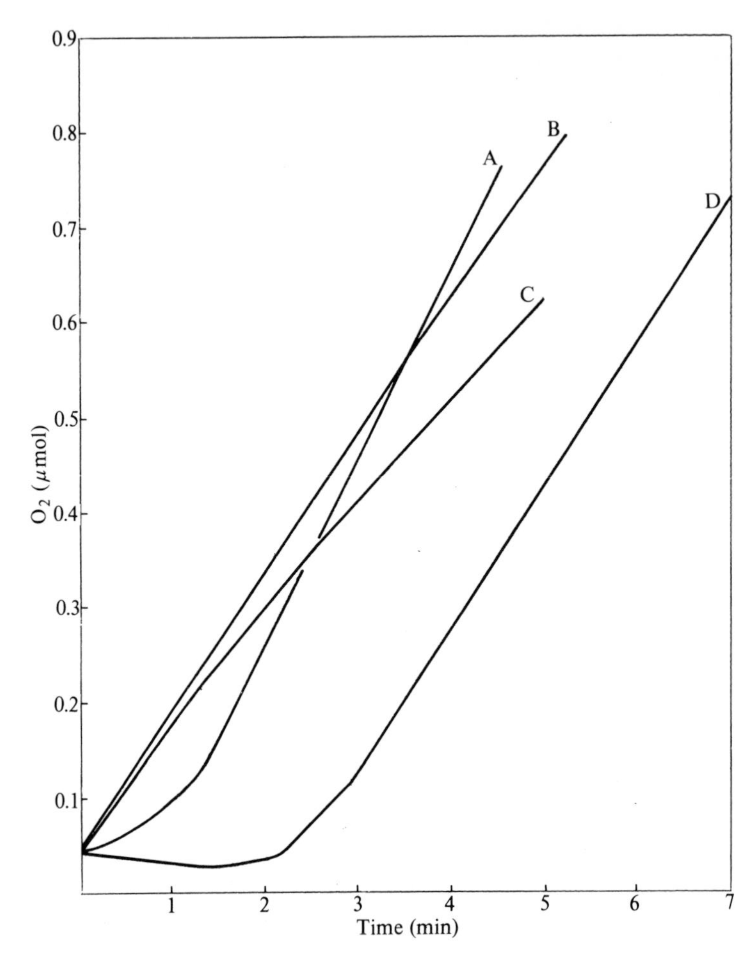

Fig. 4.1. The induction period in isolated chloroplasts. Type A spinach chloroplasts were prepared from leaves harvested after 12 h darkness and allowed to fix carbon dioxide in the light. The rate of carbon dioxide-dependent oxygen evolution is plotted against time. Curve D shows the induction period of 2–3 min. Curve A: as for D, but ribose 5-phosphate added to the preparation. Curve B: as for D, but phosphoglycerate added to the preparation. Curve C: as for D, but chloroplasts resuspended in medium in which another chloroplast sample had been illuminated for 30 min and then removed by centrifugation. The induction period is shortened by compounds released from illuminated chloroplasts in the same way as it is by Calvin cycle metabolites. Data from Walker (1976).

from those which operate when fixation is proceeding at its maximum rate.

The simplest factors which can regulate a metabolic pathway are pH, the amounts of active enzymes present, and the supply of substrates or cofactors. The substrates and cofactors of the Calvin cycle enzymes are carbon dioxide,

NADPH, ATP, and Mg^{2+}. The extent to which these factors can regulate carbon dioxide fixation *in vivo* will be examined.

4.2 Regulation by availability of carbon dioxide

The K_m values for carbon dioxide of whole leaves, intact chloroplasts, and the activated forms of isolated ribulose diphosphate carboxylase/oxygenase are approximately equal to atmospheric carbon dioxide concentrations (Chapter 3). Thus an increase in the carbon dioxide concentration around a plant should increase its rate of carbon dioxide fixation, and hence the rate of plant growth. This is a well-established phenomenon, and carbon dioxide-enrichment of the atmosphere in greenhouses has been used commercially to promote the growth of some species.

The carbonic anhydrase present in the chloroplast stroma has sometimes been suggested to 'concentrate' carbon dioxide for the carboxylase/oxygenase enzyme, but it is not at all clear how it could do this (Jacobson, Fong, and Heath 1975). However, C_4 plants (see Chapter 5) do have an efficient carbon dioxide-concentrating mechanism and are much less affected by variations in atmospheric carbon dioxide concentrations than are C_3 plants.

4.3 Regulation by change in amount of active enzymes

The initial rate of an enzyme-catalysed reaction is usually directly proportional to the concentration of enzyme present under the conditions used for kinetic studies *in vitro*. If an enzyme catalyses a rate-limiting step in a metabolic pathway, then an increase in the amount of active enzyme present should increase the flux through that pathway, and vice versa. Hence the light-'activations' and dark-'deactivations' of ribulose diphosphate carboxylase, glyceraldehyde 3-phosphate dehydrogenase, phosphoribulokinase, fructose diphosphatase, and sedoheptulose diphosphatase in leaves (Chapter 3) are potential regulatory mechanisms. Activation of all these enzymes, except ribulose diphosphate carboxylase, involves light-induced generation of membrane dithiols and/or the thioredoxin system. Probably both are involved *in vivo* and the discrepancy between the work of the groups of Anderson and of Buchanan is more apparent than real. Perhaps in pea chloroplasts the membrane-bound dithiols actually belong to a form of thioredoxin which is tightly bound to the membrane instead of being a stromal protein, as it appears to be in spinach chloroplasts. How deactivation of the enzymes in the dark is achieved remains to be established (Chapter 3).

However, the changes in enzyme activity brought about by illumination of leaves or of isolated chloroplasts take several minutes for completion and often do not exceed two- or threefold changes in amount. Substantial activity of all Calvin cycle enzymes can be detected in extracts of leaves, protoplasts, or chloroplasts in the dark. Indeed, results quoted in the preceding chapter show that at least some of the glucose 6-phosphate dehydrogenase activity of chloroplasts, supposedly inactivated by the same dithiol system in the light, is actually capable of functioning if the NADPH/NADP$^+$ ratio is lowered. It could be argued that the disruption of chloroplasts necessary to assay their enzyme content causes enzyme activity to appear by interfering with the deactivation system. However, if the Calvin cycle enzymes listed above were completely inactivated in the dark by some dithiol system, type A chloroplasts should not fix carbon dioxide in the dark under any circumstances. However, Werdan, Heldt, and Milovancev (1975) showed that they could. Fixation in the dark requires an alternative source of the ATP and NADPH usually supplied by the photosynthetic light reactions, which they achieved by adding dihydroxyacetone phosphate to the chloroplasts. This compound enters the stroma via the C_3 translocator in the envelope (see Chapter 6) and either becomes transformed into ribulose 5-phosphate or is oxidized to phosphoglycerate, yielding ATP and NADPH. The ATP can be used in the phosphoribulokinase reaction and the phosphoglycerate can leave the chloroplast in exchange for dihydroxyacetone phosphate. The NADPH formed was removed by supplying oxaloacetate to the chloroplasts to allow NADPH reoxidation by chloroplast malate dehydrogenase, also supposedly inactivated in the dark. Figure 4.2 summarizes the above

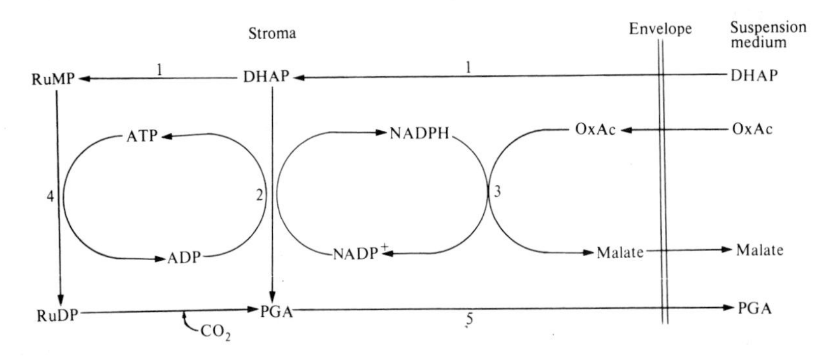

Fig. 4.2 Scheme for carbon dioxide fixation by intact chloroplasts supplied with dihydroxyacetone phosphate and oxaloacetate in the dark. Abbreviations: RuMP, ribulose monophosphate; RuDP, ribulose diphosphate; DHAP, dihydroxyacetone phosphate; PGA, phosphoglycerate; OxAc, oxaloacetate. 1. Added DHAP enters via the C_3 translocator and is converted to RuMP by aldolase, transketolase, epimerase, and fructose diphosphatase (see Fig. 3.1). 2. DHAP is isomerized to phosphoglyceraldehyde by triose phosphate isomerase and the phosphoglyceraldehyde oxidized to phosphoglycerate by NADP$^+$-glyceraldehyde 3-phosphate dehydrogenase. 3. The NADPH so produced is removed by NADP$^+$-malate dehydrogenase. Oxaloacetate and malate exchange across the envelope using the dicarboxylate translocator. 4. The ATP produced is used by phosphoribulokinase to make RuDP for the carboxylase/oxygenase enzyme. 5. PGA leaves the chloroplast in exchange for incoming DHAP.

scheme of operation, which needs ribulose diphosphate carboxylase, fructose diphosphatase, phosphoribulokinase, glyceraldehyde 3-phosphate dehydrogenase (working in the reverse direction), epimerase, and transketolase activities. Werdan *et al.* (1975) were able to demonstrate significant rates of carbon dioxide fixation, dependent on dihydroxyacetone phosphate and oxaloacetate, by chloroplasts in the dark. Hence the above enzymes, and $NADP^+$-malate dehydrogenase, *must* have been at least partially active in the intact organelles. In other words, even in isolated type A chloroplasts deactivation is not complete in the dark.

Turning off the light supplied to a preparation of type A chloroplasts fixing carbon dioxide at a steady rate causes fixation to cease almost immediately. It starts again as soon as the light is turned on, provided that the dark interval is not too long. Since the thiol-dependent activation and deactivation processes take several minutes for completion, they obviously cannot be involved in such rapid responses of the rate of carbon dioxide fixation to light intensity. There remains the possibility that the time taken for enzyme activation after a prolonged dark spell might help to account for the phenomenon of the induction period (Fig. 4.1), in addition to the mechanism involving raising the concentration of Calvin cycle intermediates (Huber 1978, 1979). However, the activity of the various Calvin cycle enzymes, even after a prolonged dark period, is great enough to achieve the rates of carbon dioxide fixation seen immediately *after* the induction period (Leegood and Walker 1980, 1981: Robinson and Walker 1980) which does indicate that light-dependent control of enzyme activity cannot be a significant regulatory mechanism.

What, then, is the function of these 'light-activation' systems? Only the reduced forms, obtained by thiol treatment, of enzymes such as fructose diphosphatase, sedoheptulose diphosphatase, and glyceraldehyde 3-phosphate dehydrogenase have the kinetic properties necessary to permit their functioning in the Calvin cycle (Chapter 3). Illuminated chloroplasts generate hydrogen peroxide (Chapters 2 and 8), which is capable of oxidizing fructose diphosphatase, and possibly the other reduced enzymes, into forms which cannot participate in the Calvin cycle (Charles and Halliwell 1980*a*). The membrane-bound dithiol groups and/or the thioredoxin system may be required in the light in order to regenerate the reduced forms of the enzymes from oxidized forms generated in this way. Thiol groups on enzymes are known to oxidize slowly in the presence of oxygen, a process catalysed by ions of transition metals such as iron that are likely to be present in the stroma (Chapter 8). In the dark, when the dithiol/thioredoxin systems do not function, the reduced Calvin cycle enzymes may become oxidized to an extent depending on the stromal composition or even on how the leaves or chloroplast preparations have been handled. This would account for the occurrence of significant 'dark activity' of most enzymes and the great variation in the extent of reported 'light activation'. In conclusion, it seems probable, according to current evidence, that the dithiol/thioredoxin systems are not themselves a regulatory mechanism, but exist to generate and protect the reduced forms of certain Calvin cycle enzymes, which are then controlled by other means. Without them, however, no carbon dioxide would be fixed at all. In agreement with this theory, the measurable fructose diphosphatase activity of illuminated type A spinach chloroplasts incubated in the presence of catalase to remove H_2O_2 was always greater than that amount necessary to account for observed rates of CO_2 fixation. If catalase was omitted from the reaction mixture, enzyme activity was lower and could sometimes limit the rate of CO_2 fixation. These enzyme assays were performed using physiological concentrations of substrate and Mg^{2+} (Charles and Halliwell 1981).

4.4 Regulation by changes in the availability of NADPH and ATP

When leaves or isolated chloroplasts are first illuminated, the electron-transport chain becomes operative. ADP is phosphorylated to ATP, and $NADP^+$ reduced to NADPH. Both the ATP/ADP and NADPH/$NADP^+$ ratios rise from less than unity to values of 2.0–3.0 (Krause and Heber 1976; Lendzian and Bassham 1975; Giersch, Heber, Kobayashi, Inoue, Shibata, and Heldt 1980). The chloroplast stroma contains an adenylate kinase enzyme, which catalyses the reaction

$$\text{ATP} + \text{AMP} \rightleftarrows 2\text{ADP} \tag{4.1}$$

and appears to be close to its equilibrium position under both light and dark conditions. The energy charge of chloroplasts, defined as the ratio

$$\frac{[\text{ATP}] + 0.5\ [\text{ADP}]}{[\text{ATP}] + [\text{ADP}] + [\text{AMP}]}$$

varies between 0.27 and 0.6 in the dark, increasing initially on illumination up to values of 0.8 (Kobayashi, Inoue, Furaya, Shibata, and Heber 1979).

Both the energy charge and the NADPH/$NADP^+$ ratios remain high during the induction period, presumably because rates of carbon dioxide fixation are low and therefore the use of ATP and NADPH is restricted. When induction is complete and carbon dioxide fixation is proceeding at a steady rate, both ratios have usually fallen from their initial light-induced values, although they are still greater than the ratios in the dark (Giersch *et al.* 1980). On the basis of experiments with isolated enzymes, the rise in NADPH/$NADP^+$ ratios has been suggested to stimulate the ribulose diphosphate carboxylase and glyceraldehyde 3-phosphate dehydrogenase enzymes, but the evidence for a physiological role of these effects is not convincing (see the preceding chapter). However, increased NADPH/$NADP^+$ ratios are probably the major mechanism by which chloroplast glucose 6-phosphate dehydrogenase is prevented from functioning in the light (Chapter 3).

The supply of ATP probably limits the rate of carbon dioxide fixation when chloroplasts are placed under sub-optimal illumination, since addition of HCO_3^- to type A chloroplast preparations under such conditions transiently decreases their internal concentration of ATP, but not that of NADPH (Heber 1973).

However, the rate of carbon dioxide fixation after the induction period in saturating light does not appear to be controlled by the ATP/ADP ratio. Indeed, high rates of carboxylation seem to be associated with lower ATP/ADP ratios. When low concentrations of NH_4Cl are added to illuminated chloroplasts, the ATP/ADP ratio falls somewhat but the rate of carbon dioxide fixation is often increased (Champigny 1978; Krause and Heber 1976).

Obviously, there must be a limit to the extent to which the ATP/ADP ratio can fall without affecting carbon dioxide fixation. If too much ADP accumulates, conversion of phosphoglycerate to 1,3-diphosphoglycerate will be inhibited since phosphoglycerate kinase catalyses a freely-reversible reaction. This has been shown clearly using a reconstituted spinach chloroplast system (Walker 1976) consisting of chloroplast stroma, thylakoids, Mg^{2+}, ADP, and $NADP^+$. On illumination, electron transport and associated oxygen release occur until all the $NADP^+$ has been reduced to NADPH. Release of oxygen can be restored by adding phosphoglycerate, which uses up ATP and NADPH by the phosphoglycerate kinase and glyceraldehyde-3-phosphate dehydrogenase reactions. Addition of ADP severely decreases phosphoglycerate-dependent oxygen evolution. Oxygen is rapidly evolved again when the ADP has been phosphorylated and the kinase is once more allowed to work in the 'forward' direction.

The K_m of spinach phosphoribulokinase for ATP (about 0.4 mM) is much lower than that of phosphoglycerate kinase (about 2 mM). Hence addition of ribulose 5-phosphate to the reconstituted chloroplast system also inhibits phosphoglycerate-dependent oxygen evolution as the phosphoribulokinase reaction 'drains away' the available ATP and makes ADP, which slows the forward reaction of the phosphoglycerate kinase (Walker 1976). Such mechanisms should help to prevent the ATP/ADP ratio from falling too far during carbon dioxide fixation in the light. They may well be the most important regulatory mechanisms *in vivo* when the light intensity is insufficient to saturate photosynthesis.

Of course, the rapid fall in ATP/ADP ratios in chloroplasts when the light is turned off will decrease the activity of both kinase enzymes in spinach chloroplasts. Pea-leaf phosphoribulokinase, which seems to be insensitive to adenylate control (see the preceding Chapter) might have its remaining dark activity after 'dark inactivation' shut off by the rise in the concentration of 6-phosphogluconate in the stroma (Anderson 1973) as glucose 6-phosphate dehydrogenase becomes active.

4.5 Regulation by changes in the pH and magnesium ion concentration in the stroma

Illumination of chloroplasts causes an uptake of protons from the stroma into the thylakoids (Chapter 2) which raises the stromal pH from about 7.0 to 8.0 (Werdan *et al.* 1975). The movement of protons is balanced by fluxes of other ions, one of these being Mg^{2+}, which leaves the thylakoids as protons enter.

This Mg^{2+} may become bound to the outside of the thylakoids or it may enter the stroma (Chapter 2). Since the envelope of the chloroplasts is impermeable to Mg^{2+}, a rise in the stromal concentration of this ion may result. It is not yet clear what concentration changes do occur: rises in the stromal Mg^{2+} concentrations in the light of between one and three mM have been reported (Portis and Heldt 1976; Krause 1977). The concentration of free Mg^{2+} in the stroma in the dark has not been conclusively established either, although some free Mg^{2+} must be present, as it is required for the action of chloroplast phosphofructokinase in the dark (Section 4.9). A value of 1-4 mM has been suggested (Ben-Hayyim and Krause 1980). A small contribution to the alkalinization of the stroma in the light is made by an exchange of protons for K^+ ions across the envelope (Chapter 2), which does not occur in the dark.

Enzymes such as fructose diphosphatase, sedoheptulose diphosphatase, ribulose diphosphate carboxylase, and phosphoribulokinase (Hurwitz *et al.* 1956) show little, if any, activity at pH 7.0 and low Mg^{2+} concentration, but are much more active at pH 8 and increased Mg^{2+} concentrations (e.g. Baier and Latzko 1975). Figure 4.3 shows some typical results for the spinach chloroplast fructose diphosphatase enzyme. Increasing the pH from 7.0 to 8.0 at, say, 3 mM Mg^{2+} increases enzyme activity from almost zero to over 100 units per mg protein. If Mg^{2+} concentrations are simultaneously increased an even greater activation can result. This would be a very simple, yet effective, means of 'switching on' the Calvin cycle upon illumination of leaves. In the dark, the stromal Mg^{2+} concentration and pH fall as Mg^{2+} ions re-enter the thylakoids in exchange for protons, which should stop the enzymes from working.

Evidence for this proposed regulation by pH and Mg^{2+} comes from the experiments of Werdan *et al.* (1975), who altered the stromal pH of isolated spinach chloroplasts by various manipulations. They found that carbon dioxide fixation could not occur at a pH of 7.2 or below, but was optimal at pH 8.1. Carbon dioxide fixation by a reconstituted chloroplast system requires not only a high Mg^{2+} concentration, but also a pH of 8.0 (Bassham *et al.* 1974).

Addition of the ionophore A23187 to type A chloroplasts allows Mg^{2+} to leave the stroma by exchange for protons across the envelope. Hence the Mg^{2+} concentration in the stroma falls and it becomes more acidic (Table 4.1), changes which inhibit carbon dioxide fixation completely and also cause a sharp rise in the concentrations of fructose and sedoheptulose diphosphates within the chloroplast, accompanied by a fall in hexose monophosphates. Such results point strongly to an inhibition of fructose and sedoheptulose diphosphatases as the primary effect of lowering pH and Mg^{2+} concentration. Restoration of Mg^{2+} to the chloroplast by including it in the incubation medium with A23187 largely reverses these changes (Table 4.1). Treatment of chloroplasts with nitrite (NO_2^-) or glyoxylate (HOOC·CHO) inhibits carbon dioxide fixation and reduces the stromal pH without affecting the Mg^{2+} concentration, and the site of inhibition is again the fructose and sedoheptulose diphosphatases (Heldt, Chon, Lilley, and

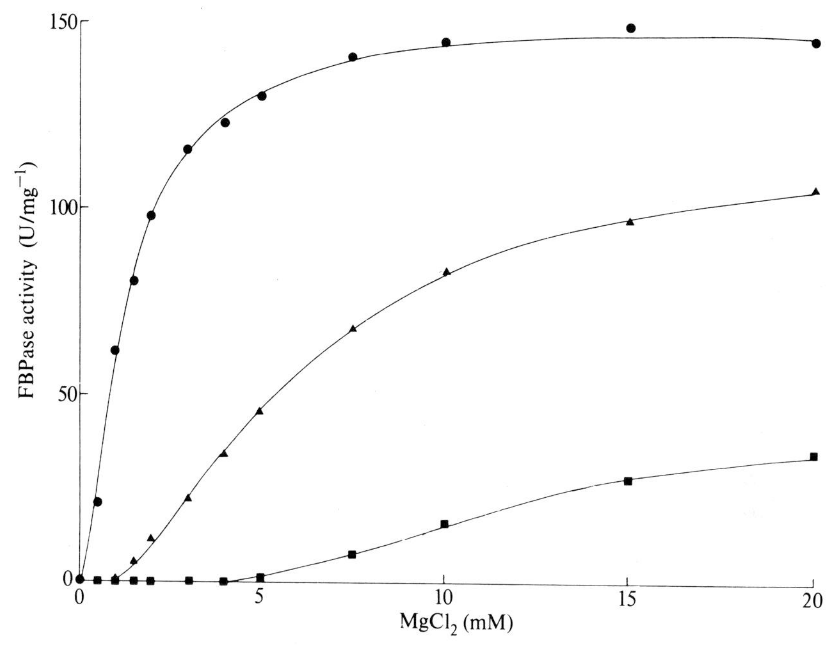

Fig. 4.3. Response of spinach chloroplast fructose diphosphatase to pH and Mg^{2+} ion concentration. The effect of Mg^{2+} ion concentration on the activity of fructose diphosphatase at three different pH values is shown. The concentration of fructose diphosphate was 0.4 mM, the stromal concentration in both light and dark (Chapter 3). Data were abstracted from Charles and Halliwell (1980*b*). ● pH 8.0; ▲ pH 7.5; ■ pH 7.0. The enzyme used had been treated with dithiothreitol to generate the 'physiological' form, which is produced *in vivo* by the action of membrane-bound dithiol groups and/or the thioredoxin system.

Portis 1978; Enser and Heber 1980).

Purified ribulose diphosphate carboxylase is activated by Mg^{2+} and carbon dioxide, and activation is speeded up at high pH values (Chapter 3). In the chloroplast fixing carbon dioxide at a steady rate, the enzyme may well be largely, if not completely, activated. Since activation and deactivation take several minutes for completion (unlike the effects of changes in Mg^{2+} concentration and pH on the fructose and sedoheptulose diphosphatases, which are virtually instantaneous) it seems unlikely that they are involved in control of carbon dioxide fixation under steady-state conditions. Table 4.1 shows that ribulose diphosphate carboxylase/oxygenase activity, as assessed by the concentration of its substrate, did not appear to be inhibited by an artificial lowering of stromal pH and Mg^{2+} concentration. However, activation of the carboxylase/oxygenase by increased stromal pH and Mg^{2+} could conceivably sometimes play a part in determining the length of the induction period if, after a prolonged period in the dark, the leaf enzyme becomes partially deactivated *in vivo*. Often,

Table 4.1. Effect of the ionophore A 23187 on carbon dioxide fixation and metabolite concentrations in isolated spinach chloroplasts.

Parameter measured	Addition to chloroplasts		
	None	2 μM A23187	2 μM A23187 + 5 mM Mg^{2+}
CO_2 fixation (μmol h^{-1} mg chl^{-1})	92	0	85
Stromal pH	7.9	7.55	7.60
Stromal metabolite concentrations (nmol mg chl^{-1})			
Ribulose diphosphate	18	18	31
Pentose monophosphates (total)	3	1	3
Fructose diphosphate	4	22	7
Sedoheptulose diphosphate	8	23	7
Hexose monophosphates (total)	58	11	39

Data from Heldt *et al.* (1978).

In chloroplasts not treated with the ionophore, a shift of stromal pH from 7.9 to 7.6 would inhibit CO_2 fixation by about 50 per cent. Hence the complete inhibition by the ionophore must at least in part be due to the loss of stromal Mg^{2+}. Addition of 5 mM Mg^{2+} together with the ionophore scarcely raised the stromal pH but restored most of the CO_2 fixation. Presumably the inhibitory effect of the lower pH was counterbalanced by an excess of Mg^{2+} (see Portis, Chon, Mosbach, and Heldt 1977; Heldt *et al.* 1978).

however, the enzyme activity in the dark is more than sufficient to account for the rate of carbon dioxide fixation after the induction period, which rules out such a suggestion (Leegood and Walker 1980, 1981).

4.6 Regulation of the Calvin cycle by other ions

The efflux of Mg^{2+} from the thylakoids is insufficient to electrically balance the uptake of protons from the stroma, so other ion movements must occur (Chapter 2), possibly movements of Cl^- and/or K^+. Calcium is also present within the chloroplast at concentrations of 15 mM (Nobel 1969) or even higher, since a concentration of 25 mM may be calculated from the data of Portis and Heldt (1976) by assuming the internal chloroplast volume to be 24 μl mg chlorophyll^{-1}. This Ca^{2+} does not appear to be involved in balancing proton movements (Chapter 2) (Hind, Nakatani, and Izawa 1974; Barber 1976) and its function is unknown, although a role in the electron transport chain has been suggested. It should be noted, however, that A23187 is a much better ionophore for Ca^{2+} than for Mg^{2+} (e.g. Weissmann, Anderson, Serhan, Samuelson, and Goodman 1980). In the presence of A23187, Ca^{2+} at concentrations of 1 mM completely inhibits carbon dioxide fixation by type A spinach chloroplasts, apparently by inhibiting fructose and sedoheptulose diphosphatases (Charles and Halliwell 1980*c*). If Ca^{2+} ions were involved in regulating the Calvin cycle they should be completely bound in the light, perhaps to the ionized carboxyl groups ($-COO^-$) present on the inner and outer surfaces of the thylakoid membranes (Chapter 2). When the light is turned off and the pH of the stroma falls, a small fraction of the groups on the outer surface will become protonated and unable to bind Ca^{2+} (Schapendonk, Hemrika-Wagner, Theuvenet, Wong Fong Sang, Vredenberg, and Kraayenhof 1980). Thus a small part of the Ca^{2+} will be released and could inhibit the Calvin cycle. Ca^{2+} ions also inhibit chloroplast protein synthesis, and could prevent this process occurring in the dark (Chapter 2).

There is no direct evidence for such a regulatory process at the moment, but it is worth investigating in view of the uncertainties concerning changes in Mg^{2+} ion concentrations. Calmodulin, a Ca^{2+}-binding protein which regulates several enzymes in animal tissues, has recently been detected in spinach leaves (Van Eldik, Grossman, Iverson, and Watterson 1980).

4.7 What regulates the Calvin cycle? A Summary

During the induction period, carbon dioxide fixation is limited by the low concentration of Calvin cycle intermediates. It might also conceivably be limited in some cases by the slow light-activation of some Calvin cycle enzymes by thiol mechanisms (e.g. if much H_2O_2 is formed – see Chapter 8) and by the slow activation of ribulose diphosphate carboxylase/oxygenase by increases in pH and stromal Mg^{2+} concentration.

Light activation and dark deactivation of enzymes is too slow and too limited in extent to account for the rapid responses of carbon dioxide fixation by chloroplasts after the induction period to changes in light intensity. Probably the actions of increased stromal pH and Mg^{2+} concentration in rapidly switching on fructose and sedoheptulose diphosphatases is the most significant regulatory mechanism here. The rate of supply of ATP and NADPH does not seem to control the rate of carbon dioxide fixation in the steady state in chloroplasts at saturating light intensity, but a fall of the ATP/ADP ratio to too low a level would shut off the phosphoglycerate kinase step, as probably happens in a light-to-dark transition. If illumination is sub-optimal, then ATP supply may control the rate of carbon dioxide fixation (Heber 1973). Finally, the overall entry of carbon into the cycle is regulated by the available carbon dioxide, which depends not only on the atmospheric carbon dioxide concentration, but also on the degree of stomatal opening of the leaf.

It is clear that there is no one means by which the Calvin cycle is regulated. The metabolic control point can shift depending on the exact conditions to which the leaf is exposed. The rate of starch synthesis and the export of metabolites to the cytoplasm, which depend on the cytoplasmic phosphate concentration, must also be balanced against the requirements of the Calvin cycle itself.

4.8 Synthesis of starch

The synthesis of the $\alpha1\rightarrow4$ linkages of starch occurs in chloroplasts by the conversion of glucose 1-phosphate into ADP-glucose (eqn (4.2)), which then reacts with a primer (eqn (4.3)). UDP-glucose does not seem to be involved in starch synthesis by chloroplasts (Preiss and Levi 1978).

$$\text{ATP} + \text{glucose 1-phosphate} \xrightarrow[\text{pyrophosphorylase}]{\text{ADP-glucose}} \text{ADP-glucose} + \text{pyrophosphate} \quad (4.2)$$

$$\underset{\text{PRIMER}}{\text{ADP-glucose} + (\text{glucopyranose } \alpha1\rightarrow4\text{-glucopyranose})_n} \xrightarrow[\text{synthetase}]{\text{starch}} \text{ADP} + (\text{glucopyranose } \alpha1\rightarrow4 \text{ glucopyranose})_{n+\frac{1}{2}} \quad (4.3)$$

Chloroplasts contain hexose phosphate isomerase (Schnarrenberger and Oeser 1974) and phosphoglucomutase (Muhlbach and Schnarrenberger 1978) activities, which convert fructose 6-phosphate into glucose 6-phosphate and then into the glucose 1-phosphate needed for reaction (4.2). Almost all of the ADP-glucose pyrophosphorylase and starch synthetase activities of leaf tissues are located within the chloroplasts (Mares, Hawker, and Possingham 1978). 'Branching enzyme' is also present to form the $\alpha1\rightarrow6$ linkages characteristic of the amylopectin component of starch (Okita, Greenberg, and Kuhn 1979). The pyrophosphate produced by the ADP-glucose pyrophosphorylase (eqn (4.2)) is hydrolysed to phosphate by the action of pyrophosphatase, a Mg^{2+}-dependent enzyme with an optimum pH of 8.2–8.6 that has been reported to be present in the chloroplast stroma (Schwenn, Lilley, and Walker 1973; Gould and Winget 1973).

The amount of ADP-glucose in chloroplasts is very small, even under conditions that greatly favour starch synthesis. The K_m of the starch synthetase enzyme for ADP-glucose is quite high, however, and so the rate of starch synthesis in leaves is probably regulated by the concentration of ADP–glucose (Heldt, Chon, Maronde, Herold, Stankovic, Walker, Kraminer, Kirk, and Heber 1977). At pH 8, the ADP-glucose pyrophosphorylase enzyme is strongly inhibited by inorganic phosphate. As the phosphate concentration in the stroma *in vivo* is about 10 mM (Chapter 3), the enzyme might be expected to be completely inhibited. However, inhibition by phosphate is reversed by adding phosphoglycerate, an accumulation of which *in vivo* would therefore 'switch on' starch synthesis by allowing formation of ADP-glucose (Kaiser and Bassham 1979*a*). The spinach chloroplast pyrophosphorylase is also inhibited by ADP, 1–2 mM concentrations giving about a 50 per cent inhibition (Ghosh and Preiss 1966). The rise in the ATP/ADP ratio on illumination of chloroplasts may also help to increase the activity of the enzyme in the light (Kaiser and Bassham, 1979*b*), although this is probably a less significant regulatory mechanism than is the effect of phosphoglycerate concentrations.

Starch synthesis by isolated illuminated chloroplasts is promoted by having a low phosphate concentration in the suspension medium (Steup, Peavey, and Gibbs 1976). Two mechanisms can account for this. Firstly, as phosphate is used up by photophosphorylation and its stromal concentration falls, its inhibitory effect on ADP–glucose pyrophosphorylase will be relieved. Secondly, loss of C_3 compounds from the chloroplasts by exchange with external phosphate through the C_3 porter (see Chapter 6) will be prevented and so the stromal concentration of phosphoglycerate should rise, again increasing pyrophosphorylase activity. Such observations are relevant *in vivo*: phosphate starvation of whole leaves causes an abnormally large accumulation of starch within them (Herold, Lewis, and Walker 1976; Heldt *et al.* 1977). The phosphate concentration in the cytoplasm of leaves can also be decreased by feeding them with mannose, which is converted in the cytoplasm to mannose 6-phosphate. Mannose phosphate cannot be further metabolized by the leaf and its formation 'drains away' inorganic phosphate. Supplying mannose to leaves causes enhanced starch synthesis (Heldt *et al.* 1977; Walker 1976). Obviously alterations in the phosphate metabolism of a plant can produce profound effects on the balance between the movement of metabolites between chloroplast and cytoplasm, starch synthesis, and the operation of the Calvin cycle.

4.9 Breakdown of starch

The starch formed in chloroplasts in the light is broken down again in the dark. The present data do not rule out the possibility that it might be undergoing continuous synthesis and breakdown in the light as well.

There are two potential pathways by which starch could be degraded (Fig. 4.4). The action of amylases and maltase could release glucose, which might be phosphorylated by hexokinase and then converted into other intermediates. Two types of amylase activity have been described in plant tissues. α-Amylase attacks $\alpha 1 \rightarrow 4$ glucosidic linkages at random points in the amylose and amylopectin molecules, eventually resulting in almost complete hydrolysis to maltose. β-Amylase also hydrolyses $\alpha 1 \rightarrow 4$ linkages, but does so from the non-reducing end of the chains, cleaving off maltose units. Both enzymes must operate in conjunction with a 'de-branching' enzyme to remove $\alpha 1 \rightarrow 6$ linkages (Street and Cockburn 1972). Alternatively, starch could be degraded to glucose 1-phosphate by the action of a phosphorylase enzyme (Fig. 4.4).

In isolated type A pea (*Pisum sativum*) shoot chloroplasts, no α-amylase, β-amylase, or maltase activities can be detected and phosphate is required for

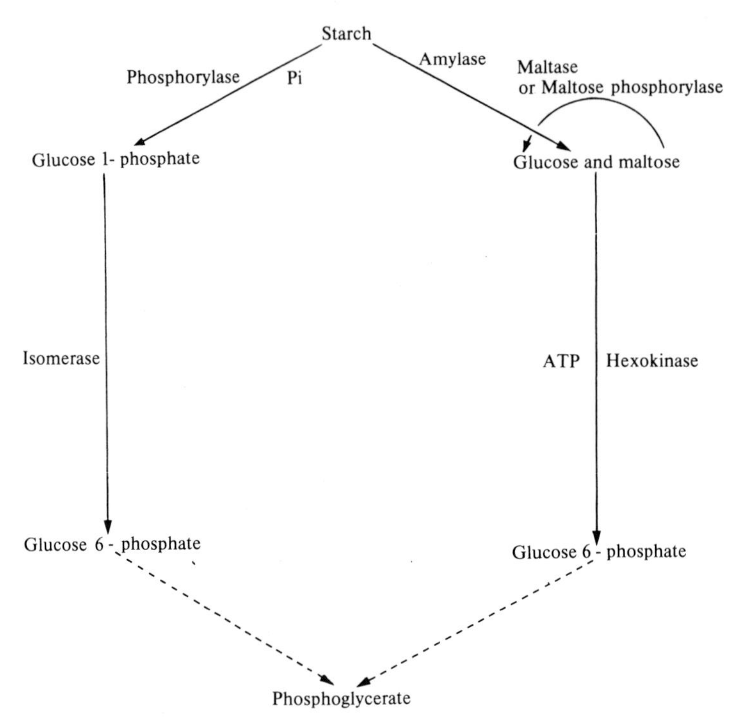

Fig. 4.4. Possible pathways of starch degradation in chloroplasts.

starch degradation to occur. Hence only a phosphorylase pathway seems to operate (Stitt, Bulpin, and Ap Rees 1978; Levi and Preiss, 1978; Stitt and Ap Rees 1980). Since plant phosphorylases have K_m values for inorganic phosphate in the millimolar range, the stromal phosphate concentration might play a part in regulating their activity *in vivo* (Steup *et al.* 1976; Steup, Schachtele, and Latzko 1980; Levi and Preiss 1978). There is no evidence for regulation of plant phosphorylases by phosphorylation/dephosphorylation cycles of the type known to occur with the animal enzymes.

The situation for spinach leaf chloroplasts is more complicated. Evidence for both the phosphorylase and the amylase routes of degradation has been reported, although the relative significance of these two pathways may depend on the conditions under which the plant is grown (Pongratz and Beck 1978). There is no reason why both pathways should not operate simultaneously, provided that the amylase pathway receives a supply of ATP for the action of hexokinase (Fig. 4.4) by the operation of later stages of the glycolytic pathway (see below). A de-branching enzyme has also been reported in spinach chloroplasts (Okita *et al.* 1979).

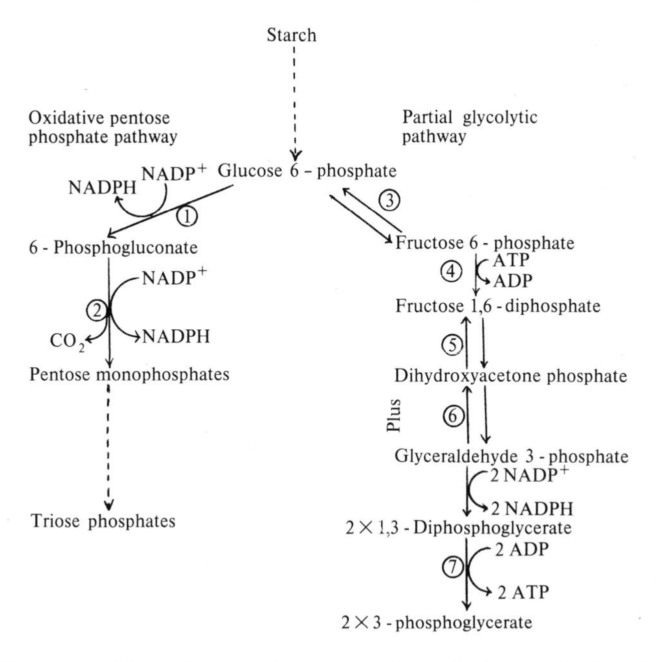

Fig. 4.5. Two possible pathways for conversion of glucose 6-phosphate into phosphoglycerate by chloroplasts in the dark. Conversion of glyceraldehyde 3-phosphate into 1,3-diphosphoglycerate might occur by reversal of the action of glyceraldehyde 3-phosphate dehydrogenase, generating NADPH. Alternatively, the enzyme might use NAD^+ and generate NADH in this reaction. Enzymes required: (1) glucose 6-phosphate dehydrogenase; (2) 6-phosphogluconate dehydrogenase; (3) hexose phosphate isomerase; (4) phosphofructokinase; (5) aldolase; (6) triose phosphate isomerase; (7) phosphoglycerate kinase.

Degradation of starch by chloroplasts in the dark eventually results in the formation of C_3 compounds, some of which can be exported from the chloroplast. Formation of phosphoglycerate from glucose 6-phosphate can be accounted for by two possible pathways, a partial glycolytic pathway and the oxidative pentose phosphate pathway (Fig. 4.5). Experiments on chloroplast extracts indicate that both pathways occur and that chloroplast 'glycolysis' probably supplies ATP to the chloroplast in the dark (Kaiser and Bassham 1979*c*; Stitt and Ap Rees 1979; 1980). Further metabolism of phosphoglycerate by the glycolytic pathway in chloroplasts cannot occur, however, as enolase, phosphoglyceromutase and/or pyruvate kinase activities are low or absent (Stitt and Ap Rees 1979). Small amounts of phosphofructokinase activity are present to convert fructose-6-phosphate to fructose 1,6-diphosphate as shown in Fig. 4.5 (Kelly and Latzko 1977).

4.10 Synthesis of sucrose

Sucrose is the main form of carbohydrate exported from the leaves of higher plants and it is rapidly labelled when $^{14}CO_2$ is supplied to illuminated leaves. The major enzymes involved in its synthesis in leaves are UDP-glucose pyrophosphorylase (eqn (4.4)), sucrose phosphate synthetase (eqn (4.5)), and sucrose phosphatase (eqn (4.6)) (Preiss and Kosuge 1970).

$$\text{glucose 1-phosphate} + \text{UTP} \rightarrow \text{UDP-glucose} + \text{pyrophosphate} \qquad (4.4)$$

$$\text{UDP-glucose} + \text{fructose 6-phosphate} \rightarrow \text{sucrose phosphate} + \text{UDP} \qquad (4.5)$$

$$\text{sucrose phosphate} + H_2O \rightarrow \text{sucrose} + \text{phosphate} \qquad (4.6)$$

However, isolated intact chloroplasts contain little, if any, UDP-glucose pyrophosphorylase or sucrose phosphate synthetase (Bird, Cornelius, Keys, and Whittingham 1974; Mares, Hawker, and Possingham 1978) and so they lack the capacity to convert carbon dioxide into sucrose. Hence synthesis of labelled sucrose from $^{14}CO_2$ *in vivo* by illuminated leaves must occur in the cytoplasm using products of carbon dioxide fixation exported from the chloroplast (Robinson and Walker 1979). This view is supported by the results of Larsson and Albertson (1974), who showed that the only isolated leaf chloroplast preparations that could make sucrose were those that had a substantial degree of cytoplasmic contamination. Sucrose phosphate synthetase is strongly inhibited by its product, UDP, and by inorganic phosphate (C. H. Foyer, personal communication).

References

Anderson, L. E. (1973). *Biochim. biophys. Acta* **321**, 484–8.
Baier, D. and Latzko, E. (1975). *Biochim. biophys. Acta* **396**, 141–7.
Barber, J. (1976) in *The intact chloroplast, Topics in photosynthesis,* Vol. 1 (ed. J. Barber) pp. 89–134. Elsevier, Amsterdam.
Bassham, J. A. (1979) in *Photosynthesis II, Encyclopaedia of plant physiology,* Vol. 6 (eds. M. Gibbs and E. Latzko) pp. 9–30 Springer, Berlin.
— Levine, G., and Forger, J. (1974). *Plant Sci. Lett.* **2**, 15–21.
Ben Hayyim, G. and Krause, G. H. (1980). *Archs Biochem. Biophys.* **202**, 546–57.
Bird, I. F., Cornelius, M. J., Keys, A. J., and Whittingham, C. P. (1974). *Phytochemistry* **13**, 59–64.
Champigny, M. L. (1978) in *Photosynthesis '77* (eds. D. O. Hall, J. Coombs, and T. W. Goodwin) pp. 479–88. Biochemical Society, London.
Charles, S. A. and Halliwell, B. (1980*a*). *Biochem. J.* **189**, 373–6.
— — (1980*b*). *Biochem. J.* **185**, 689–93.
— — (1980*c*). *Biochem. J.* **188**, 775–9.
— — (1981). *Planta* **151**, 242–6.
Enser, U. and Heber, U. (1980). *Biochim. biophys. Acta* **592**, 577–91.
Ghosh, H. P. and Preiss, J. (1966). *J. biol. Chem.* **244**, 4491–504.

Giersch, C., Heber, U., Kobayashi, Y., Inoue, Y., Shibata, K. and Heldt, H. W. (1980). *Biochim. biophys. Acta* **590**, 59–73.
Gould, J. M. and Winget, G. D. (1973). *Archs Biochem. Biophys.* **154**, 606–13.
Heber, U. (1973). *Biochim. biophys. Acta* **305**, 140–52.
Heldt, H. W., Chon, C. J., Lilley, R. McC, and Portis, A. (1978) in *Photosynthesis '77* (eds. D. O. Hall, J. Coombs, and T. W. Goodwin) pp. 469–78. Biochemical Society, London.
— — Maronde, D., Herold, A., Stankovic, Z. S., Walker, D. A., Kraminer, A., Kirk, M. R., and Heber, U. (1977). *Pl. Physiol., Lancaster* **59**, 1146–55.
Herold, A., Lewis, D. H., and Walker, D. A. (1976). *New Phytol.* **76**, 397–407.
Hind, G., Nakatani, H. Y. and Izawa, S. (1974). *Proc. natn. Acad. Sci. U.S.A.* **71**, 1484-8.
Huber, S. (1978). *FEBS Lett.* **92**, 12–14.
— (1979). *Biochim. biophys. Acta* **545**, 131–40.
Hurwitz, J., Weissbach, R., Horecker, B. L., and Smyrniotis, P. Z. (1956). *J. biol. Chem.* **218**, 769–83.
Jacobson, B. S., Fong, F., and Heath, R. L. (1975). *Pl. Physiol., Lancaster* **55**, 468–74.
Kaiser, W. M. and Bassham, J. A. (1979*a*). *Pl. Physiol., Lancaster* **63**, 105–8.
— — (1979*b*). *Pl. Physiol., Lancaster* **63**, 109–13.
— — (1979*c*). *Planta* **144**, 193-200.
Kelly, G. J. and Latzko, E. (1977). *Pl. Physiol., Lancaster* **60**, 290–4.
Kobayashi, Y., Inoue, Y., Furaya, F., Shibata, K. and Heber, U. (1979). *Planta* **147**, 69–75.
Krause, G. H. (1977). *Biochim. biophys. Acta* **460**, 500–10.
— and Heber, U. (1976) in *The intact chloroplast, Topics in photosynthesis* Vol. 1 (ed. J. Barber) pp. 171–214. Elsevier, Amsterdam.
Larsson, C. and Albertsson, P. (1974). *Biochim. biophys. Acta* **357**, 412–19.
Leegood, R. C. and Walker, D. A. (1980). *Archs Biochem. Biophys.* **200**, 575–82.
— — (1981). *Plant Cell Envir.* **4**, 59–66.
Lendzian, K. and Bassham, J. A. (1975). *Biochim. biophys. Acta* **430**, 478–89.
Levi, C. and Preiss, J. (1978). *Pl. Physiol., Lancaster* **61**, 218–20.
Mares, D. J., Hawker, J. S. and Possingham, J. V. (1978). *J. exp. Bot.* **29**, 829–35.
Muhlback, H. and Schnarrenberger, G. (1978). *Planta* **141**, 65–70.
Nobel, P. S. (1969). *Biochim. biophys. Acta* **172**, 134–43.
Okita, T. W., Greenberg, E., Kuhn, D. N., and Preiss, J. (1979). *Pl. Physiol., Lancaster* **64**, 187–92.
Pongratz, P. and Beck, E. (1978). *Plant Physiol.* **62**, 687–9.
Portis, A. R. and Heldt, H. W. (1976). *Biochim. biophys. Acta* **449**, 434–46.
— Chon, C. J., Mosbach, A., and Heldt, H. W. (1977). *Biochim. biophys. Acta* **461**, 313–25.
Preiss, J. and Kosuge, T. (1970). *A. Rev. Pl. Physiol.* **21**, 433–66.
— and Levi, C. (1978) in *Photosynthesis '77* (eds. D. O. Hall, J. Coombs, and T. W. Goodwin) pp. 457–68. Biochemical Society, London.
Robinson, S. P. and Walker, D. A. (1979). *FEBS Lett.* **107**, 295–9.
— (1980). *Archs Biochem. Biophys.* **202**, 617–23.
Schapendonk, A. H. C. M., Hemrika-Wagner, A. M., Theuvenet, A. P. R., Wong Fong Sang, H. W., Vredenberg, W. J., and Kraayenhof, R. (1980). *Biochemistry* **19**, 1922–7.
Schnarrenberger, C. and Oeser, A. (1974). *Eur. J. Biochem.* **45**, 77–82.

Schwenn, J. D., Lilley, R. McC., and Walker, D. A. (1973). *Biochim. biophys. Acta* **325**, 586–95.
Steup, M., Peavey, D. G., and Gibbs, M. (1976). *Biochem. Biophys. Res. Commun.* **72**, 1554–61.
— Schachtele, C., and Latzko, E. (1980). *Z. Pflanzenphysiol.* **96**, 365–74.
Stitt, M. and Ap Rees, T. (1979). *Phytochemistry.* **18**, 1905–11.
— — (1980). *Biochim. biophys. Acta* **627**, 131–43.
— Bulpin, P. V., and Ap Rees, T. (1978). *Biochim. biophys. Acta* **544**, 200–14.
Stokes, D. M. and Walker, D. A. (1971). *Pl. Physiol., Lancaster* **48**, 163–5.
Street, H. E. and Cockburn, W. (1972). *Plant metabolism*, 2nd edn. Pergamon Press, Oxford.
Van Eldik, L. J., Grossman, A. R., Iverson, D. B. and Watterson, D. M. (1980). *Proc. natn. Acad. Sci. U.S.A.* **77**, 1912–16.
Walker, D. A. (1976) in *The intact chloroplast, Topics in photosynthesis,* Vol. 1 (ed. J. Barber) pp. 235–78. Elsevier, Amsterdam.
Weissman, G., Anderson, P., Serhan, C., Samuelsson, E., and Goodman, E. (1980). *Proc natn Acad.* Sci. *U.S.A.* **77**, 1506–10.
Wedan, K., Heldt, H. W., and Milovancev, M. (1975). *Biochim. biophys. Acta* **396**, 276–92.

5 ENZYMES OF CARBON DIOXIDE FIXATION. THE C_4 PATHWAYS AND CRASSULACEAN ACID METABOLISM

5.1 Introduction

The Calvin cycle is the only pathway of *net* photosynthetic carbon dioxide fixation in all green plants and in some plants, such as soybean, wheat, alfalfa, lettuce, peas, and spinach, it is the only pathway by which carbon dioxide is fixed in the light. Such plants are often referred to as 'C_3 plants' since the three-carbon compound phosphoglycerate is the first product of carbon dioxide fixation. In some plants, usually referred to as 'C_4 plants', carbon dioxide entering the leaf is at first fixed into four-carbon acids (malate, aspartate, and oxaloacetate). These compounds are transported to a different part of the leaf and there decarboxylated to yield carbon dioxide, which is fixed by the operation of the Calvin cycle. Hence the C_4 pathways themselves do not achieve a *net* fixation of carbon dioxide.

C_4 plants are present in at least ten major plant families, including both mono- and dicotyledons. Well-known species include sugar-cane, crabgrass, maize, and sorghum. Some genera contain both C_3 and C_4 species, e.g. *Atriplex* (Bjorkman 1976).

The leaves of C_4 plants differ from those of C_3 plants in several ways, a major difference being in leaf anatomy. In C_4 leaves the vascular tissue is usually surrounded by a layer of 'bundle-sheath' cells containing many chloroplasts, which is in turn surrounded by layers of mesophyll cells, also rich in chloroplasts. This 'Kranz' (from the German for 'halo' or 'wreath') type of leaf anatomy is not seen in C_3 plants. Figures 5.1 and 5.2 show typical examples. Although some C_3 plants do have a bundle-sheath, it rarely contains the large number of chloroplasts characteristic of a C_4 bundle-sheath and the mesophyll is not arranged concentrically around it (Laetsch 1974). A few C_4 plants, such as *Suaeda monioca* (Shomer-Ilan, Neumann-Ganmore, and Waisel 1979) have a different leaf anatomy, but there are still two or more anatomically and functionally different cell layers.

A second difference between C_3 and C_4 leaves is in the carbon dioxide compensation point. This is defined as that concentration of carbon dioxide which allows photosynthesis to proceed at such a rate that it just balances the carbon dioxide released by the leaf, so that there is no net gain of carbon by the plant. At 25 °C and 21 per cent oxygen, C_3 leaves have compensation points in the range of 40–50 μl of carbon dioxide per litre, whereas the values for C_4 leaves are much lower, often close to zero (Krenzer, Moss, and Crookston 1975). Hence C_4 leaves are capable of net carbon dioxide fixation at much lower atmospheric carbon dioxide concentrations than are C_3 leaves. Photo-

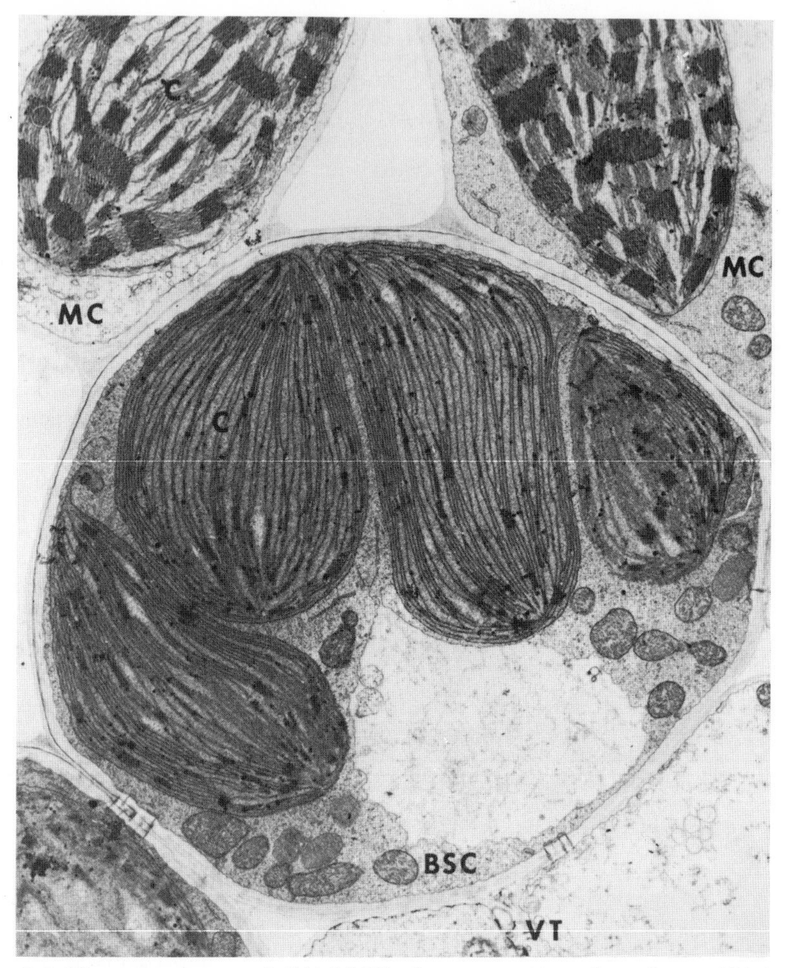

Fig. 5.1. Electron micrograph (X16 500) of a cross-section of a crabgrass (*Digitaria sanguinalis*) leaf. BSC, bundle sheath cell; C, chloroplast; M, mitochondrion: MC, mesophyll cells; P, peroxisome; VT, vascular tissues.

synthetic carbon dioxide fixation in C_4 plants is much less inhibited by elevated oxygen concentrations than it is in C_3 plants (Chapters 7 and 8).

If C_3 and C_4 plants are supplied with carbon dioxide containing the carbon isotope ^{13}C, they fix the $^{13}CO_2$ less well than ordinary carbon dioxide ($^{12}CO_2$) because the carbon dioxide-fixing enzymes discriminate against the slightly-heavier carbon atoms (Deleens, Lerman, Nato, and Moyse 1975). However, the discrimination is less in the case of C_4 plants. Since about 1 per cent of the total atmospheric carbon dioxide is $^{13}CO_2$ some of it will be picked up by plant tissues, and the $^{13}C/^{12}C$ ratio of the carbon fixed by the plant is therefore

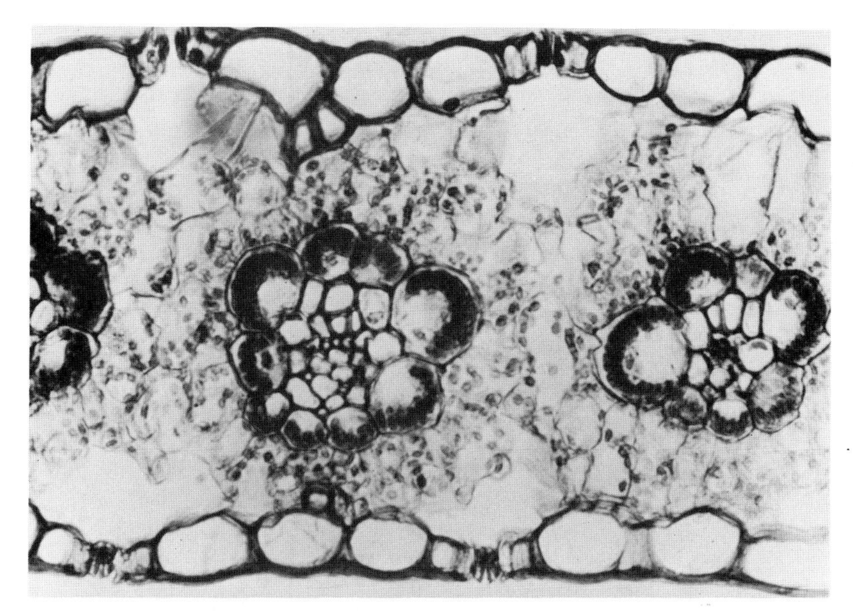

Fig. 5.2. Light photomicrograph (X185) of a maize (*Zea mays*) leaf in cross-section showing Kranz leaf anatomy. BSC, bundle-sheath cells; MC, mesophyll cells; VT, vascular tissues.

greater in the case of C_4 plants than it is for C_3 plants (Troughton 1979). This is usually expressed as a $\delta^{13}C$ ratio in parts per thousand (‰) by comparison with a standard mineral specimen.

$$\delta^{13}C\,(‰) = \left[\frac{{}^{13}C/{}^{12}C \text{ sample tested} - {}^{13}C/{}^{12}C \text{ standard}}{{}^{13}C/{}^{12}C \text{ standard}}\right] \times 1000 \quad (5.1)$$

Values of $\delta^{13}C$ are negative because of the isotope discrimination, but less so in the case of C_4 plants (Table 5.1).

The existence of plants with characteristics intermediate between C_3 and C_4 has been claimed. For example, the genus *Atriplex* contains both C_3 and C_4 species, and artificially-created hybrids between them show some intermediate properties (Bjorkman 1976). Some naturally-occurring *Panicum* species, such as *Panicum milioides*, also seem to be intermediates as illustrated in Table 5.1 (Rathnam and Chollet 1980; Morgan, Brown, and Reger 1980). Raghavendra, Rajendrudo, and Das (1978) presented evidence which suggests that on a single plant of *Mollugo nudicaulis* the young leaves have anatomy and carbon dioxide-fixation patterns of the C_3 type, whereas the older leaves are of the C_4 type. Similar phenomena have been reported in other C_4 plants (e.g. Crespo, Frean,

Table 5.1. Intermediate photosynthetic characteristics of *Panicum milioides*

Parameter	Typical C_3 plant	*P. milioides*	Typical C_4 plant
CO_2 compensation point ($\mu l\ litre^{-1}$)	40–60	16	0–5
Oxygen inhibition of photosynthesis (21% vs 2% $[O_2]$)	30%	15–20%	0%
Leaf $\delta\,^{13}C/^{12}C$ ratio (‰)	−22 to −40	−26	−9 to −19
PEP carboxylase activity ($\mu mol\ mg\ chlorophyll^{-1}\ h^{-1}$)	27	147	>1000
Leaf anatomy	non-Kranz	Bundle-sheath cells arranged around vascular tissue but no concentric arrangement of mesophyll cells	Kranz type (see text)

Mesophyll and bundle-sheath cells of *P. milioides* both contain PEP carboxylase and Calvin cycle enzymes, unlike C_4 plants. However, pyruvate-phosphate dikinase is exclusively in the mesophyll and NAD-malic enzyme, the only C_4 acid decarboxylase detected, is in the bundle-sheath cells. Data were abstracted from Rathnam and Chollet (1980) and from Troughton (1979).

Cresswell, and Tew 1979). Submerged aquatic angiosperms show compensation points and carbon dioxide-fixation patterns characteristic of both C_3 and C_4 species (e.g. *Hydrilla verticillata* seem to change its carbon dioxide fixation mechanism at different times of the year) but they do not have a Kranz-type leaf anatomy (Bowes, Holaday, Van, and Haller 1978). While it could be argued that 'pure' C_3 and C_4 plant types are the extremes of a broad spectrum, the number of well-characterized intermediate species reported to date is small.

5.2 Biochemistry of carbon dioxide fixation in C_4 plants

It has been well established (for reviews of earlier work see Coombs 1976; Rathnam 1978) that the primary carboxylation reaction in all C_4 plants is the conversion of phosphoenolpyruvate (PEP) to oxaloacetate by the action of PEP carboxylase (eqn (5.2)), an enzyme that is probably mainly, if not exclusively, located in the cytoplasm of the mesophyll cells (Rathnam and Chollet 1980). The true substrate of this enzyme is HCO_3^- rather than carbon dioxide, and its affinity for HCO_3^- is high (Coombs 1976).

$$\text{PEP} + HCO_3^- \xrightarrow[\text{PEP carboxylase}]{Mg^{2+}} \text{oxaloacetate} + \text{phosphate} \qquad (5.2)$$

Mesophyll cells contain high activities of carbonic anhydrase, presumably to facilitate rapid conversion of dissolved carbon dioxide into HCO_3^- (Rathnam 1978). PEP carboxylase is also present in C_3 leaves, but activities are much lower than in C_4 leaves (Table 5.1). PEP carboxylase is not present in significant amounts in bundle-sheath cells. In contrast, ribulose diphosphate carboxylase/oxygenase is present at high activities in bundle-sheath cells, but can scarcely be detected in the mesophyll of C_4 leaves.

The subsequent fate of oxaloacetate produced by PEP carboxylase depends on the C_4 plant being studied. Three groups of plants can be distinguished, as summarized in Table 5.2.

5.2.1 Malate formers

This group includes plants such as maize, sorghum, and crabgrass (*Digitaria sanguinalis*). The term 'malate former' is slightly misleading, since it refers to malate as being the major labelled compound first detected upon feeding $^{14}CO_2$ to leaves. However, 'malate formers' fix significant amounts of carbon dioxide into aspartate (e.g. Chapman and Hatch 1979) and similarly 'aspartate formers' produce variable amounts of malate, although aspartate is the major labelled product (Table 5.2).

The mesophyll chloroplasts of malate formers contain high activities of an $NADP^+$-dependent malate dehydrogenase. Oxaloacetate produced by the action

Table 5.2. Current views of the mechanisms of C_4 photosynthesis

Group of plants	CO_2 is fixed in the mesophyll by:	CO_2 is translocated from mesophyll to bundle-sheath mainly as:	CO_2 is released in the bundle-sheath mainly by:	CO_2 release in the bundle-sheath cells may be largely inhibited by:
'Malate formers' (e.g. maize, sorghum, crabgrass)	PEP carboxylase	L-Malic acid	$NADP^+$-malic enzyme L-malate + $NADP^+$ $\rightleftarrows$ NADPH + CO_2 + pyruvate	Oxalic acid
'Aspartate formers' type 1 (e.g. *Panicum maximum*)	PEP carboxylase	Aspartic acid	PEP carboxykinase oxaloacetate + ATP $\rightleftarrows$ PEP + CO_2 + ADP	3-Mercaptopicolinic acid
'Aspartate formers' type 2 (e.g. *Atriplex spongiosa*)	PEP carboxylase	Aspartic acid	NAD^+-malic enzyme L-malate + NAD^+ $\rightleftarrows$ NADH + CO_2 + pyruvate	No specific inhibitor of NAD^+-malic enzyme known

of PEP carboxylase in the cytoplasm is believed to enter the mesophyll chloroplasts and become reduced to malate by the $NADP^+$-malate dehydrogenase, using NADPH produced by the photosynthetic light reactions (eqn (5.3)).

$$\text{oxaloacetate} + \text{NADPH} + \text{H}^+ \rightarrow \text{L-malate} + \text{NADP}^+ \qquad (5.3)$$

The malate is then believed to leave the mesophyll cells and travel to the bundle-sheath. The only significant enzyme activity capable of decarboxylating malate that has been found to date in the bundle-sheath cells is an $NADP^+$-malic enzyme (Tables 5.2 and 5.3). Inhibition of this enzyme by adding oxalic acid prevents malate decarboxylation by isolated bundle-sheath cells (Rathnam 1978). Hence it is believed that carbon dioxide is released from malate by the malic enzyme, and fixed into the Calvin cycle by the operation of ribulose diphosphate carboxylase/oxygenase.

The pyruvate (Table 5.2) might then be expected to return to the mesophyll cells to maintain the carbon balance of the leaf. The mesophyll chloroplasts contain high activities of a pyruvate-phosphate dikinase enzyme, which presumably converts pyruvate into PEP *in vivo* using ATP produced by photophosphorylation (eqn (5.4)).

$$\text{ATP} + \text{pyruvate} + \text{Pi} \xrightarrow[\text{dikinase}]{\text{pyruvate-Pi}} \text{AMP} + \text{PEP} + \text{PPi} \qquad (5.4)$$

The above reactions are summarized in Fig. 5.3.

In 'malate formers' the bundle-sheath chloroplasts have poorly-developed grana (e.g. Fig. 5.1) and the isolated chloroplasts often show low rates of photoreduction of $NADP^+$ with water as electron donor. They appear to be partially deficient in photosystem II activity, although the extent of this deficiency is controversial (Laetsch 1974; Bishop, Anderson, and Smillie 1972; Gutierrez, Kanai, Huber, Ku, and Edwards 1974; Gregory, Droppa, Horvath, and Evans 1979; Walker and Izawa 1980). Fixation of the carbon dioxide released from malate into the Calvin cycle needs two NADPH and three ATP molecules per carbon dioxide molecule fixed. One NADPH molecule can be provided by the action of $NADP^+$-malic enzyme (Fig. 5.3) but it is possible that the photosystem II activity in the bundle-sheath chloroplasts of malate formers may sometimes be insufficient to provide the rest of the NADPH, and all of the ATP required. The NADPH problem could be overcome if some phosphoglycerate were shuttled back to the mesophyll for reduction there to phosphoglyceraldehyde in the mesophyll chloroplasts, which have normal PSII activity. The phosphoglyceraldehyde could then return to the bundle-sheath cells. Whereas most Calvin cycle enzymes can scarcely be detected in the mesophyll cells, high activities of phosphoglycerate kinase and $NADP^+$-glyceraldehyde 3-phosphate

Table 5.3. Activity and localization of photosynthetic enzymes in bundle-sheath cells of the three types of C_4 plant

Enzyme	Plant type					
	Malate former		Aspartate former Type 1		Aspartate former Type 2	
	Activity	Location	Activity	Location	Activity	Location
$NADP^+$-malic enzyme	622	Chp	Trace	–	Trace	–
PEP carboxykinase	Trace	–	714	Chp/cyto*	Trace	–
NAD^+-malic enzyme	149	Mitos	243	Mitos	399	Mitos
Aspartate aminotransferase	457	Mitos	1078	Mitos	1347	Mitos
Alanine aminotransferase	193	Cyto	620	Cyto	1219	Cyto
NAD^+-malate dehydrogenase	2850	Mitos	8325	Chp, mitos	8270	Mitos, chp
RuDP carboxylase	240	Chp	281	Chp	305	Chp

Enzyme activities are expressed as μmol mg chlorophyll^{-1} h^{-1}. Data are taken from Rathnam (1978). The species used as representative of each of the three subgroups were *Zea mays*, *Panicum maximum*, and *Panicum miliaceum*. Abbreviations of sites of enzyme location: Chp, chloroplasts; cyto, cytoplasm; mitos, mitochondria.

*Localization may depend on species.

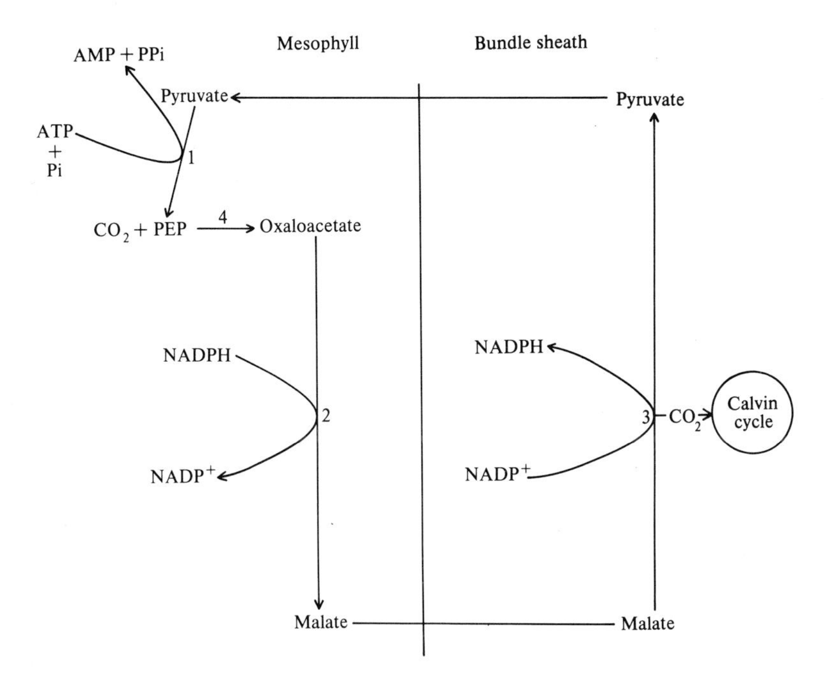

Fig. 5.3. Carbon dioxide fixation by a C_4 'malate former'. Enzymes: 1, pyruvate-phosphate dikinase; 2, $NADP^+$-malate dehydrogenase; 3, $NADP^+$-malic enzyme; 4, PEP carboxylase.

dehydrogenase are present in mesophyll as well as bundle-sheath cells, which is consistent with the occurrence of such a shuttle (Hatch 1971). There is no direct experimental evidence for the shuttle, however. It is possible that cyclic photophosphorylation might provide some of the ATP needed (Chapman, Berry, and Hatch 1980).

The scheme of Fig. 5.3 was originally based on differential localization of enzymes between mesophyll and bundle-sheath cells (Hatch 1971). Some objections to these studies were raised on the grounds that inactivation of enzymes such as ribulose diphosphate carboxylase/oxygenase by phenolic compounds during fractionation of the leaf was responsible for the apparent lack of most Calvin cycle enzymes in mesophyll cells. However, these objections have not been substantiated (Chollet and Ogren 1975). For example, Huber, Hall, and Edwards (1976) did not detect ribulose diphosphate carboxylase/oxygenase in mesophyll cell extracts by gel electrophoresis, which would have revealed even an inactivated protein if one had been present. Fluorescent antibody labelling of sections of C_4 leaves using antibodies raised against ribulose diphosphate carboxylase/oxygenase almost exclusively stains the bundle-sheath cells

(Hattersley, Watson, and Osmond 1976). Hence the data in Tables 5.2 and 5.3 are now generally held to be correct (Rathnam 1978).

Perhaps the most striking evidence for the scheme of Fig. 5.3 comes from work with isolated mesophyll and bundle-sheath cells (Rathnam and Chollet 1980). For example, illuminated crabgrass mesophyll cells fix carbon dioxide into malate, but not into Calvin cycle intermediates (Salin, Campbell, and Black 1973). If phosphoglycerate is supplied, it is reduced in the light by the combined operation of the phosphoglycerate kinase and glyceraldehyde 3-phosphate dehydrogenase reactions, but again no significant amount of carbon dioxide is fixed into Calvin cycle intermediates (Salin and Black 1974). Comparable results were obtained with mesophyll chloroplasts from *Atriplex spongiosa* (Kagawa and Hatch 1974*a*, *b*).

When pyruvate is supplied to isolated crabgrass mesophyll cells, a low rate of carbon dioxide fixation is observed in the light (Huber and Edwards 1975). Presumably the dikinase enzyme converts pyruvate into PEP (eqn (5.4)), providing substrate for PEP carboxylase. The mesophyll chloroplasts would generate the ATP used by the dikinase enzyme, but they would have no means of removing NADPH produced by non-cyclic photophosphorylation. Providing the cells with oxaloacetate as well as with pyruvate, as should happen *in vivo* (Fig. 5.3), greatly enhances the rate of carbon dioxide fixation, since NADPH can then be oxidized by the malate dehydrogenase reaction, leading to increased electron transport and generation of more ATP for the dikinase reaction. This increased rate of carbon dioxide fixation is inhibited by DCMU, but the low rate seen in the presence of pyruvate alone is insensitive to DCMU. Huber and Edwards (1975) therefore suggested that the ATP needed in the latter case was being produced by cyclic photophosphorylation, which can be demonstrated under appropriate conditions in isolated mesophyll chloroplasts (Huber and Edwards, 1976). The physiological significance of cyclic photophosphorylation has not been established in either C_3 or C_4 plants (Chapter 2), but it could be relevant *in vivo* if, after a dark period, PEP has to 'build up' from pyruvate in the mesophyll cells when the light is first turned on. ATP would initially be required without NADPH, which would not be used up until oxaloacetate had been produced.

Isolated bundle-sheath cells fix carbon dioxide into intermediates of the Calvin cycle. Rates of carbon dioxide fixation by cells from many malate-formers are low, possibly because of the partial deficiency in photosystem II activity, but they can be raised by supplying appropriate Calvin cycle intermediates. In aspartate formers, the bundle-sheath chloroplasts show normal grana and photosynthetic activities (Laetsch 1974; Gutierrez *et al.* 1974) and isolated bundle-sheath cells can fix carbon dioxide into the Calvin cycle at high rates, e.g. 100 μmol h^{-1} mg chlorophyll^{-1} for *Panicum capillare* cells (Gutierrez *et al.* 1974) and even higher rates for cells from *Atriplex spongiosa* (Kagawa and Hatch 1974*a*, *b*). If malate is supplied to illuminated bundle-sheath cells, carbon is rapidly lost from position four of the molecule and appears in Calvin cycle intermediates (Rathnam 1978).

5.2.2 Aspartate formers of type 1

Extracts made from the leaves of such C_4 plants as *Panicum maximum*, *Chloris gayana*, and *Sporobolus fimbriatus* contain only low activities of $NADP^+$-malic enzyme, but high activities of aspartate and alanine aminotransferases, and of PEP carboxykinase (Table 5.3) which catalyses the reaction

$$\text{oxaloacetate} + \text{ATP} \rightarrow \text{PEP} + CO_2 + \text{ADP} \qquad (5.5)$$

Oxaloacetate formed by the action of PEP carboxylase in the mesophyll cytoplasm seems to be mainly converted into aspartate by transamination reactions using glutamate and alanine as amino-donors. Aspartate passes to the bundle-sheath cells and is converted back to oxaloacetate, which is decarboxylated by PEP carboxykinase (eqn (5.5)) to release carbon dioxide for fixation into the Calvin cycle (Fig. 5.4). Evidence for this role of PEP carboxykinase is provided by the observation that treatment of *Panicum maximum* leaves with 3-mercaptopicolinic acid, a specific inhibitor of PEP carboxykinase, raises the compensation point to a level characteristic of that of a C_3 plant (Ray and Black 1976). Similarly, this inhibitor did not affect carbon dioxide fixation into the Calvin cycle by isolated bundle-sheath strands from *Eriochloa borumensis*, but it caused a 70–75 per cent inhibition of aspartate-dependent oxygen evolution from the cells, presumably by preventing the internal release of carbon dioxide from aspartate (Rathnam and Edwards 1977).

The fate of the PEP produced by reaction (5.5) in the bundle-sheath cells is unknown. It might return to the mesophyll as such, or be converted into pyruvate which could then be transaminated to alanine. Nitrogen must be returned from bundle-sheath to mesophyll in some form to allow continued operation of the transaminases, and alanine or glutamate would be likely candidates for a return journey (Fig. 5.4). Isolated bundle-sheath strands from *Eriochloa* released both PEP and alanine into the surrounding solution when supplied with aspartate and 2-ketoglutarate in the light (Rathnam and Edwards 1977), although in *Chloris gayana* PEP is probably not a significant transport metabolite (Hatch 1979).

Inspection of Fig. 5.4 reveals a second question raised by the current scheme for carbon dioxide fixation in aspartate formers of type 1, i.e. the function of the mesophyll chloroplasts. Conversion of pyruvate into PEP by pyruvate-phosphate dikinase in these chloroplasts (the extent of which will depend on how much, if any, PEP returns from bundle-sheath to mesophyll *in vivo* – see Fig. 5.4) consumes light-generated ATP, but there is no apparent role for the NADPH also generated during non-cyclic electron flow. It might be argued that the mesophyll chloroplasts carry out only cyclic or pseudocyclic photo-

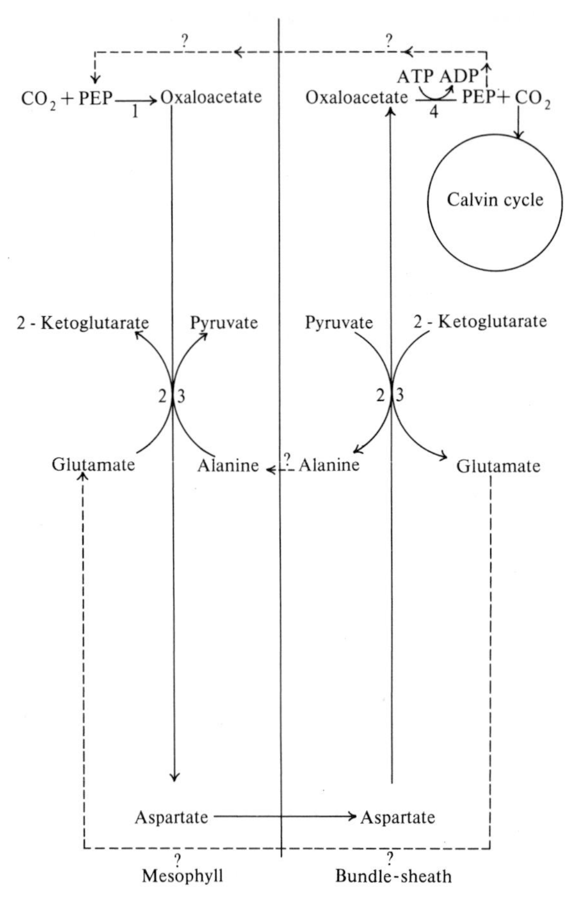

Fig. 5.4. Carbon dioxide fixation by an aspartate former of type 1. Enzymes: 1, PEP carboxylase; 2, aspartate aminotransferase; 3, alanine aminotransferase; 4, PEP carboxykinase. ?---- indicates suggested, but unproven, pathways of metabolite flow.

phosphorylation (Edwards and Huber 1979). However, cyclic photophosphorylation is of debatable physiological significance in the chloroplasts of higher plants (Chapter 2). A high rate of pseudocyclic electron flow seems unlikely because of the damaging effects of the hydrogen peroxide and superoxide radicals that this process generates (see Chapters 2 and 8), but it is, of course, possible that the mesophyll chloroplasts are efficiently protected against these species. Alternatively, the NADPH generated by non-cyclic photophosphorylation could be consumed in processes other than carbon dioxide fixation. Nitrate reduction, which absorbs considerable reducing power, occurs largely in the mesophyll cells of C_4 leaves (Chapter 10). Phosphoglycerate might be shuttled back from the bundle-sheath cells for reduction to phosphoglyceraldehyde in the mesophyll chloroplasts, as is believed to happen with malate-formers. However, the bundle-sheath cells of aspartate formers are perfectly capable of themselves providing sufficient NADPH to operate the Calvin cycle, since they fix carbon dioxide into the cycle at high rates *in vitro* (see Section 5.2.1), and so a triose phosphate shuttle would not seem to be necessary. The problem remains to be resolved.

5.2.3 Aspartate formers of type 2

In leaves of C_4 plants such as *Atriplex spongiosa*, *Amaranthus edulis*, and *Panicum miliaceum*, the activities of both $NADP^+$-malic enzyme and PEP carboxykinase are low, but there are large amounts of aspartate and alanine aminotransferases in both mesophyll and bundle-sheath cells (Table 5.3). Most carbon is transferred from mesophyll to bundle-sheath as aspartate, which must then give rise to carbon dioxide for fixation into the Calvin cycle. The only decarboxylating enzyme so far detected in large amounts in the bundle-sheath cells of type 2 aspartate formers is an NAD^+-malic enzyme, located in the mitochondria. It has therefore been suggested that aspartate entering the bundle-sheath cells is transaminated back into oxaloacetate, which is then reduced to malate by an NAD^+-linked malate dehydrogenase. Malate is then decarboxylated by the NAD^+-malic enzyme and the resulting pyruvate converted into alanine by transamination for return to the mesophyll. This rather speculative scheme is shown in Fig. 5.5. Unfortunately, no specific inhibitor of the NAD^+-malic enzyme is yet available so it is not possible to check the proposed role of this enzyme *in vivo*. Some circumstantial evidence does support the occurrence of the scheme shown in Fig. 5.5, however. According to this scheme, aspartate decarboxylation requires no energy input and should therefore be independent of light, a phenomenon which has been confirmed experimentally for isolated bundle-sheath cells (Rathnam 1978). In contrast, bundle-sheath cells from type 1 aspartate formers require light in order to release carbon dioxide from aspartate, presumably because of the ATP requirement of PEP carboxylase (Fig. 5.4). In addition to the NAD^+-malic enzyme, the major isoenzyme of aspartate aminotransferase and the malate dehydrogenase activities of bundle-sheath cells are located in the mitochondria (Table 5.3), which might explain why aspartate formers of type 2 have larger and/or more numerous mitochondria in their bundle-sheath cells than do other C_4 plants (Hatch, Kagawa, and Craig 1975; Kagawa and Hatch 1975). Isolated bundle-sheath mitochondria have been shown to catalyse aspartate decarboxylation (Kagawa and Hatch 1975; Rathnam 1978). The carbon dioxide released *in vivo* must diffuse into the bundle-sheath chloroplasts for fixation by the Calvin cycle. Examination of Fig. 5.5 shows that, as for type 1 aspartate formers, the fate of any NADPH generated by mesophyll chloroplasts is unaccounted for.

The bundle-sheath cells of malate formers contain no significant PEP carboxykinase activity, but some NAD^+-malic enzyme is present (Table 5.3). Similarly,

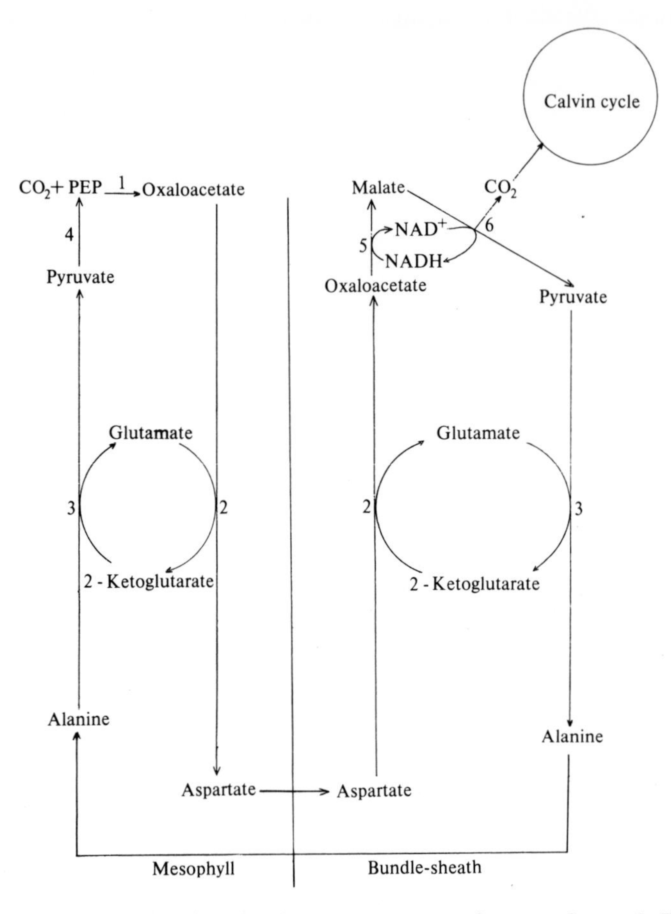

Fig. 5.5. Carbon dioxide fixation by a C_4 aspartate former of type 2. Enzymes: 1, PEP carboxylase; 2, aspartate aminotransferase; 3, alanine aminotransferase; 4, pyruvate-Pi dikinase; 5, NAD^+–malate dehydrogenase; 6, NAD^+-malic enzyme.

aspartate formers of type 1 possess NAD^+-malic enzyme in their bundle-sheath mitochondria. It is possible that this enzyme could play some part in carbon dioxide-releasing reactions in the bundle-sheath cells of these two types of plant. However, use of the specific inhibitors oxalate and 3-mercaptopicolinate (Table 5.2) has shown that the major carbon dioxide-releasing reactions are as in Figs. 5.3 and 5.4, and that any contribution made by NAD^+-malic enzyme is minor (Rathnam 1978).

5.3 Synthesis of starch and sucrose in C_4 plants

Most starch formed by C_4 leaves appears in bundle-sheath chloroplasts, although mesophyll chloroplasts often show some starch formation (Laetsch 1974). Leaf fractionation studies show that enzymes for the synthesis of sucrose and of starch, and for starch degradation, are present in both mesophyll and bundle-sheath cells (Chen, Dittrich, Campbell, and Black 1974; Mbaku *et al.* 1978). Presumably the pathways of starch and sucrose formation in the bundle-sheath cells are identical to those in C_3 plants (see Chapter 4), but how starch could be formed in mesophyll chloroplasts is not at all understood. Perhaps there is a shuttle of hexose sugars from bundle-sheath to mesophyll.

5.4 Regulation of carbon dioxide fixation in C_4 plants

The activities of pyruvate-phosphate dikinase and (in malate formers) $NADP^+$-malate dehydrogenase in C_4 leaves or in isolated mesophyll chloroplasts are increased on illumination and decrease again in the dark. *In vitro*, deactivation of the dikinase is achieved by incubation with ADP plus a 'protein factor', and it is reversed by inorganic phosphate (Hatch 1981). By contrast, activation of $NADP^+$-malate dehydrogenase appears to involve a dithiol system such as the thioredoxin mechanism (Kagawa and Hatch 1977). Activation and deactivation are fairly slow processes and are, by themselves, insufficient to regulate ongoing carbon dioxide fixation. Metabolic regulation of PEP carboxylase by adenylate control or by inhibition by C_4 acids has been proposed, but the kinetic complexity of this enzyme makes such proposals difficult to evaluate (Coombs 1976; Rathnam and Chollet 1980). The NAD^+-malic enzyme in the mitochondria of aspartate formers of type 2 is competitively inhibited by HCO_3^- and activated by low concentrations of fructose diphosphate (Hatch 1981), which could be involved in co-ordinating the rate of decarboxylation and the activity of the Calvin cycle (Chapman and Hatch 1977). Presumably the Calvin cycle enzymes in the bundle-sheath are regulated by mechanisms similar to those in C_3 plants, although the light-activation of Calvin cycle enzymes is not as marked as in C_3 plants (Steiger, Ziegler, and Ziegler 1971). The $NADP^+$-malic enzyme of maize bundle-sheath cells may be controlled by light-induced changes in pH and Mg^{2+} concentration (Asami, Inoue, and Akazawa 1979).

An area to which little attention has been given is the mechanism by which metabolites enter and leave the mesophyll and bundle-sheath cells. Such studies are essential before the shuttle systems depicted in Figs. 5.3-5 can be regarded as finally established. Numerous plasmodesmata connecting adjacent mesophyll and bundle-sheath cells have been observed in cross-sections of C_4 leaves, and these may represent the route of transfer. Since PEP carboxylase in C_4 plants is a cytoplasmic enzyme, whereas pyruvate-phosphate dikinase is located in the mesophyll chloroplasts, there must be some means for import of pyruvate and export of PEP across the envelope of such chloroplasts. Indeed, Huber and Edwards (1977*a*, *b*) detected both a pyruvate carrier and an exchange porter for PEP and inorganic phosphate in mesophyll chloroplasts from crab-grass. Neither system could be detected in chloroplasts of the C_3 plant, spinach.

5.5 The function of the C_4 pathways

What is the function of C_4 photosynthesis: why do some plants fix carbon dioxide into four-carbon acids only to release it again within the leaf for fixation by ribulose diphosphate carboxylase/oxygenase? The answer seems to lie in the low compensation point. At low carbon dioxide concentrations, C_4 plants can trap much more of the available carbon than C_3 plants, and so they should have a selective advantage. This is due to the low K_m of PEP carboxylase for HCO_3^-. For example, the C_4 plant *Atriplex sabulosa* can continue growing at carbon dioxide concentrations that are unusable by the C_3 plant, *Atriplex glabriuscula* (Bjorkman 1976). C_4 plants are at their greatest advantage if photosynthesis is operating at high light intensities and temperatures, when the carbon dioxide in the air around plants tends to become depleted and the stomata are often partially closed to reduce water loss, therefore hindering carbon dioxide penetration into the leaf (Bjorkman 1976). Under temperate conditions, the advantage of C_4 over C_3 plants would be less marked, or absent (Moore 1979).

When the C_4 pathways were first discovered, it was proposed that the C_4 reactions act as a 'carbon dioxide-concentrating mechanism': decarboxylation of aspartate or malate within the bundle-sheath cells would raise the carbon dioxide concentration in the vicinity of their ribulose diphosphate carboxylase/oxygenase and hence promote carbon dioxide fixation because of the high K_m for carbon dioxide of this enzyme. However, it is now known that if proper activation procedures are used, the K_m of ribulose diphosphate carboxylase/oxygenase from C_4 plants is decreased and the V_{max} is increased in the same way as for the C_3 enzyme (see Chapter 3 and Yeoh, Badger, and Watson 1980). The K_m values of the 'activated' C_4 carboxylase/oxygenases are still higher than atmospheric carbon dioxide concentrations, however, so they will increase their activity in response to increased carbon dioxide and the argument is still valid. Any carbon dioxide which escapes from bundle-sheath cells will be absorbed again as it passes through the surrounding mesophyll, and recycled (Rathnam and Chollet 1978). It has also been suggested that C_4 plants utilize nitrogen from the soil more efficiently than C_3 plants (Ray and Black 1979; Bolton and Brown 1980) since they need synthesize less ribulose diphosphate carboxylase, the major soluble leaf protein, to fix a given amount of carbon dioxide. The advantages of C_4 metabolism will be discussed further in Chapter 7.

Because of the greater growth rate of C_4 plants under certain environmental conditions, it has sometimes been suggested that it might be beneficial to breed C_4 characteristics into C_3 plants of agricultural importance. Some initial experiments along these lines were carried out by Bjorkman (1976). He crossed the C_3 species *Atriplex triangularis* with its near relative, the C_4 plant *Atriplex rosea*. The hybrids so produced had a leaf anatomy intermediate between that of C_3 and C_4 plants. On feeding $^{14}CO_2$ to the hybrid leaves, approximately half of the radioactivity went initially into malate and aspartate and the other half into phosphoglycerate. However, the transfer of radioactivity from the C_4 acids into Calvin cycle intermediates was greatly impaired. It seemed that the bulk of the carbon dioxide fixed by PEP carboxylase was released in the wrong part of the leaf, so that it was not picked up by ribulose diphosphate carboxylase/ oxygenase. Presumably this was a consequence of the intermediate leaf anatomy, but as far as carbon dioxide compensation point was concerned, it meant that the hybrid plant had a value similar to that of its C_3 parent. Such results illustrate the potential difficulties of the breeding approach.

5.6 Carbon dioxide fixation by leaf epidermis

Epidermis stripped from the leaves of both C_3 and C_4 plants has been found to contain PEP carboxylase activity and to fix $^{14}CO_2$ into malate and aspartate in the light and, to a lesser extent, in the dark. Epidermal carbon dioxide fixation seems to be associated particularly with guard cells, and it might be involved in the regulation of stomatal opening although the details have yet to be established (Raschke and Dittrich 1977; Thorpe, Brady, and Milthorpe 1979). In at least one C_3 plant, *Vicia faba*, the guard cell chloroplasts lack Calvin cycle enzymes (Outlaw, Manchester, Dicamelli, Randall, Rapp, and Veith 1979).

5.7 Crassulacean acid metabolism

Crassulacean acid metabolism (CAM) refers to a variant of photosynthetic carbon metabolism carried out by some succulent plants living in arid conditions. Many such plants have a striking appearance (e.g. Figs 5.6–5.8). Well-known CAM plants are *Bryophyllum*, *Sedum*, and *Kalanchoe* species and an extensive list of such plants has been given by Szarek and Ting (1977). CAM plants show the following features in their normal environments (Kluge 1978; Kluge and Ting 1978).

(1) A diurnal alteration of the organic acid content of the photosynthesizing tissues, especially malic acid. Malate accumulates during the night and is used up during the day. This malate is stored largely in cell vacuoles, as has been shown by direct measurement of the malate content of vacuoles isolated from CAM leaves at different times (Kenyon, Kringstad, and Black 1978). Accumulation of malate in the vacuole may be driven by an active co-transport of protons (Luttge and Ball 1980).

(2) Diurnal alteration of the starch content of the leaves inversely to the malic acid content, i.e. starch is synthesized during the day and used up at night.

Fig. 5.6. The giant saguaro plant (*Carnegia gigantea*) found in the southwestern United States is one of the largest and most impressive CAM plants.

(3) A large net uptake of carbon dioxide from the atmosphere occurs at night, whereas carbon dioxide uptake is depressed during the day. The stomata tend to be open at night and more-or-less closed during the day. Figure 5.9 illustrates some of these features for the plant *Kalanchoe daigremontiana*.

Plants living in hot, arid environments have evolved many mechanisms to reduce water loss. Accumulation of water in succulent leaves is one. CAM represents another such mechanism: decarboxylation of accumulated malate during the day provides carbon dioxide for fixation into the Calvin cycle when the stomata are closed. Indeed, it is possible that the increased internal carbon dioxide caused by malate decarboxylation in the light is the signal for stomatal closure (Cockburn, Ting, and Sternberg 1979). Hence CAM plants can shut their stomata during the day, to decrease water loss, and still continue to use

Fig. 5.7. *Sedum prealtum*, a typical CAM plant. Note the fleshy leaves.

light energy to 'fix CO_2' derived from malate, i.e. they might be considered to have achieved a temporal separation of primary carbon dioxide fixation from the photosynthetic light reactions. This is done at a heavy cost, however. The energy needed to fix carbon dioxide into malate in the dark is provided by degradation of stored carbohydrate, which is replaced in the light by the action of the Calvin cycle. The *net* energy gain to the plant of these metabolic processes is very small, resulting in very slow growth. This may be contrasted with the situation in C_4 plants, where light provides the energy to drive *both* the formation of C_4 acids *and* the Calvin cycle simultaneously.

The extent to which CAM operates in a plant depends to a large extent on its environment. For example, provision of an ample water supply often causes stomatal opening and direct carbon dioxide fixation by CAM leaves during the day (Osmond 1978; Kluge 1978) which enables the plant to grow somewhat faster. Temperature and day length also have an effect. In plants such as *Kalanchoe* sp the enzymic capacity for CAM remains even when the plant is placed under non-arid conditions. In others, called 'inducible' CAM plants by Osmond (1978), the enzymes of CAM metabolism only appear when the plant is placed under water-stress or exposed to certain photoperiods. *Mesembryanthemum crystallinum* is an example of an inducible CAM plant (Winter and Luttge 1979).

5.8 Carboxylation and decarboxylation in CAM plants

Dark fixation of carbon dioxide in the leaves of CAM plants is catalysed by the enzyme PEP carboxylase, which has been reported to be located in the leaf cytoplasm in some plants (Spalding, Schmitt, Ku, and Edwards 1979) but in the chloroplasts of others (Schnarrenberger, Gross, Burkhard, and Herbert 1980), although latency experiments of the type needed to rule out non-specific absorption of enzyme onto the chloroplast envelope (Chapter 1) were not performed in the latter study.

The PEP required by the carboxylase could be provided by breakdown of stored carbohydrate through the glycolytic pathway. Alternatively, if the ribulose diphosphate carboxylase/oxygenase enzyme were active in the dark, the phosphoglycerate produced by its action could give rise to PEP (see Fig. 5.10 and Bradbeer 1975). The ribulose diphosphate needed would then have to be generated from stored hexose sugars by the oxidative pentose-phosphate pathway and the action of phosphoribulokinase.

These two possibilities can be distinguished by examining the labelling pattern in malate produced during $^{14}CO_2$ fixation in the dark (Fig. 5.10). Most experiments of this type carried out on CAM leaves favour the primary carboxy-

Fig. 5.8. Typical CAM plants. (a) *Kalanchoe daigremontiana*, a broad-leaved succulent of the Crassulaceae.

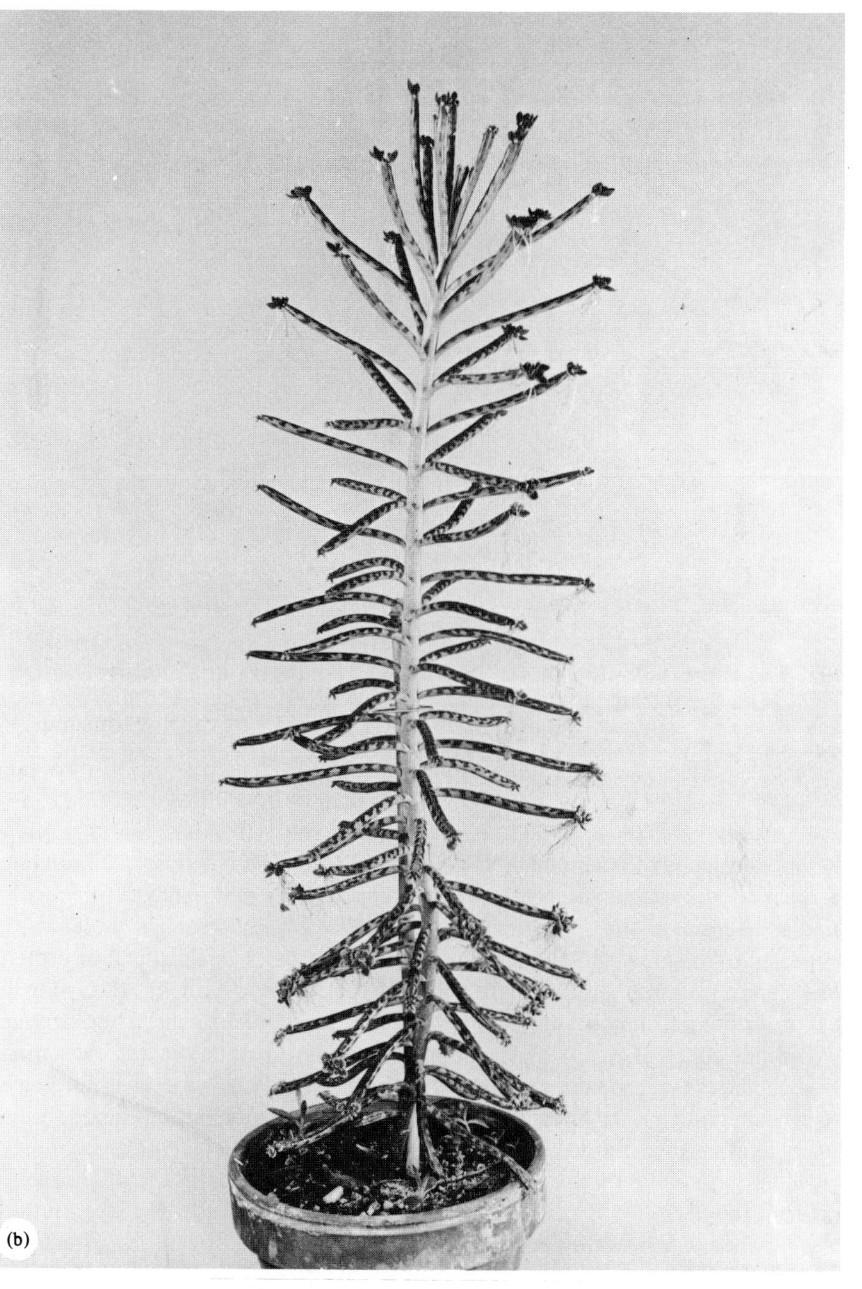

Fig. 5.8 (b) *Kalanchoe tubiflora*.

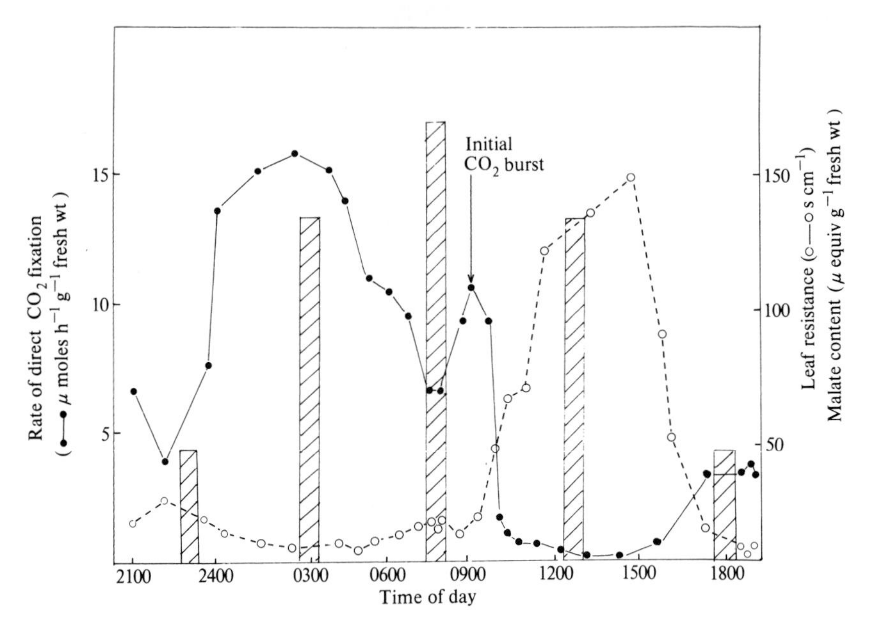

Fig. 5.9. Diurnal rhythm of carbon dioxide assimilation in *K. daigremontiana*. Light intensity during growth and assay was 27×10^3 erg cm^{-1} s^{-1} (400–700 nm) and day and night air temperatures were 27 and 15°C. CO_2 fixation rate = ●——●; leaf resistance to water vapour exchange = ○——○; and malic acid content at various times is shown by vertical bars. Data from Osmond (1978).

lation mechanism (Osmond 1978) (Fig. 5.10). The ribulose diphosphate carboxylase/oxygenase of CAM plants has activation and catalytic properties similar to those of the C_3 enzyme (Badger *et al.* 1975) and it might therefore be expected to have low activity in chloroplasts in the dark (see Chapter 3). Further, PEP carboxylase has a much higher affinity for carbon dioxide (as HCO_3^-) than has the ribulose diphosphate carboxylase/oxygenase, and so the latter enzyme would not compete effectively at normal atmospheric carbon dioxide concentrations. Hence the primary carboxylation mechanism would seem in any case to be inherently more likely. If higher concentrations of carbon dioxide were used, fixation by both enzymes might occur, which could explain the secondary labelling patterns detected in a few experiments (Bradbeer 1975). It is usually difficult to tell from published papers exactly what carbon dioxide concentration was supplied to the leaf.

Many CAM plants show an initial burst of carbon dioxide fixation at the beginning of the light period (Fig. 5.9). Recent studies on *Kalanchoe daigremontiana* show that the malate synthesized during this period has a double-labelled pattern, i.e. both carboxylases are active to some extent. Presumably

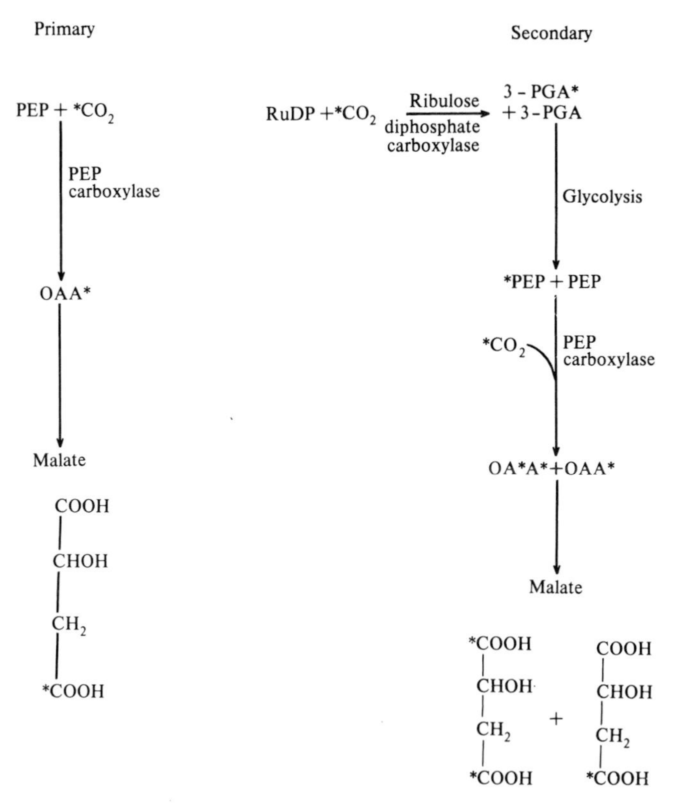

Fig. 5.10. Predicted distribution of ^{14}C in malate labelled by primary and secondary carboxylation of PEP, in short-term experiments with $^{14}CO_2$. Asterisks show the number and position of labelled atoms of intermediates. Note that the primary sequence would result in 100 per cent of labelled carbon appearing in the C-4 carboxyl of malate, compared with 66.7 per cent for the secondary sequence.

illumination increases the activity of ribulose diphosphate carboxylase rapidly, but reduces PEP carboxylase activity more slowly. How PEP carboxylase is prevented from fixing carbon dioxide after the 'burst' is not clearly established. It has been reported that PEP carboxylase activity in leaf extracts is lower in illuminated than in darkened leaves (Queiroz 1974) but the differences in activity are only 3-4 fold, or less in some CAM plants (e.g. see Pierre and Queiroz 1979), so this cannot be the sole explanation. PEP carboxylase in CAM plants is inhibited by malic acid at millimolar concentrations (e.g. K_i for malate inhibition of the *Sedum* and *Kalanchoe* enzymes is 3.6 mM – Kluge and Ting 1978). Malic acid formed in the cytoplasm of CAM leaves at night is transported into the vacuoles. As the vacuoles 'fill up' towards the end of the dark period, the malic acid concentration in the cytoplasm might rise sufficiently to reduce the activity of PEP carboxylase. In the light, transport of malate out of the vacuole for decarboxylation might keep the cytoplasmic concentration high enough to severely inhibit PEP carboxylase. As the malate is used up, inhibition will gradually be relieved so that by the end of the light period, the carboxylase will be completely functional once again (Kluge and Ting 1978; Kluge, Bocher, and Jungnickel 1980).

Malate decarboxylation by CAM plants in the light appears to be catalysed in some species by $NADP^+$-malic enzyme and in others by conversion to oxaloacetate followed by the action of PEP carboxykinase (Dittrich, Campbell, and Black 1973). Significant amounts of NAD^+-malic enzyme have also been detected in the leaves or leaf mitochondria of some CAM plants (e.g. Spalding, Arron, and Edwards 1980). If CAM is an evolutionary offshoot of C_4 metabolism, as seems likely, then one would probably expect to find the three different decarboxylation mechanisms shown in Table 5.2 in different CAM species.

References

Asami, S., Inoue, K., and Akazawa, T. (1979). *Archs Biochem. Biophys.* **196**, 581-7.

Badger, M. R., Andrews, T. J., and Osmond, C. B. (1975) in *Proceedings of the third international congress on photosynthesis* (ed. M. Avron), pp. 1421–9. Elsevier, Amsterdam.

Bishop, D. G., Anderson, K. S., and Smillie, R. M. (1972). *Pl. Physiol., Lancaster* **50**, 774–7.

Bjorkman, Q. (1976) in *CO_2 metabolism and plant productivity* (eds. R. H. Burris and C. C. Black) pp. 287–309. University Park Press, Baltimore, MD.

Bolton, J. K. and Brown, R. H. (1980). *Pl. Physiol., Lancaster* **66**, 97–100.

Bowes, G., Holaday, A. S., Van, T. K., and Haller, W. T. (1978) in *Photosynthesis '77* (eds. D. O. Hall, J. Coombs, and T. W. Goodwin) pp. 289–98. Biochemical Society, London.

Bradbeer, J. W., Cockburn, W., and Ranson, S. L. (1975) in *Environmental and biological control of photosynthesis* (ed. R. Marcelle) pp. 265–72. Junk, The Hague.

Chapman, K. S. R. and Hatch, M. D. (1977). *Archs Biochem. Biophys.* **184**, 298–306.

— — (1979). *Biochem. Biophys. Res. Commun.* **86**, 1274–80.

— Berry, J. A. and Hatch, M. D. (1980). *Archs Biochem. Biophys.* **202**, 330–41.

Chen, T. M., Dittrich, P., Campbell, W. H., and Black, C. C. (1974). *Archs Biochem. Biophys.* **163**, 246–62.

Chollet, R. and Ogren, W. L. (1975). *Bot. Rev.* **41**, 137–79.

Cockburn, W., Ting, I. P., and Sternberg, L. O. (1979). *Pl. Physiol., Lancaster* **63**, 1029–32.

Coombs, J. (1976) in *The intact chloroplast, Topics in photosynthesis,* Vol. 1 (ed. J. Barber) pp. 279–313. Elsevier, Amsterdam.

Crespo, H. M., Frean, M., Cresswell, C. F., and Tew, J. (1979). *Planta* **147**, 257–63.

Deleens, E., Lerman, J. C., Nato, A., and Moyse, A. (1975) in *Proceedings of the third international congress on photosynthesis* (ed. M. Avron) pp. 1267–76. Elsevier, Amsterdam.

Dittrich, P., Campbell, W. H., and Black, C. C. (1973). *Pl. Physiol., Lancaster* **52**, 357–61.

Edwards, G. E. and Huber, S. C. (1979) in *Photosynthesis II, Encyclopaedia of plant physiology,* Vol. 6 (eds. M. Gibbs and E. Latzko) pp. 102–12. Springer, Berlin.

Gregory, R. P. F., Droppa, M., Horvath, G., and Evans, E. H. (1979). *Biochem. J.* **180**, 253–6.

Gutierrez, M., Kanai, R., Huber, S. C., Ku, S. B., and Edwards, G. E. (1974). *Z. Pflazenphysiol.* **72**, 305–19.

Hatch, M. D. (1971) in *Photosynthesis and photorespiration* (eds. M. D. Hatch, C. B. Osmond, and R. O. Slatyer) Wiley, New York.

— (1979). *Archs Biochem. Biophys.* **194**, 117–27.

— (1981) in *Proceedings of the Fifth International Congress on Photosynthesis.* In press.

— Kagawa, T., and Craig, S. (1975). *Aust. J. Pl. Physiol.* **2**, 111–28.

Hattersley, P. W., Watson, L. and Osmond, C. B. (1976) in *Transport and transfer processes in plants* (eds. I. F. Wardale and J. Passiouriou). Academic Press, New York.

Huber, S. C. and Edwards, G. E. (1975). *Pl. Physiol., Lancaster* **55**, 835–44.

— — (1976). *Biochim. biophys. Acta* **449**, 420–33.

— — (1977*a*). *Biochim. biophys. Acta* **462**, 583–602.

— — (1977*b*). *Biochim. biophys. Acta* **462**, 603–12.

— Hall, T. C., and Edwards, G. E. (1976). *Pl. Physiol., Lancaster* **57**, 730–3.

Kagawa, T. and Hatch, M. D. (1974*a*). *Aust. J. Pl. Physiol.* **1**, 51–64.

— — (1974*b*). *Biochem. Biophys. Res. Commun.* **59**, 1326–31.

— — (1975). *Archs Biochem. Biophys.* **167**, 687–96.

— — (1977). *Archs Biochem. Biophys.* **184**, 290–7.

Kenyon, W. H., Kringstad, R., and Black, C. C. (1978). *FEBS Lett.* **94**, 281–3.

Kluge, M. (1978) in *Photosynthesis '77* (eds. D. O. Hall, J. Coombs, and T. W. Goodwin) pp. 335–45. Biochemical Society, London.

— and Ting, I. P. (1978). *Crassulacean acid metabolism*, Ecological Studies No. 30, Springer, Berlin.

— Bocher, M., and Jungnickel, G. (1980). *Z. Pflanzenphysiol.* **97**, 197–204.

Krenzer, E. G., Moss, D. L., and Crookston, R. K. (1975). *Pl. Physiol., Lancaster* **56**, 194–206.

Laetsch, W. M. (1974). *A. Rev. Pl. Physiol.* **25**, 27–52.

Luttge, U. and Ball, E. (1980). *Pl. Cell Envir.* **3**, 195–200.

Mbaku, S. B., Fritz, G. J., and Bowes, G. (1978). *Pl. Physiol., Lancaster* **62**, 510–15.

Moore, P. D. (1979). *Nature, Lond.* **280**, 193–4.

Morgan, J. A., Brown, R. H., and Reger, B. J. (1980). *Pl. Physiol., Lancaster* **65**, 156–9.

Osmond, C. B. (1978) in *CO_2 metabolism and plant productivity* (eds. R. H. Burris, and C. C. Black) pp. 217–34. University Park Press, Baltimore, MD.

Outlaw, W. H., Manchester, J., Dicamelli, C. A., Randall, D. D., Rapp, B., and Veith, G. M. (1979). *Proc. natn. Acad. Sci. U.S.A.* **76**, 6371–5.

Pierre, G. N. and Queiroz, O. (1979). *Planta* **144**, 143–51.

Queiroz, O. (1974). *A. Rev. Pl. Physiol.* **25**, 115–34.

Raghavendra, A. S., Rajendrudu, G. and Das, V. R. S. (1978). *Nature, Lond.* **273**, 143-4.
Raschke, K. and Dittrich, P. (1977). *Planta* **134**, 69-75.
Rathnam, C. K. M. (1978). *Sci. Prog.* **65**, 409-35.
— and Chollet, R. (1978). *Biochem. Biophys. Res. Commun.* **85**, 801-8.
— — (1980). *Progress in phytochemistry,* Vol. 6 (eds. L. Reinhold, J. B. Harborne, and T. Swain) pp. 1-48. Pergamon Press, Oxford.
— and Edwards, G. E. (1977). *Planta* **133**, 135-44.
Ray, T. B. and Black, C. C. (1976). *J. biol. Chem.* **251**, 5824-6.
— — (1979) in *Photosynthesis II, Encyclopaedia of plant physiology,* Vol. 6 (eds. M. Gibbs and E. Latzko) pp. 77-101. Springer, Berlin.
Salin, M. L. and Black, C. C. (1974). *Plant Sci. Lett.* **2**, 303-8.
— Campbell, W. H., and Black, C. C. (1973). *Proc. natn. Acad. Sci. U.S.A.* **70**, 3730-3.
Schnarrenberger, R. C., Gross, D., Burkhard, Ch., and Herbert, M. (1980). *Planta* **147**, 477-84.
Shomer-Ilan, A., Neumann-Ganmore, R., and Waisel, Y. (1979). *Pl. Physiol., Lancaster* **64**, 963-5.
Spalding, M. H., Schmitt, M. R., Ku, S. B., and Edwards, G. E. (1979). *Pl. Physiol., Lancaster* **63**, 738-43.
— Arron, G. P., and Edwards, G. E. (1980). *Archs Biochem. Biophys.* **199**, 448-56.
Steiger, E., Ziegler, J., and Ziegler, H. (1971). *Planta* **96**, 109-18.
Szarek, S. R. and Ting, I. P. (1977). *Photosynthetica* **11**, 330-42.
Thorpe, N., Brady, C. J., and Milthorpe, F. L. (1979). *Aust. J. Pl. Physiol.* **6**, 409-16.
Troughton, J. H. (1979) in *Photosynthesis II, Encyclopaedia of plant physiology,* Vol. 6 (eds. M. Gibbs and E. Latzko) pp. 140-9. Springer, Berlin.
Walker, G. H. and Izawa, S. (1980). *Pl. Physiol., Lancaster* **65**, 685-90.
Winter, K. and Luttge, U. (1979). *Ber. dt. Bot. Ges.* **92**, 117-32.
Yeoh, H. H., Badger, M. R., and Watson, L. (1980). *Pl. Physiol., Lancaster* **66**, 1110-12.

6 IMPORT AND EXPORT ACROSS THE CHLOROPLAST ENVELOPE IN C_3 PLANTS

6.1 Introduction

During photosynthesis chloroplasts fix carbon dioxide in order to provide the leaf with the substances required for cell growth and renewal. Both carbon dioxide and phosphate must enter the chloroplast, and carbon compounds must be exported from it to allow sucrose synthesis in the cytosol (Chapter 4). Illumination of leaves raises the ATP/ADP ratio, not only in the chloroplast but also in the cytoplasm (Krause and Heber 1976), so there must be some mechanism for ATP transfer between these compartments. Chloroplasts provide the cytoplasm with reducing power for processes such as nitrate reduction (Chapter 10) and so NAD(P)H must also be exported.

The chloroplast envelope consists of two lipoprotein membranes. Because of their chemical nature, such membranes will allow the passage of uncharged, non-polar molecules, but they will not normally be permeable to charged species such as phosphate, Mg^{2+}, ATP, NADP(H), or sugar phosphates. Table 6.1 lists some common metabolites which do not appear to cross the envelope at significant rates. Other charged substances, such as phosphoglycerate, can cross the envelope only because of the presence of specific transport systems. Our knowledge of such transport systems has come from a variety of experimental approaches, which are briefly indicated below. It must be pointed out, however, that most studies have been carried out on chloroplasts from mature spinach leaves, and that chloroplasts from younger plants or other species might have different permeability properties.

6.2 Methods to study metabolite transfer across the envelope

6.2.1 Determination of metabolite distribution in vivo *or* in vitro

In many early studies of metabolite transfer, the distribution of radioactively-labelled metabolites between chloroplast and cytoplasm was determined after feeding $^{14}CO_2$ to leaves. In order to prevent the redistribution of water-soluble compounds between the various subcellular fractions during isolation, non-aqueous fractionation techniques were usually employed. Initially, radioactivity was detected only within the chloroplast fraction, but it rapidly appeared in the cytoplasm in the form of sucrose, phosphoglycerate, fructose 6-phosphate, glucose 6-phosphate, fructose diphosphate, and UDP-glucose. In contrast, labelled ribulose and sedoheptulose diphosphates remained confined to the chloroplast and did not enter the cytoplasm, which suggests that they cannot cross the envelope. However, the appearance of the other metabolites in the cytoplasm does not mean that they left the chloroplast as such. For example, there is evidence showing (Section 6.2.3) that the labelled fructose diphosphate found in the cytoplasm is probably synthesized there from C_3 compounds

Table 6.1. Permeability properties of the chloroplast envelope of C_3 plants

(1) Compounds which do not enter or leave at significant rates

NAD^+	$NADP^+$	Phosphoenolpyruvate	Sedoheptulose diphosphate
NADH	NADPH	Sedoheptulose-7-phosphate	Ribulose diphosphate
Sorbitol	Pyrophosphate	Fructose diphosphate	Sucrose
Glycine	Alanine	Serine	Small cations (Na^+, Mg^{2+}, H^+)
		Acetyl-coenzyme A	(there may be a small, light-dependent flux of H^+ – see Chapter 2)

(2) Compounds which can penetrate, probably by simple diffusion of the unionized form

CO_2	Formate	Methanol	Glycollate
Acetate	Glycerol	Ethanol	Glycerate
Glyoxylate	Pyruvate		

(3) Compounds for which specific carrier systems exist, but the maximal rate of transport is low

Carrier system	*Compounds transported*	*Comments*
Adenine nucleotide translocator	ATP, ADP, AMP	Most ATP exported from chloroplasts indirectly, probably as C_3 compounds (see text). Rate of translocator <5 μmol h^{-1} mg chlorophyll^{-1} in chloroplasts from mature spinach leaves.
Sugar carrier	D-glucose, D-mannose, D-xylose, D-ribose, L-arabinose, maltose.	Inhibited by phloretin. Rates <7 μmol h^{-1} mg chlorophyll^{-1}. Physiological significance unclear.
Amino acid carrier	L-leucine, L-isoleucine	Found in pea chloroplasts only (McLaren & Barber 1977). Rates ~4 μmol h^{-1} mg chlorophyll^{-1}. Similar rates found in spinach chloroplasts but attributed to simple diffusion (Heldt 1976). Glycine not transported.
Purine and pyrimidine carriers	Adenine Cytosine	Reported in pea chloroplasts only (Barber and Thurman 1978a). Very low rates. Physiological significance unknown.

(4) Compounds for which specific carrier systems exist, with high rates of transport

Carrier system	*Compounds transported*	*Comments*
C_3 translocator	Phosphoglycerate Inorganic phosphate Dihydroxyacetone phosphate	Exchange system. Allows import of phosphate into chloroplast and export of C_3 compounds for sucrose synthesis in cytoplasm, and possibly for generation of ATP, NADH, and NADPH (see text).
Dicarboxylate translocator	Malate, succinate Aspartate Glutamate Oxaloacetate	Exchange system, but a small amount of unidirectional flow possible (Heldt 1976). Required for import of oxaloacetate for aspartate synthesis in chloroplasts and export of glutamate formed by glutamate synthetase (Chapter 10). May have a role in export of reducing power to cytoplasm.

(5) Compounds apparently transported at significant rates for which no specific mechanism is yet known

ribose 5-phosphate; ribulose 5-phosphate; xylulose 5-phosphate

---

Unless otherwise stated, conclusions are based on the reviews by Walker (1974), by Heldt (1976) and by Krause and Heber (1976). The permeability barrier of chloroplasts is the inner envelope membrane. Almost all studies have been carried out on chloroplasts from mature spinach leaves.

exported from the chloroplast, rather than being exported directly from the chloroplast itself. Hence the *in vivo* approach can give misleading results.

Related studies have been carried out using isolated type A chloroplasts. During carbon dioxide fixation by such preparations, certain metabolites are released into the medium and an examination of them should give some idea of which compounds can cross the envelope. For example, ribulose and sedoheptulose diphosphates rarely appear in the medium, whereas phosphoglycerate does so rapidly. Provided that great care is taken to allow for the presence of 'leaky' (type B) chloroplasts in the type A preparation, this approach can give useful results.

6.2.2 Effect of metabolites on the photosynthetic activity of chloroplasts

If a compound which cannot penetrate the envelope is added to type A chloroplast preparations, it should have no effect on their photosynthetic activity, nor should it be acted upon by an enzyme located within the chloroplast. Conversely, compounds that do have effects may be assumed to have penetrated (but see below).

Isolated type A spinach chloroplasts will not photoreduce $NADP^+$, and their rate of photophosphorylation using added ADP is slow. NAD(H) or NADP(H)-dependent enzymes present in the stroma cannot be assayed in intact chloroplasts using external cofactors (e.g. see Table 1.6). It may be concluded that the envelope is impermeable to NAD(H) and NADP(H) and only slightly permeable to ADP. Addition of phosphoglycerate is effective in decreasing the length of the induction period of carbon dioxide fixation in chloroplasts isolated from leaves kept in the dark (Chapter 4), but sedoheptulose diphosphate is not. Hence phosphoglycerate seems to penetrate the envelope whereas sedoheptulose diphosphate cannot.

It must be remembered, however, that type A chloroplast fractions are often contaminated by cytoplasmic enzymes (Chapter 1). Hence an added metabolite which has an effect on the chloroplasts might have been converted by extra-chloroplast enzymes into something else, which is the true effector. Fructose diphosphate, which poorly penetrates the envelope, can easily be converted by contaminating aldolase activity into phosphoglyceraldehyde and dihydroxy-acetone phosphate, which penetrate easily. Calvin cycle enzymes released from the variable percentage of broken chloroplasts always present in a type A preparation can also bring about such transformations. It is possible that such metabolic transformations account for some of the reported penetration of pentose monophosphates into chloroplasts (Table 6.1).

6.2.3 Measurement of uptake into chloroplasts

If a compound can penetrate the envelope, it should be found inside chloroplasts that have been incubated in its presence. Such uptake is usually studied by the 'silicone oil' method. A centrifuge tube containing a denaturing agent such as perchloric acid is taken, and on top of the acid is placed a layer of light silicone oil, followed by the chloroplast suspension on top of this (Heldt 1980). The radioactively-labelled substance whose penetration is to be studied is added to the chloroplast suspension, and after the required incubation time the tube is rapidly centrifuged. The chloroplasts spin through the oil layer, which brushes off most of the medium adhering to them, and are destroyed by the denaturant, which can then be analysed for radioactivity. The total sedimentation time can be less than 2 s, and the technique is often used for measurements of the kinetics of uptake (Heldt 1976, 1980).

Experiments of this type showed that part of the chloroplast volume was accessible to almost all low molecular-weight compounds tested, such as sucrose or sugar phosphates, but not to proteins or to dextran. The size of the 'sucrose permeable' space varied with the osmolarity of the medium surrounding the chloroplasts and was found to correlate with the size of the space between the two envelope membranes as determined by electron microscopy. For example, in a hypotonic medium the intermembrane space is small but its size increases as tonicity is increased, and the size of the sucrose-permeable space follows a similar pattern. It is therefore believed that the sucrose-permeable space is identical with the intermembrane space, which means that the outer envelope membrane must be unspecifically permeable to small molecules. If so, it follows that the discrimination between penetrant and non-penetrant molecules is achieved by the inner envelope membrane. A similar situation exists in mitochondria from both plant and animal tissues.

6.2.4 Studies on isolated envelope membranes

Isolated envelope membrane vesicles have been shown to exhibit permeability properties similar to those of intact chloroplasts, and they should therefore prove a useful tool for checking results obtained by other methods, such as the silicone oil technique (Poincelot 1975).

6.2.5 Osmotic response of isolated chloroplasts

Isolated spinach chloroplasts suspended in sucrose solutions behave as perfect osmometers, undergoing swelling or shrinkage to an extent dependent on the osmolarity of the sucrose solution (Nobel and Wang 1970). This confirms the impermeability of the inner envelope membrane to sucrose. If a metabolite can penetrate the envelope, then the osmotic responses will be smaller when the chloroplast is placed in a solution of that metabolite than they would be in a sucrose solution. Although this technique has been employed to study the entry of amino acids (Nobel and Wang 1970) and of alcohols (Wang and Nobel 1971) into pea chloroplasts, it is not widely used because it is difficult to relate the degree of swelling to the rate of penetration into the chloroplast. Also, the osmotic swelling technique would not detect uptake that occurs by exchange with internal solutes, since this would not result in an osmotic water flux.

6.3 The permeability systems detected

6.3.1 Simple diffusion of carboxylic acids

The inner envelope membrane is permeable to certain monocarboxylic acids, e.g. acetic acid, formic acid, glyceric acid, glycollic acid, and glyoxylic acid, which are non-polar enough to cross the membrane in their unionized form without the aid of specific transport mechanisms. Their distribution between chloroplast and cytoplasm will be determined by simple concentration gradient effects, e.g. formation of glycollate in chloroplasts during photorespiration (Chapter 7) will lead to its diffusion out into the cytoplasm. The presence of a pH gradient between the chloroplast stroma and the cytoplasm in the light will also influence the distribution of these molecules.

6.3.2 Movement of carbon dioxide

When HCO_3^- is supplied to isolated chloroplasts or to envelope vesicles, it appears to penetrate at a rate that greatly exceeds the maximum rate of carbon dioxide fixation by chloroplasts *in vitro*. Heldt (1976) has suggested that the compound that actually crosses the inner membrane is not HCO_3^- but carbon dioxide, which is known to traverse biological membranes rapidly. If this is so, then HCO_3^- added to the medium would equilibrate with carbon dioxide, which would cross the envelope and re-equilibrate with HCO_3^- in the stroma. Indeed, the rate of isotope uptake by chloroplasts illuminated in the presence of low concentrations of $H^{14}CO_3^-$ is increased by addition of carbonic anhydrase to the suspension medium (Shiraiwa and Miyachi 1978). *In vivo*, carbonic anhydrase is present in both the chloroplasts and the cytoplasm of leaf tissues (Read 1979).

If the rate of carbon dioxide diffusion is assumed not to be a rate-limiting step, then the carbon dioxide concentrations on each side of the envelope should be equal. Hence the distribution of HCO_3^- between suspension medium and stroma would be given by the following equation.

$$\log_{10} \frac{[HCO_3^-]_{stroma}}{[HCO_3^-]_{medium}} = pH_{stroma} - pH_{medium} \qquad (6.1)$$

Heldt (1976) found that this equation was obeyed in the case of isolated spinach chloroplasts, which suggests that the above assumption of rapid penetration of carbon dioxide is true. On illumination of chloroplasts the stroma becomes more alkaline (Chapter 2), the value of ($pH_{stroma} - pH_{medium}$) increases and so the ratio $[HCO_3^-]_{stroma}/[HCO_3^-]_{medium}$ must also increase, i.e. relatively more HCO_3^- is present in the stroma. However, the action of the more alkaline pH on the reactions catalysed by stromal carbonic anhydrase (eqn (6.2))

$$HCO_3^- + H^+ \rightleftarrows H_2CO_3 \rightleftarrows CO_2 + H_2O \quad (6.2)$$

would be to decrease the CO_2/HCO_3^- ratio, i.e. the increased stromal $[HCO_3^-]$ does *not* correspond to an increased supply of carbon dioxide, the true substrate for the ribulose diphosphate carboxylase/oxygenase enzyme (Chapter 3).

6.3.3 Movement of inorganic phosphate and of triose phosphates

The silicone-oil technique (Heldt 1976) and studies on envelope vesicles (Poincelot 1975) have shown that 3-phosphoglycerate, dihydroxyacetone phosphate, and inorganic phosphate can penetrate the chloroplast envelope rapidly. In contrast, glucose 6-phosphate, fructose 6-phosphate, ribulose diphosphate, sedoheptulose diphosphate, and fructose diphosphate are scarcely taken up at all.

Unlike the uptake of molecules which enter by simple diffusion, uptake of the above triose phosphates shows saturation kinetics, from which K_m and V_{max} values can be determined. V_{max} values are greater than 100 μmol of C_3 compound transported h^{-1} mg chlorophyll^{-1} for spinach chloroplasts. Phosphoglycerate, dihydroxyacetone phosphate and inorganic phosphate each competitively inhibits the uptake of the others. The K_i values for inhibition of transport are identical with the K_m values for transport of the metabolite by itself. Hence it has been proposed that all these compounds are transported by the same 'facilitated diffusion' carrier, which has also been shown to catalyse a counter-exchange process. For example, uptake of phosphoglycerate into the chloroplasts is matched by release of an equivalent amount of phosphate or of dihydroxyacetone phosphate. The likely significance of this shuttle system *in vivo* is to allow import of phosphate into the chloroplast for photophosphorylation in exchange for the triose sugars needed for biosynthetic reactions (e.g. of sucrose) in the cytoplasm. Thiol groups are involved in the functioning of the C_3 translocator, since it is inhibited by reagents that bind to such groups. Pyrophosphate inhibits the C_3 translocator, but it is not itself transported at significant rates in chloroplasts from mature spinach leaves (Flugge, Freisl, and Heldt 1980).

6.3.4 Movement of ATP, ADP, and AMP

On illumination of leaves, the ATP/ADP ratio in the cytoplasm rises rapidly and so ATP must be quickly exported from chloroplasts. An exchange porter system for adenine nucleotides (ATP, ADP, AMP) has been detected in the inner envelope membranes of spinach chloroplasts but its V_{max} values are low, usually less than 5 μmol h^{-1} mg chlorophyll^{-1} at 20 °C. Also, the porter works best when supplied with *external* ATP rather than ADP, so it seems unlikely that it contributes much to the rapid supply of ATP to the cytoplasm in illuminated leaves. Indeed, it might be argued that it is better adapted to transport cytoplasmic ATP into chloroplasts for biosynthetic reactions in the dark. The relationship, if any, of the adenine nucleotide translocator to the ATPase and adenylate kinase activities detected on chloroplast envelope membranes (Chapter 1) remains to be discovered.

So how is ATP exported from the chloroplast in the light? An efficient system would be the operation of a phosphoglycerate/dihydroxyacetone phosphate shuttle, as shown in Fig. 6.1. Dihydroxyacetone phosphate from the Calvin cycle could be exported from the chloroplast through the 'phosphate translocator'. In the cytoplasm it could yield ATP by a partial glycolytic pathway and the resulting phosphoglycerate could enter the chloroplast in exchange for more dihydroxyacetone phosphate. Hence both ATP and NADH would be generated in the cytoplasm. This shuttle would presumably operate alongside the phosphate–triose phosphate exchange that provides carbon for sucrose synthesis in the cytoplasm. It can be demonstrated to occur at high rates when appropriate glycolytic enzymes are added to chloroplasts *in vitro* (Krause and Heber 1976) and it seems to be a perfectly feasible transfer mechanism. Direct proof that it operates *in vivo* is lacking, however. It should also be noted that

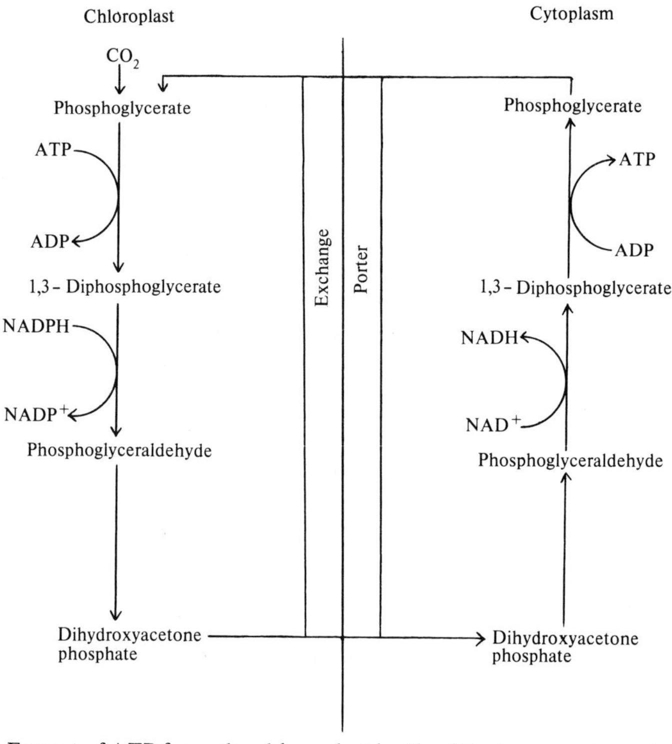

Fig. 6.1. Export of ATP from the chloroplast by the dihydroxyacetone phosphate/phosphoglycerate shuttle. The shuttle is written as a closed cycle, but some triose phosphates will be retained in the cytoplasm for biosynthesis and the balance required by the exchange porter maintained by import of phosphate from the cytoplasm.

current views of the mechanism of photorespiration (Chapter 7) show that metabolism of the glycine produced in the cytoplasm from glycollate generated by the chloroplast gives rise to large amounts of ATP. Hence the requirement for the triose phosphate shuttle (Fig. 6.1) might be small under conditions of rapid photorespiration. The envelopes of chloroplasts isolated from young leaves of certain plants, e.g. pea, show a significantly greater permeability to adenine nucleotides than do the envelopes of mature spinach chloroplasts, and shuttle systems will not necessarily operate in these tissues.

6.3.5 Movement of dicarboxylic acids

The inner envelope membrane has been shown to contain a porter system capable of translocating several dicarboxylic acids at high rates. Again, it seems to operate mainly as an exchange system. The preferred substrates of the dicarboxylate translocator are oxaloacetate and L-malate, but fumarate, succinate, 2-ketoglutarate, and L-aspartate are also translocated at significant rates (Heldt 1976). L-Glutamate and, to a lesser extent, glutamine are more slowly translocated by this system (Barber and Thurman 1978*a*, *b*), but tricarboxylic acids or phosphate are not transported at all.

The dicarboxylate porter is required for import of oxaloacetate into chloroplasts to allow aspartate synthesis in the stroma and for the export of glutamate produced by the glutamate synthetase reaction (Chapter 10). Since malate dehydrogenases are present in both chloroplast and cytoplasm (Chapter 3), it has also been suggested that the dicarboxylate translocator could be involved in export of reducing power from the chloroplast. NADPH lost from the chloroplast would appear as NADH in the cytoplasm (Fig. 6.2). Such shuttles would presumably be needed because of the impermeability of the envelope to NAD(H)

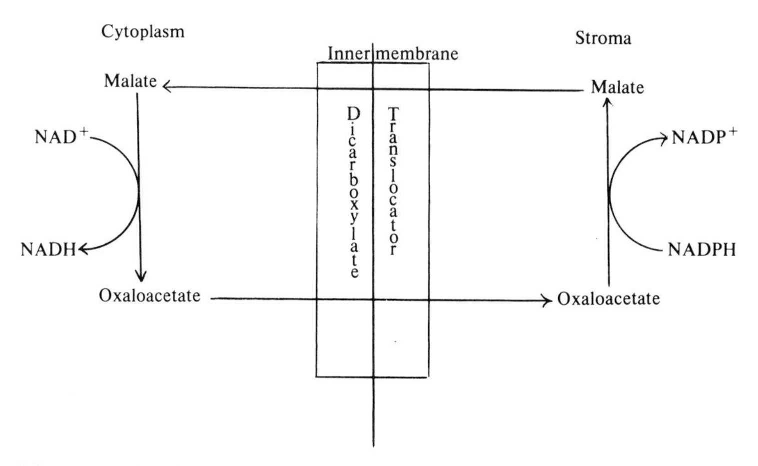

Fig. 6.2. A 'malate/oxaloacetate' shuttle for export of reducing power from illuminated chloroplasts, or its import into chloroplasts.

and to NADP(H). The malate/oxaloacetate shuttle can be made to occur *in vitro* by adding NAD^+-malate dehydrogenase and its substrates to type A chloroplasts (Krause and Heber 1976; Anderson and House 1979), but the rate *in vivo* might be limited because the concentration of oxaloacetate in plant cells is so low that its rate of transport by the dicarboxylate translocator would be slow (K_m of translocator for oxaloacetate is about 0.26 mM: cytoplasmic concentration is less than 0.02 mM – Krause and Heber 1976). Since isoenzymes of aspartate aminotransferase are present in chloroplasts and cytoplasm, and cellular concentrations of amino acids are much greater than those of oxaloacetate, it has been suggested that a more complex shuttle might operate *in vivo* to export reducing power from the chloroplast (Fig. 6.3). It must be realized, however, that these shuttle systems are to some extent 'paper chemistry' in that there is no direct evidence for their operation *in vivo*. Export of reducing power to the cytoplasm, e.g. for nitrate reduction, can also be achieved by the phosphoglycerate shuttle (Fig. 6.1). If NADH produced as in Fig. 6.1 is not used up quickly enough in the cytoplasm, it is possible that the malate/aspartate/oxaloacetate shuttle systems could operate in reverse and drive reducing power back into the chloroplast (Giersch, Heber, Kaiser, Walker, and Robinson 1980).

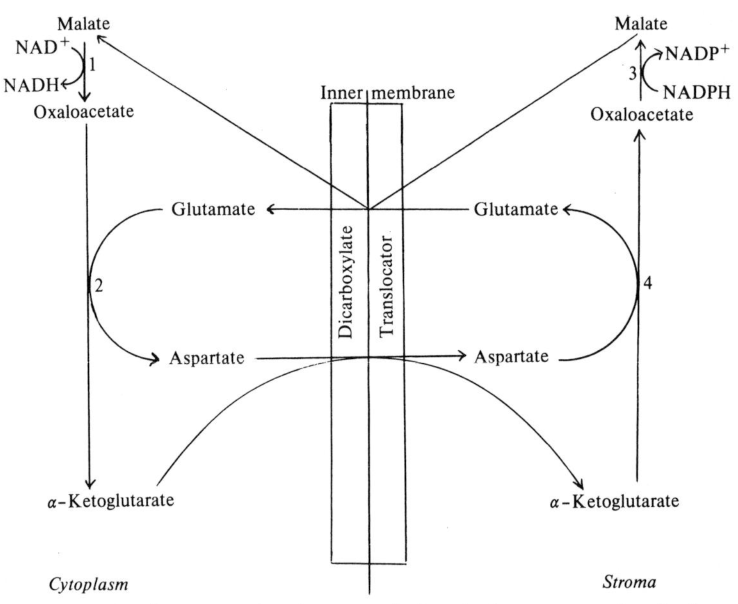

Fig. 6.3. A 'malate/aspartate/oxaloacetate' shuttle for export of reducing power from illuminated chloroplasts or its import into chloroplasts. Since the concentration of aspartate in plant tissues is higher than that of oxaloacetate, it has been suggested that the above shuttle could achieve higher rates of NADH formation in the cytoplasm than could the malate/oxaloacetate shuttle (Fig. 6.2). Enzymes: 1-cytoplasmic NAD^+-malate dehydrogenase; 2-cytoplasmic isoenzyme of aspartate aminotransferase; 3-chloroplast $NADP^+$-malate dehydrogenase; 4-chloroplast isoenzyme of aspartate aminotransferase.

Cytoplasmic NADPH is required for biosynthetic purposes, but the extent to which it is supplied by the operation of the oxidative pentose phosphate pathway in the cytoplasm of illuminated leaves is not clear. This pathway can certainly operate in the light, although the rates seem to be lower than those in the cytoplasm of darkened leaves (Raven 1972; Anderson and Nehrlich 1977). Leaf cytoplasm contains a non-reversible $NADP^+$-dependent glyceraldehyde 3-phosphate dehydrogenase (Bamberger, Ehrlich, and Gibbs 1975) which could function in a shuttle mechanism to generate NADPH in the cytoplasm (Fig. 6.4). This shuttle, if it occurs, would have to be integrated with that shown in Fig. 6.1.

6.3.6 *Movement of sugars*

The silicone-oil technique has detected the presence of a sugar-transporting carrier system in the envelope of spinach chloroplasts (Schafer, Heber, and Heldt 1977). It can transport D-glucose, D-mannose, D-xylose, D-ribulose, and L-arabinose by an exchange system, it is inhibited by phloretin and it will not transport L-glucose, in the latter two respects being similar to the glucose

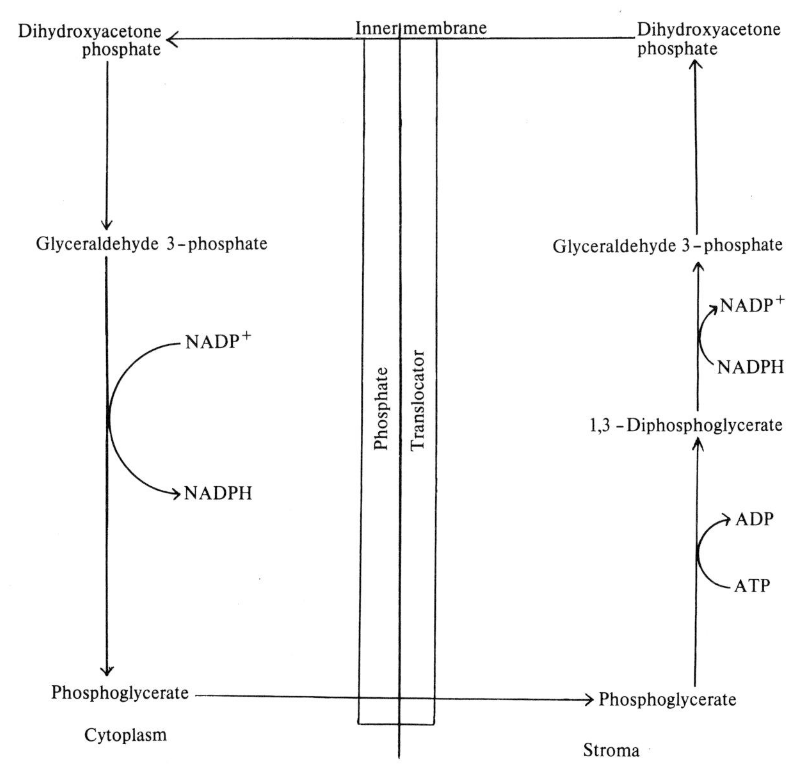

Fig. 6.4. Proposed shuttle for export of NADPH from chloroplast to cytoplasm.

transport systems found in animal tissues. The maximum rates of sugar transport are, however, very low (less than 7 μmol h^{-1} mg $chlorophyll^{-1}$) and the physiological significance of the porter is unkown. Although it has been suggested that glucose formed by the amylase pathway of starch degradation (Chapter 4) might have to be transported into the cytoplasm for breakdown by glycolysis if hexokinase is not present in chloroplasts (Schafer *et al.* 1977), this pathway is not a major route of starch degradation in spinach chloroplasts. Maltose, another product of starch degradation, can also be translocated by this sugar carrier (Herold, Leegood, McNeil, and Robinson 1981). Pea chloroplasts also have a sugar transport system in their envelopes (Wang and Nobel 1971).

6.3.7 Movement of amino acids

Experiments on pea chloroplasts using the osmotic swelling technique suggested that the envelopes were freely permeable to all amino acids (Nobel and Cheung 1972), but this has been disputed (McLaren and Barber 1977). When the question was investigated on spinach chloroplasts using either the swelling technique (Gimmler, Schafer, Kraminer, and Heber 1974) or the silicone oil technique, the rates of transport were found to be greatest for hydrophobic amino acids such as isoleucine and phenylalanine, which were suggested to enter by simple diffusion. Even for these amino acids, the rates of uptake at physiological amino acid concentrations were less than 4 μmol mg $chlorophyll^{-1}$ h^{-1} (Heldt 1976). More polar amino acids such as glycine, serine, and L-alanine, were not transported at significant rates when supplied to chloroplasts at physiological concentrations. McLaren and Barber (1977) found similar low rates with pea chloroplasts, but concluded that leucine transport was mediated by a carrier of low activity (Table 6.1). However, glutamate and aspartate are rapidly translocated by the dicarboxylate carrier.

References

Anderson, J. W. and House, C. M. (1979). *Pl. Physiol., Lancaster* **64**, 1064–9.

Anderson, L. E. and Nehrlich, S. C. (1977). *FEBS Lett.* **76**, 64–6.

Bamberger, E. S., Ehrlich, B. A., and Gibbs, M. (1975). *Pl. Physiol., Lancaster* **55**, 1023–30.

Barber, D. J. and Thurman, D. A. (1978*a*). *Plant Cell Env.* **1**, 305–6.

— — (1978*b*). *Plant Cell Env.* **1**, 297–303.

Flugge, U. I., Freisl, M., and Heldt, H. W. (1980). *Pl. Physiol. Lancaster* **65**, 574–7.

Giersch, C., Heber, U., Kaiser, G., Walker, D. A., and Robinson, S. P. (1980). *Archs Biochem. Biophys.* **205**, 246–59.

Gimmler, H., Schafer, H., Kraminer, H., and Heber, U. (1974). *Planta* **120**, 47–61.

Heldt, H. W. (1976) in *The intact chloroplast, Topics in photosynthesis* Vol. 1 (ed. J. Barber) pp. 215–34. Elsevier, Amsterdam.

— (1980). *Meth. Enzymol.* **69C**, 604–13.

Herold, A., Leegood, R. C., McNeil, P. H., and Robinson, S. P. (1981). *Pl. Physiol., Lancaster* **67**, 85–8.

Krause, G. H. and Heber, U. (1976) in *The intact chloroplast, Topics in photosynthesis,* Vol. 1 (ed. J. Barber) pp. 171–214. Elsevier, Amsterdam.
McLaren, J. S. and Barber, D. J. (1977). *Planta* **136**, 147–51.
Nobel, P. S. and Cheung, Y. N. S. (1972). *Nature New Biol.* **237**, 207–8.
Nobel, P. S. and Wang, C. T. (1970). *Biochim. biophys. Acta* **211**, 79–87.
Poincelot, R. P. (1975). *Pl. Physiol., Lancaster* **55**, 849–52.
Raven, J. A. (1972). *New Phytol.* **71**, 227–47.
Read, M. L. (1979). *Pl. Physiol., Lancaster* **63**, 216–17.
Schafer, G., Heber, U., and Heldt, H. W. (1977). *Pl. Physiol., Lancaster* **60**, 286–9.
Shiraiwa, Y. and Miyachi, S. (1978). *FEBS Lett.* **95**, 207–10.
Walker, D. A. (1974) in *MTP International Review of Science,* Vol. 11 (ed. D. H. Northcote) pp. 1–49. Butterworths, London.
Wang, C. T. and Nobel, P. S. (1971). *Biochim. biophys. Acta* **241**, 200–12.

7 PHOTORESPIRATION

7.1 Introduction

In the dark, green leaf cells obtain their energy by the oxidation of stored food reserves, principally starch. They therefore rely heavily on the metabolic pathways of glycolysis, the Krebs cycle, and oxidative phosphorylation to provide ATP, and the rate of these pathways may easily be measured as carbon dioxide evolution. Carbon dioxide can also be released by the action of the oxidative pentose-phosphate pathway in both chloroplasts and cytoplasm (Chapter 4) operating to provide NADPH and pentose sugars. The rate of carbon dioxide release in the dark, which will be referred to as 'dark respiration', reaches a maximum at concentrations of oxygen as low as 2 per cent ($^v/_v$), presumably because of the high affinity of leaf mitochondria for oxygen.

The above pathways may continue to operate to some extent in the light (see Section 7.2) but in C_3 plants the rate of respiration in the light, measured as uptake of oxygen or release of carbon dioxide, is often much greater than the rate in the dark. This light-dependent extra respiration, which is often known as 'photorespiration', is the main reason why C_3 plants have such high carbon dioxide compensation points. As explained in Chapter 5, the carbon dioxide compensation point is defined as that concentration of carbon dioxide which allows photosynthesis to proceed at such a rate that carbon dioxide fixation just balances the carbon dioxide released by the leaf. At 25°C and atmospheric concentrations of oxygen (21 per cent), C_3 leaves usually have compensation points in the range 40–50 μl carbon dioxide per litre. In contrast the values for C_4 leaves are much lower, often being close to or equal to zero. Hence C_4 leaves are capable of net carbon dioxide fixation at much lower atmospheric concentrations than are C_3 leaves because they show little, if any, photorespiratory carbon dioxide release. This could be because photorespiration does not occur in C_4 leaves, or because any carbon dioxide released is immediately refixed by phosphoenolpyruvate carboxylase in the mesophyll cells (Chapter 5), a point which is discussed further in Section 7.5.

The rates of carbon dioxide release by photorespiration and by 'dark respiration' in the light are extremely difficult to measure, since the net flux of carbon dioxide is usually into the leaf due to the action of the Calvin cycle. Unlike dark respiration, however, the rate of photorespiration increases as the oxygen concentration around the leaf is increased, and it is not saturated even at 100 per cent oxygen. As a consequence, the net rate of carbon dioxide fixation by C_3 leaves (that is carbon dioxide fixed by the Calvin cycle minus carbon dioxide lost by photorespiration minus carbon dioxide lost by dark respiration) is greater when they are surrounded by an atmosphere containing only 2 per cent oxygen than it is in normal air (21 per cent oxygen), whereas oxygen concentration has a much smaller effect on carbon dioxide fixation by C_4 leaves. This 'oxygen effect' can be regarded as a diagnostic test for photorespiration: any gas-exchange phenomenon that one wishes to attribute to photorespiration must be shown to be stimulated by raising oxygen concentration in the range 2-21 per cent. For example, the carbon dioxide compensation point of C_3 leaves increases almost linearly with oxygen concentration.

A second difference between photorespiration and dark respiration is in the substrates that give rise to the carbon dioxide released. Isotopic labelling studies have shown that photorespiratory carbon dioxide is derived from early products of photosynthesis, clearly distinct from the fuel reserves metabolized by the 'dark' respiratory processes. Indeed, photorespiration can be attributed to the oxidative decarboxylation of the two carbon compound glycollic acid, formed during the operation of the Calvin cycle (Section 7.2.8). Hence for the purposes of this Chapter photorespiration will be defined as follows:

> Photorespiration is a light-dependent carbon dioxide release by leaves that is due to the oxidative decarboxylation of glycollate derived from the Calvin cycle and is inhibited at low oxygen concentrations.

7.2 The measurement of photorespiration

If the rate at which carbon dioxide is fixed into the Calvin cycle in an illuminated green leaf is x μmol h^{-1} mg chlorophyll^{-1}, the rate of photorespiration is y μmol CO_2 released h^{-1} mg chlorophyll^{-1} and the rate of carbon dioxide release by the Krebs cycle plus oxidative pentose phosphate pathway is z μmol h^{-1} mg chlorophyll^{-1} , then the photosynthetic rate actually measured (net photosynthesis) would be given by

$$\text{net photosynthesis} = x - (y + z) \tag{7.1}$$

This illustrates the first problem in measuring photorespiration (y): we need to know the value of z in the light. A second problem arises because a proportion of the carbon dioxide evolved during photorespiration and dark respiration will be fixed into the Calvin cycle before it escapes from the leaf, i.e. any determination of $(y + z)$ that is made by gas exchange measurements will be an underestimate of the true value. The percentage of carbon dioxide refixed has been estimated as 17-34 per cent of total carbon dioxide production in the light (e.g. Canvin 1979) but it might well vary with environmental conditions. Carbon dioxide released within leaf cells has to cross various membranes and exit through the leaf stomata, and the barrier presented by these systems to the diffusion of carbon dioxide must also be considered in interpreting gas exchange measurements (e.g. Jackson and Volk 1970; Canvin 1979).

7.2.1 What is the rate of dark respiration in the light?

There is considerable export of ATP and reducing power from chloroplast to cytoplasm in the light (Chapter 6) and inhibition of the oxidative pentose-phosphate pathway in the cytoplasm by increased $NADPH/NADP^+$ ratios and of mitochondrial oxidative phosphorylation by increased ATP/ADP ratios might be expected. A rise in the cytoplasmic $NADH/NAD^+$ ratio should decrease the turnover of the Krebs cycle. Hence one would expect a drop in the rate of 'dark respiration' in the light and there is much evidence in favour of this, although the extent of the drop may well depend on such factors as the light intensity, which can control the supply of ATP and reducing power to the cytoplasm. The oxidative pentose-phosphate pathway is prevented from operating in chloroplasts in the light (Chapter 4) and the rate of the pathway in the cytoplasm is probably reduced, although to what extent is difficult to evaluate (Raven 1972*a*). Measurement of the carbon dioxide compensation point of C_3 leaves as a function of the oxygen concentration in the surrounding atmosphere produces a graph which can often (but not always) be extrapolated through the origin, implying that, within the errors of the method, dark respiration ceases in the light (Chollet and Ogren 1975). Photosynthetically-assimilated $^{14}CO_2$ does not rapidly enter the organic acids of the Krebs cycle nor its associated amino acids (glutamate and aspartate) in the leaves of several plants. Since triose phosphates are rapidly exported from chloroplasts in the light (Chapter 6) this implies a block either in the later stages of glycolysis, at the pyruvate dehydrogenase step, or in the cycle itself. However, the complete inhibition of the cycle is unlikely since the feeding of labelled amino acids or organic acids to illuminated leaves produces a rapid entry of isotope into Krebs cycle intermediates. Raven (1972*b*) concluded that the rate of carbon dioxide production in the illuminated green plant cell by 'dark respiration' can be 25–100 per cent of that in the dark, depending on the tissue used and the conditions to which it is exposed. In young leaves, for example, the Krebs cycle may turn over rapidly in the light to provide bio-synthetic intermediates for growth (Graham and Chapman 1979). In summary, a definitive answer to the question posed at the head of this section is not yet possible.

7.2.2 Measurement of photorespiration as a 'post-illumination burst' (PIB)

In 1955, Decker discovered that illuminated leaves of several C_3 plants show a rapid rate of carbon dioxide release on switching off the light. It continues for 1-2 min, gradually falling to the rate characteristic of dark respiration. The size of this 'burst' of carbon dioxide release is increased by increasing the light intensity or temperature before darkening the leaf, and decreased at low oxygen concentrations, and it was suggested that the post-illumination burst is a continuation of photorespiration for a brief period in the dark. Hence the size of the burst could be used as a measure of the photorespiration rate. In agreement with this, the work of Canvin (1979) and his group showed that the specific activity of $^{14}CO_2$ evolved in a post-illumination burst was closely similar to that of the $^{14}CO_2$ evolved in photorespiration in the light. A post-illumination burst cannot be detected with leaves of the C_4 plant, maize, although it has been observed in some other C_4 plants, but in many cases such bursts are not eliminated at low oxygen concentrations and therefore cannot represent photorespiration as defined above (Canvin 1979). Perhaps they are due to carbon dioxide release from accumulated malate or aspartate (see Chapter 5 and Rathnam 1978).

The PIB assay is quite easy to perform and the results seem to correlate reasonably well with estimates of photorespiration rates by other methods. Obviously there will be some contribution to the carbon dioxide release from the dark respiratory pathways, and the possibility of changes in stomatal resistance to carbon dioxide movement during a light-dark transition must be considered. A 'low oxygen' control must be included to eliminate artefacts arising from other sources of carbon dioxide (see above).

There is a serious theoretical objection to the use of the PIB as a means of measuring photorespiration, however. The PIB presumably arises by the metabolism of residual glycollate and compounds derived from it (e.g. glycine and serine) to carbon dioxide. The PIB would thus depend on the amount of these intermediates present in the leaf (their 'pool size'), which is not necessarily related to the rate at which carbon is passing through these intermediates. The pool size of glycollate is usually very small (Section 7.4.1) but glycine and serine derived from it can sometimes accumulate in significant amounts.

7.2.3 Compensation point and carbon dioxide-extrapolation methods

As already explained, the size of the carbon dioxide compensation point for a leaf gives an indication of the photorespiratory capacity of a plant, although it will include a contribution from dark respiration if this process is not completely suppressed in the light. The size of the compensation point increases with oxygen concentration and with temperature, but it is independent of light intensity above a low value. The compensation points for C_3 leaves at 25°C under 21 per cent oxygen are usually in the range of 40–50 $\mu l\ CO_2\ l^{-1}$.

Photorespiration might also be assessed by plotting the rate of photosynthesis as a function of carbon dioxide concentration and extrapolating the resulting curve to zero carbon dioxide. The 'negative carbon dioxide uptake' thereby obtained should be a measure of photorespiration (Jackson and Volk 1970). This method would also detect dark respiration and does not overcome the reassimilation problem.

7.2.4 Measurement of photorespiration as carbon dioxide efflux into carbon dioxide free air

If a leaf is placed in a closed chamber, illuminated, and carbon-dioxide-free air passed over it, an infrared gas analyser can be used to measure the concentration of carbon dioxide in the issuing gas stream. From this value and the known rate of flow, the rate of carbon dioxide release by the leaf can be calculated as an estimate of photorespiration. Obviously, the occurrence of dark respiration and reassimilation of photorespiratory carbon dioxide are still problems. It must also be assumed that the carbon-dioxide-free air does not itself alter the rate of photorespiration measured, although in support of this, Decker (1955, 1959) found that the post-illumination burst was unaltered whether a leaf was maintained at the carbon dioxide compensation point or in normal air.

7.2.5 The 'leaf disc' method

A variation of the carbon dioxide efflux method of measuring photorespiration was developed by Goldsworthy (1966) and by Zelitch (1968, 1979*a*). In essence, they increased the sensitivity of the method by using radioactive carbon dioxide ($^{14}CO_2$). Leaf discs are allowed to photosynthesize in a closed vessel containing $^{14}CO_2$ until the compensation point is reached, and illumination is continued for a further period of about 30 min. The discs are then swept with carbon dioxide-free air either in light or in darkness, the $^{14}CO_2$ released trapped in alkali and the radioactivity of the resulting solution measured by liquid scintillation counting (Zelitch 1979*a*). The ratio of $^{14}CO_2$ evolved in the light to that evolved in the dark is taken as a measure of the rate of photorespiration, and changes in this ratio are attributed to increases or decreases in photorespiration rate. Photorespiration as determined by this method was decreased at low oxygen concentration and increased at high temperatures, in agreement with the results of other methods.

There is an important theoretical objection to the technique, however, which arises from the fact that only the total radioactivity of the $^{14}CO_2$ evolved is measured, and not the specific radioactivity. The specific radioactivity of $^{14}CO_2$ released in the light can often be greater than that in the dark because dark respiration uses stored fuel reserves which might not have become fully labelled during preillumination, whereas photorespiration uses recent products of photosynthesis. Further, the specific activities of the carbon dioxide evolved can change with time in both light and darkness. Hence the measured radioactivity of the carbon dioxide trapped cannot be simply related to the *actual amounts* of carbon dioxide released in the light or dark. Thus the technique cannot properly measure photorespiration rates nor the effect of environmental factors on them unless these specific activities are known. Such information is generally not available. The results will also be affected by the internal reassimilation of evolved carbon dioxide.

Data obtained by the leaf-disc assay are often at variance with those of other assays for photorespiration, both in terms of the high rates of photorespiration detected (Canvin 1979) and in showing the effects of potential inhibitors of photorespiration. For example, both Zelitch (1979*b*) and Chollet (1978) found that glycidic acid and glyoxylic acid inhibited photorespiration as measured by this assay in discs taken from tobacco leaves, but Chollet (1978) found that these inhibitors had no effect on photorespiration rates measured by several other methods.

7.2.6 Measurement of photorespiration by the oxygen inhibition method

It is generally accepted that photorespiration is suppressed at low oxygen concentrations. Indeed, 30–50 per cent increases in the net photosynthetic carbon dioxide assimilation rates of C_3 leaves are observed when the surrounding oxygen concentration is decreased from 21 to 1-3 per cent. This effect is not seen in C_4 leaves. The size of the increase could be taken as a measure of photorespiration. However, the oxygen inhibition of net carbon dioxide fixation is not only due to photorespiratory carbon dioxide release: it can be partially attributed to a direct competitive inhibition of ribulose diphosphate carboxylase by oxygen (Chapter 3) and to an increased generation of oxygen free-radicals within the chloroplast (Chapter 8). Hence the increase in carbon dioxide uptake obtained at low oxygen would be an overestimate of photo-respiration. In contrast, the increase could be an underestimate if photo-respiration were not completely suppressed at low oxygen. The generation of oxygen in illuminated chloroplasts means that their internal oxygen concentration would often be higher than that in the atmosphere surrounding the leaf. Indeed, isolated illuminated chloroplasts can produce glycollate at very low external oxygen concentrations (Section 7.3). Perhaps these two opposing factors might sometimes combine to produce the right answer, but one should not depend on this!

7.2.7 Measurement of photorespiration by the $^{14}CO_2$ uptake method

This method was designed by Canvin's group to overcome the problem of internal refixation of photorespiratory carbon dioxide. A leaf is allowed to photosynthesize in unlabelled carbon dioxide ($^{12}CO_2$) until a steady state is reached, at which time $^{14}CO_2$ is rapidly introduced. Initially $^{14}CO_2$ will be fixed into the Calvin cycle but not released by photorespiration, since the isotope will take time to move through the Calvin cycle and the glycollate pathway. Hence the difference between the initial rate of $^{14}CO_2$ uptake into the Calvin cycle and the previous rate of $^{12}CO_2$ uptake should be equal to the rate of photorespiration, although there will still be a contribution from dark respiration if it occurs in the light. Since recently-fixed $^{14}CO_2$ can be released and recycled within 15–45 s in leaf cells (e.g. Raven 1972b) measurements must be made more quickly than this. Times as low as 15 s have been used, although this introduces technical difficulties in measurement, especially at high carbon dioxide concentrations and low specific radioactivities. Also, $^{14}CO_2$ presented to the leaf enters and mixes with $^{12}CO_2$ in the gas spaces beneath the stomata, causing a dilution of its specific activity which will become more significant the shorter the time allowed for $^{14}CO_2$ assimilation.

7.2.8 Measurement of photorespiration as the rate of the glycollate pathway

Since photorespiration is largely due to the oxidative decarboxylation of glycine derived from glycollate (Section 7.4.4), then a measurement of the rate of synthesis and loss of glycollate, glycine, or its product serine in illuminated leaves should give an estimate of the true rate of photorespiratory carbon dioxide release. Such a measurement would be unaffected by carbon dioxide reassimilation and would not contain any contribution from dark respiration. The steady-state concentration of glycollate in illuminated leaves is extremely low, however, which makes it difficult to extract and measure. Hence glycine and serine are usually examined and are frequently determined together because of problems in separation. Unfortunately, the kinetics of labelling of serine by $^{14}CO_2$ supplied to leaves are extremely complicated and inconsistent with a single origin (Canvin, Fock, and Lloyd 1978), i.e. serine can be formed from sources other than glycollate. Indeed, metabolic pathways leading to serine from phosphoglycerate and glycerate are known to exist (Section 7.4.3). Hence the results of this technique are difficult to interpret unless the flux through glycine alone is measured. The amounts of carbon flowing through glycine and serine in illuminated leaves are sufficient, or more than sufficient, to account for observed rates of photorespiratory carbon dioxide release.

7.2.9 Overall assessment of methods used to measure photorespiration

The difficulties encountered in measuring photorespiration outlined above and summarized in Table 7.1 have been discussed more extensively in reviews by Goldsworthy (1970), Jackson and Volk (1970), Raven (1972b), Zelitch (1979*a*, *b*), Chollet and Ogren (1975), and Canvin (1979). It may be concluded that none of the methods available to measure the rate of photorespiration gives an accurate result, although determination of the rate of carbon flow through glycine seems to be potentially useful. Hence no report of photorespiration rates, nor of the effect of any compound or environmental parameter on such rates, can be accepted unless it has been demonstrated by at least two different techniques, preferably more (Martin, Ozbun, and Wallace 1972). However, several techniques do give comparable results, which are probably therefore near to the truth. Using the methods of carbon dioxide evolution into carbon dioxide-free air, extrapolation of the carbon dioxide response curve and $^{14}CO_2$ uptake, the rate of photorespiration by C_3 leaves in normal air (300 $\mu l\ l^{-1}$ of CO_2 and 21 per cent O_2) at 25 °C seems to be about 15–20 per cent of the rate of net photosynthesis, exceeding dark respiration by 1.2- to 4-fold (Canvin 1979; Chollet and Ogren 1975). Results obtained by the leaf disc method are larger, however, which casts doubt on the validity of this method (Section 7.2.5). It is generally agreed that the rate of photorespiration increases with increases in oxygen concentration. It also increases with temperature in the range 15–35 °C, faster than does photosynthesis (e.g. Keys, Sampaio, Cornelius, and Bird 1977; Fock, Klug and Canvin

Table 7.1. Methods adopted for measuring photorespiration (PR)

Name of method	Essential features	Problems
Post-illumination burst	Measure rate of CO_2 release immediately on darkening leaves	Essential to show inhibition at low $[O_2]$. Some contribution from dark respiration. Must check stomatal resistance. Not a good method on theoretical grounds.
Compensation point	Size gives indication of occurrence of photorespiration (e.g. C_3 leaves 40–50 $\mu l\ CO_2\ l^{-1}$; C_4 0–10 $\mu l\ CO_2\ l^{-1}$ at 25 °C and atmospheric $[O_2]$)	Contribution from dark respiration. Cannot be used to study effect of $[CO_2]$ on photorespiration.
Extrapolation method	Extrapolate graph of rate of photosynthesis against $[CO_2]$ to zero CO_2	Contribution from dark respiration could lead to overestimate of PR: reassimilation of CO_2 will underestimate PR
CO_2 efflux into CO_2-free air	Measure CO_2 release from leaf by IR gas analyser	Effects of dark respiration and reassimilation must be considered. Changes in stomatal aperture must be checked for during the 'sweep' with CO_2-free air. Assumes CO_2-free air does not change PR rate
Leaf disc method	Measure $^{14}CO_2$ release from pre-labelled leaf discs in light and dark	Radioactivity of $^{14}CO_2$ measured not necessarily related to amount of CO_2 evolved unless assumptions made about constancy of specific activity (which are unlikely to hold). Reassimilation of CO_2 will underestimate PR
O_2 inhibition	Compare rates of CO_2 uptake at 21% and 1–3% atmospheric $[O_2]$	PR not necessarily inhibited at low $[O_2]$ because of generation of O_2 inside chloroplast. O_2 inhibition of CO_2 uptake due to other factors besides PR
$^{14}CO_2$ uptake method	Initial rate of $^{14}CO_2$ uptake measures true rate of CO_2 fixation, uptake of unlabelled CO_2 measures net rate	Technically difficult ($^{14}CO_2$ enters glycollate pathway very quickly). Can be dilution problem with $^{12}CO_2$ in leaf spaces.
Glycollate pathway method	Measure rate of labelling of glycine and serine by $^{14}CO_2$ or $^{18}O_2$	Not all serine in leaves derived from glycollate. Assumes only one pathway of CO_2 release during PR

1979). At low light intensities the rate of carbon dioxide evolution by C_3 leaves in the light is less than that in darkness, but as the light intensity increases photorespiration increases in step with photosynthesis up to saturation. Hence the compensation point is not usually affected by an increase in light intensity above very low intensities, although it increases at higher temperatures or oxygen concentrations (Jackson and Volk 1970).

There is less agreement over the effects of carbon dioxide concentration on photorespiration rates. The PIB is decreased in bean, wheat, and sunflower leaves as carbon dioxide concentration increases, being eliminated at concentrations of 1200 $\mu l\ CO_2\ l^{-1}$, although there is often little effect of carbon dioxide concentrations in the range 0–300 $\mu l\ l^{-1}$ (an assumption made in using the technique of efflux into carbon-dioxide-free air, Section 7.2.4). However, Bravdo and Canvin (1979) using the $^{14}CO_2$ uptake method concluded that photorespiration rates did *not* decrease at high carbon dioxide concentrations in sunflower leaves (range used 20–1150 $\mu l\ CO_2$ $litre^{-1}$). Snyder and Tolbert (1974) found that the amount of carbon incorporated into the glycine–serine pool by tobacco leaves was constant in the range of atmospheric carbon dioxide concentrations from 300 $\mu l\ CO_2\ l^{-1}$ up to 1000 $\mu l\ CO_2\ l^{-1}$. Of course, since the rate of the Calvin cycle increases with carbon dioxide concentration (Chapter 4), even a constant rate of photorespiration becomes less significant when expressed as a percentage of net carbon dioxide fixation.

The importance of photorespiration in limiting the net carbon dioxide fixation, and hence the growth, of C_3 plants is illustrated by the observation that lowering the oxygen concentration around plants such as beans or soyabeans (but not C_4 plants) greatly increases the vegetative yield (Bjorkman, Hiesey, Nobs, Nicholson, and Hart 1968; Parkinson, Penman, and Tregunna 1974; Quebedeaux and Hardy 1975). Of course, part of the effect can be attributed to the other actions of oxygen on plant tissues (Section 7.2.6 and Chapter 8). Increasing the carbon dioxide/oxygen ratio in the air surrounding soyabean plants increases symbiotic nitrogen fixation by their root nodules: apparently the supply of fixed carbon to the roots, which is decreased at high rates of photorespiration and low rates of photosynthesis, is the factor limiting nitrogen-fixation (Quebedeaux, Harelka, Livak, and Hardy 1975). Hence the occurrence of photorespiration is of potentially great importance in relation to crop productivity: elimination of this process by chemical means or by selective breeding might increase crop yields significantly. Attempts have been made to breed C_4 characteristics into C_3 plants, but they have so far failed to yield progeny with low compensation points (Chapter 5). A slow growing tobacco mutant was reported to have much higher rates of photorespiration and thus lower rates of net photosynthesis than the wild-type strain (Zelitch 1979*a*, *b*) but these results were obtained using the leaf-disc assay method of photorespiration, the validity of which has been questioned (Section 7.2.5). In general in C_3 plants the rate of photorespiration seems to be positively correlated with the rate of photosynthesis (e.g. McCashin and Canvin 1979), i.e. C_3 leaves with high rates of photosynthesis also show high rates of photorespiration. Perhaps this is not surprising, since glycollate is derived from the Calvin cycle.

The design of a chemical inhibitor of photorespiration requires a knowledge of the biochemistry of glycollate synthesis and degradation, which will be considered in the next Sections.

7.3 Synthesis of glycollic acid, the substrate for photorespiratory carbon dioxide release

It is now generally accepted that photorespiration is due to the formation by the chloroplast of the two-carbon compound glycollic acid (2-hydroxyethanoic acid, $HO{\cdot}CH_2{\cdot}COOH$) and its subsequent oxidative decarboxylation in the cytoplasm. Glycollate has long been known to be an early product of photosynthesis: the glycollate derived from fixation of $^{14}CO_2$ by illuminated leaves is labelled only slightly less rapidly than the sugar phosphates of the Calvin cycle, and the kinetics of labelling show that glycollate is derived from one or more intermediates of the cycle. Glycollate does not generally accumulate in illuminated leaf tissues, which means that its rate of metabolism to carbon dioxide (and hence the rate of photorespiration) are determined by the rate of its synthesis. In general, glycollate production by leaves or by isolated chloroplasts is decreased by carbon dioxide concentrations above 300 μl litre^{-1} and increased by raising oxygen concentrations, although the relationships are not simple. For example, spinach chloroplasts produce some glycollate even at very high carbon dioxide concentrations (Robinson and Gibbs 1974; Robinson, Gibbs, and Cotler 1977; Beck 1979). Using isolated spinach chloroplasts, Eickenbusch and Beck (1973) found that at saturating carbon dioxide concentrations the percentage of the carbon dioxide entering the Calvin cycle that appeared as glycollate was increased almost proportionally by concentrations of oxygen greater than 21 per cent, but lower oxygen concentrations had no effect on this percentage. Hence a biphasic curve results when the percentage is plotted against oxygen concentrations (Fig. 7.1). The position of the 'bend' depends directly on the carbon dioxide concentrations, e.g. at very low carbon dioxide concentrations the biphasic curve becomes a straight line and glycollate formation depends on the oxygen concentration alone (Beck 1979). Oliver (1979*a*) studied the rate of glycollate synthesis in illuminated tobacco-leaf discs treated with an inhibitor of glycollate metabolism. He found that the amount of glycollate accumulated increased to a maximum as carbon dioxide concentration was increased, after which further rises in carbon dioxide concentration decreased glycollate synthesis. The carbon dioxide concentration at which glycollate synthesis reached its maximum was increased if the oxygen concentration around the leaf was increased.

Such complex results are perhaps not too surprising. If a leaf is exposed to

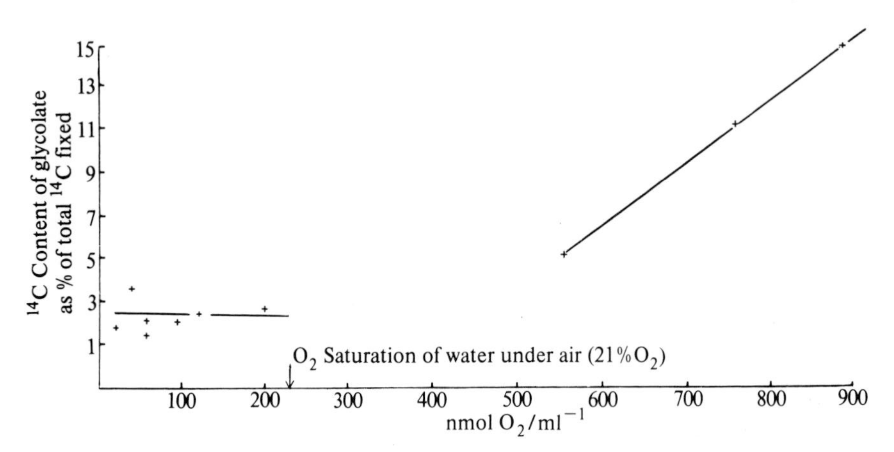

Fig. 7.1. Effect of oxygen concentration on incorporation of $^{14}CO_2$ into glycollate by illuminated spinach chloroplasts. (From Eichenbusch and Beck 1973.) The total amount of carbon dioxide fixed was approximately constant at oxygen concentrations between 78 and 564 nmol ml^{-1} but decreased at higher oxygen concentrations, i.e. although a higher percentage of the carbon fixed enters glycollate the actual *amount* of glycollate formed does not increase so much.

high oxygen and low carbon dioxide concentrations the rate of operation of the Calvin cycle will be low because of the high K_m of ribulose diphosphate carboxylase for carbon dioxide and the competitive inhibition of carboxylation by oxygen (see the legend to Fig. 7.1 and Chapter 3). Although glycollate synthesis might well be favoured by these conditions in the sense that a large percentage of the carbon dioxide fixed goes to form glycollate, the actual amount of glycollate formed is probably limited by the low concentrations of those Calvin cycle intermediates that are its precursors. If the carbon dioxide concentration is raised there will be more carbon in the Calvin cycle to act as a precursor of glycollate: even though the higher carbon dioxide level might decrease the *percentage* of that carbon incorporated into glycollate, the actual amount synthesized could be equal to or even greater than the amount synthesized at low carbon dioxide concentrations.

It is now widely believed that most, if not all, of the glycollate formed by illuminated leaves derives from the oxygenase activity of ribulose diphosphate carboxylase (see Fig. 7.2 and Chapter 3), which causes formation of phosphoglycollate. Oxygenase activity is at least adequate to account for observed rates of glycollate synthesis in C_3 leaves. Consistent with this theory is the observation that the chloroplast stroma contains high activities of a Mg^{2+}-dependent phosphatase that is specific for phosphoglycollate (Kerr 1976; Christeller and Tolbert 1978), so that this compound is hydrolysed to glycollate as soon as it is formed and does not accumulate. Such rapid removal is necessary because phospho-

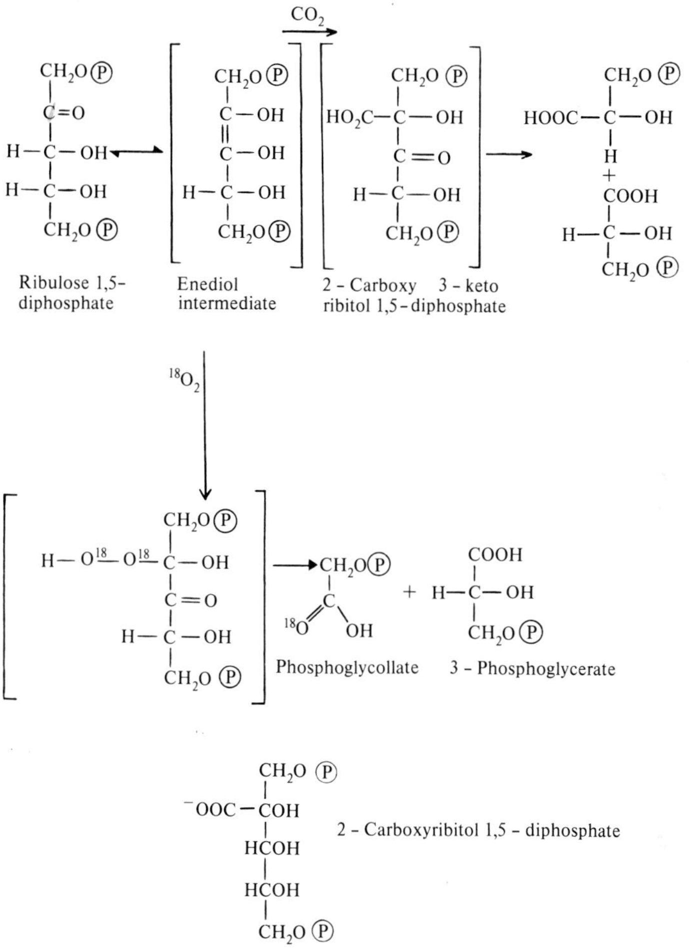

Fig. 7.2. Proposed mechanism for carboxylase and oxygenase activities of ribulose diphosphate carboxylase.

glycollate is a powerful inhibitor of the Calvin cycle enzyme triose phosphate isomerase (Wolfenden 1970). A mutant strain of the flowering plant *Arabidopsis thaliana* that had greatly decreased phosphoglycollate phosphatase activity in its leaves accumulated substantial radioactivity in phosphoglycollate when supplied with $^{14}CO_2$ in the light, unlike the wild-type strain. The mutant grew well at low oxygen concentrations, but if it was exposed to air containing 21 per cent oxygen, its rate of $^{14}CO_2$ fixation was at first high but then fell rapidly to zero, presumably due to inhibition of the Calvin cycle by accumulated phosphoglycollate (Somerville and Ogren 1979). The mutant also appeared to show abnormally low rates of photorespiration, although this could be alternatively explained as due to inhibition of the Calvin cycle and hence of glycollate synthesis under the conditions that would favour photorespiration.

There are several other lines of evidence that support the assertion that the oxygenase activity of ribulose diphosphate carboxylase is responsible for glycollate synthesis. Since oxygen and carbon dioxide are competitive inhibitors of each others action on the enzyme, this would explain why glycollate synthesis is in general favoured at high O_2/CO_2 ratios and disfavoured at low ratios, although one cannot expect the kinetics of glycollate synthesis *in vivo* to follow exactly those predicted by studies of the isolated enzyme, for the reasons discussed above. The ratio of oxygenase to carboxylase activities of the enzyme increases with temperature, which could account for the stimulatory effects of temperature on photorespiration rates (Laing, Ogren, and Hageman 1974). An additional factor that plays some part, however, is the differential effect of temperature on the solubilities of oxygen and carbon dioxide in water (Ku and Edwards 1977). As temperature increases, both oxygen and carbon dioxide become less soluble in water, but the solubility of oxygen decreases less rapidly, so that the O_2/CO_2 ratio supplied to the enzyme in solution from a fixed O_2/CO_2 ratio in the gas phase will increase with temperature and thus promote oxygenase activity.

Incorporation of $^{14}CO_2$ into glycollate by some algae, type A spinach chloroplasts and soybean leaf cells appears to be increased at high pH values (Tolbert 1973; 1979) although the pH optima for oxygenase and carboxylase activities are similar (Chapter 3). Such results can be explained by the fact that increased pH decreases the CO_2/HCO_3^- ratio in solution, i.e. the effective carbon dioxide concentration drops and oxygenase activity is increased relative to carboxylation (Heber, Andrews, and Boardman 1976; Kirk and Heber 1976; Servaites and Ogren 1977a).

When $^{18}O_2$ is supplied to ribulose diphosphate carboxylase/oxygenase *in vitro*, the oxygen isotope is incorporated only into the carboxyl group of phosphoglycollate (Fig. 7.2). Thus any glycollate produced from phosphoglycollate in illuminated leaves exposed to $^{18}O_2$ will be labelled only in the carboxyl groups, as will the glycine and serine to which it gives rise (Section 7.4). When $^{18}O_2$ was supplied to type A spinach chloroplasts illuminated in the presence of 1 mM carbon dioxide, the oxygen isotope was found in the predicted position (Fig. 7.2) in the glycollate produced and the isotope enrichment of the glycollate rapidly reached about 80 per cent of that of the $^{18}O_2$ supplied over oxygen concentrations in the surrounding gas mixture ranging from 25 to 43 per cent (Lorimer, Krause, and Berry 1977). Similar $^{18}O_2$ labelling patterns and specific activities were found in the glycollate synthesized by the green alga *Chlorella* (Lorimer, Osmond, Akazawa, and Asami 1978). Since the $^{18}O_2$ supplied in either case would become diluted by $^{16}O_2$ produced during photosynthesis it follows that more than 80 per cent of the glycollate must have been produced by the oxygenase mechanism.

Oxygen-isotope techniques have been extended to intact leaf tissues. When illuminated leaves of the C_3 plants spinach, *Atriplex hastata* or sunflower were exposed to $^{18}O_2$ at their carbon dioxide compensation points, the isotope enrichment of the glycollate, glycine, and serine that could be extracted from them reached an average value of 70 per cent of that of the oxygen supplied (Berry, Osmond, and Lorimer 1978). Similar enrichments were observed in glycine and serine when spinach leaves were exposed to an atmosphere of 100 per cent $^{18}O_2$ (Andrews, Lorimer, and Tolbert 1971). Given the dilution of $^{18}O_2$ by unlabelled oxygen, the figure of 70 per cent represents a minimum value for the amount of glycollate produced by the oxygenase pathway.

Nevertheless, these figures could be interpreted to suggest that a small percentage (<30 per cent) of the glycollate arises by a mechanism that does not incorporate $^{18}O_2$, i.e. by a means other than oxygenase activity. Such an alternative mechanism is perhaps suggested by the observation of glycollate synthesis and photorespiration at high carbon dioxide concentration, when oxygenase activity should be suppressed, and it would be interesting to perform $^{18}O_2$ labelling experiments with leaves under these conditions. The data of Eichenbusch and Beck (1973 – Fig. 1) were interpreted as suggesting two different pathways of glycollate synthesis (one low-rate pathway unaffected by oxygen concentration and a higher rate pathway dependent on oxygen concentration and due to oxygenase activity) although they are capable of alternative explanations (e.g. Christeller and Laing 1979). It is therefore worthwhile to examine some alternative sources of glycollate.

One such alternative involves the enzyme transketolase (Coombs and Whittingham 1966; Beck 1979) which catalyses the transfer of two-carbon fragments between sugar phosphates during operation of the Calvin cycle (Chapter 3). These two carbon fragments are attached to thiamin pyrophosphate (TPP) at the active site of the enzyme, and the enzyme-bound dihydroxyethyl-TPP can be oxidized non-enzymically to glycollate and TPP *in vitro* by oxidants with a redox potential greater than 120 mV, such as ferricyanide. Oxidants generated by illuminated chloroplasts that would be potentially capable of oxidizing the transketolase-TPP–C_2 complex include the superoxide radical, $O_2\cdot^-$, and H_2O_2 (Asami and Akazawa 1977; Beck 1979). However, the high concentration of a $O_2\cdot^-$-removing enzyme, superoxide dismutase, in chloroplasts (Chapter 8) makes such a role for $O_2\cdot^-$ *in vivo* most unlikely. Hydrogen peroxide, which is known to be formed in illuminated chloroplasts in significant amounts (Chapter 8) is a more likely candidate. Oxidation of dihydroxyethyl-TPP by $H_2{}^{18}O_2$ formed from $^{18}O_2$ supplied to chloroplasts would not be expected to incorporate $^{18}O_2$ into the carboxyl groups of glycollate. A reconstituted spinach chloroplast system appeared to form glycollate by the transketolase mechanism at rates of 10 μmol mg chlorophyll^{-1} h^{-1} (Shain and Gibbs 1971), although other workers have obtained much lower rates using comparable systems (Asami and Akazawa 1977; Lorimer *et al.* 1978; Christen and Gasser 1980). No direct evidence has yet been obtained for the occurrence of such reactions in isolated chloroplasts capable of fixing carbon dioxide at high rates (Kirk and Heber 1976; Krause, Thorne, and Lorimer 1977).

Nevertheless, the occurrence of the transketolase mechanism cannot be excluded under certain circumstances, perhaps using a light-generated chloroplast oxidant that has not yet been identified. Glycollate might also be formed by the oxidation of other Calvin cycle intermediates (Beck 1979), but such possibilities remain to be explored in plant tissues. For example, dihydroxyacetone phosphate bound to the active site of rabbit muscle aldolase is oxidized by H_2O_2 into hydroxymethylglyoxal phosphate, which is then further oxidized to give phosphoglycollate and formate (Christen, Anderson, and Healy 1974).

7.4 The release of carbon dioxide from glycollic acid, the source of photorespiratory carbon dioxide

7.4.1 Oxidation of glycollate to glyoxylate

The first stage in the metabolism of glycollate is its oxidation to glyoxylate (eqn (7.2)).

$$\underset{\text{glycollate}}{HOOC{\cdot}CH_2OH} + O_2 \rightarrow \underset{\text{glyoxylate}}{HOOC{\cdot}CHO} + H_2O_2 \qquad (7.2)$$

In higher plants this oxidation is catalysed by a flavoprotein (FMN) enzyme which produces hydrogen peroxide as shown in eqn (7.2). Glycollate oxidase has been shown by both cytochemical (e.g. Burke and Trelease 1975) and biochemical (Tolbert 1971) techniques to be located in leaf peroxisomes. Peroxisomes are organelles bounded by a single membrane that are rich in catalase and are often seen to be closely associated with chloroplasts in electron micrographs of leaf sections. Attachment of peroxisomes to the outside of chloroplasts *in vitro* and possibly *in vivo* as well is achieved by the presence of 10–20 mM concentrations of inorganic phosphate in the suspension medium (Schnarrenberger and Burkhard 1977). Glycollate oxidase will also oxidise L-lactic acid to pyruvate and it is not inhibited by cyanide.

Many algae, however, achieve conversion of glycollate into glyoxylate by using a different enzyme, known as glycollate dehydrogenase. It is strongly inhibited by cyanide and uses D-lactic acid (rather than the L-form) as an alternative substrate to glycollate. Glycollate oxidation by the dehydrogenase in cell extracts is not linked to uptake of oxygen but requires the presence of an artificial electron acceptor such as dichlorophenol–indophenol. In green algae the dehydrogenase is found in both peroxisomes and mitochondria, the mitochondrial glycollate oxidation being linked to the electron-transport chain and coupled to ATP synthesis with a P/O ratio that approaches a value of two (Collins, Brown, and Merrett 1975).

Leaf peroxisomes contain a high activity of catalase, and it is frequently stated that this enzyme will immediately destroy all the hydrogen peroxide generated by glycollate oxidase activity (eqn (7.2)), a statement which reveals some misunderstanding of the properties of catalase. Figure 7.3 shows that the breakdown of hydrogen peroxide by catalase requires the impact of *two* molecules of hydrogen peroxide upon the active site of a single catalase molecule, i.e. a high catalase/hydrogen peroxide ratio tends to make breakdown less efficient. At steady rates of hydrogen peroxide generation catalase compound I reaches an equilibrium with catalase and a low concentration of free hydrogen peroxide (Chance 1952). Thus catalase is very inefficient at destroying low concentrations of hydrogen peroxide. To put it another way, although the V_{max} of catalase is enormous the hydrogen peroxide concentration needed to achieve half-maximum

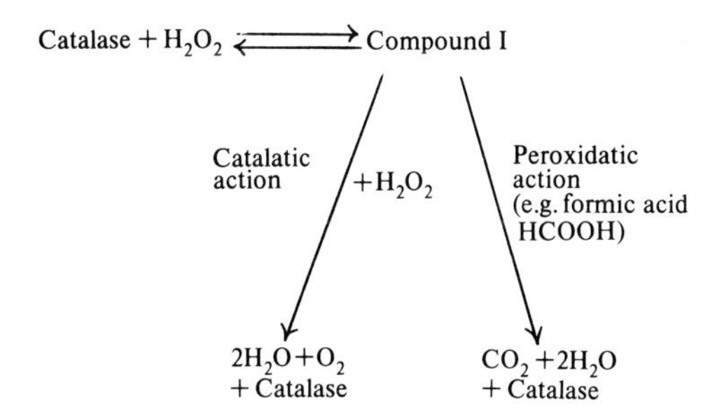

Fig. 7.3. Mode of action of catalase. Catalase is an Fe^{3+}-porphyrin enzyme. The overall breakdown of hydrogen peroxide ('catalatic action') may be represented by the equation

$$2H_2O_2 \longrightarrow 2H_2O + O_2$$

At low hydrogen peroxide concentrations compound I remains in equilibrium with catalase and a small amount of free hydrogen peroxide. Catalase can also act as a peroxidase, compound I bringing about the oxidation of, for example, formic acid to carbon dioxide. (For further details, see Chapter 8.)

velocity (K_m) is also very large: typical values for catalases from different sources are 47 mmol l^{-1} and 1100 mmol l^{-1} (Scandalios, Liu, and Campeau 1972; Jones and Suggett 1968). Because of this high K_m other reactions can compete with catalase for the available hydrogen peroxide. For example, when isolated rat liver peroxisomes, which have a high catalase activity, oxidise substrates to produce hydrogen peroxide, significant amounts of this hydrogen peroxide escape from the peroxisomes both *in vitro* (Boveris, Oshino, and Chance 1972) and *in vivo* (Chance, Sies, and Boveris 1979).

Glyoxylate undergoes a rapid non-enzymic reaction with hydrogen peroxide (eqn (7.3))

$$HOO\overset{*}{C}{\cdot}CHO + H_2O_2 \rightarrow \overset{*}{C}O_2 + H{\cdot}COOH + H_2O \qquad (7.3)$$

When radioactive glycollate is supplied to leaf tissues, the carbon dioxide evolved from it comes mainly from the carboxyl group (Zelitch 1966) although there is some release from the hydroxymethyl group. Since oxidation of glyoxylate by hydrogen peroxide releases carbon dioxide from the carboxyl group (eqn (7.3)) this non-enzymic reaction could be a source of carbon dioxide during photorespiration if sufficient of the hydrogen peroxide generated by glycollate oxidase escapes catalase action. Experiments on isolated spinach-leaf peroxisomes show that this can occur *in vitro* (Halliwell and Butt 1974; Grodzinski and Butt 1976) but even the maximum rates of glyoxylate decarboxylation achieved cannot account for more than 10–20 per cent of the rate of photorespiratory carbon dioxide release. It does not seem to be the major mechanism of carbon dioxide release. Glyoxylate decarboxylation, apparently by this mechanism, has been found to occur in isolated leaf cells (Oliver 1981) and in leaf tissues (Grodzinski and Woodrow 1981; Cresswell and Amory 1981).

A small part of any glycollate or glyoxylate fed to illuminated leaf tissues becomes oxidized to give oxalate (e.g. Seal and Sen 1970). This occurs because glycollate oxidase is capable of using glyoxylate as a substrate (eqn (7.4)).

$$\underset{\text{glyoxylate}}{HOOC{\cdot}CHO} + O_2 \rightarrow \underset{\text{oxalate}}{HOOC{\cdot}COOH} + H_2O_2 \qquad (7.4)$$

However, the K_m of the enzyme for glyoxylate is about ten times greater than that for glycollate (Richardson and Tolbert 1961). When isolated spinach-leaf peroxisomes were incubated with glycollate, only about 10 per cent of the glyoxylate formed was further converted into oxalate (Halliwell and Butt 1974). Carbon dioxide can be released from oxalate by the action of an oxalate oxidase enzyme, but the activity of this enzyme in leaf tissues is generally low (Leek, Halliwell, and Butt 1972) and oxalate metabolism is unrelated to photorespiration (Zindler-Frank 1976).

Glycollate rarely accumulates in illuminated leaf tissues and so the amount of glycollate oxidase present is sufficient or more than sufficient to cope with normal rates of glycollate production by the chloroplasts (Tolbert 1979). Accumulation of this compound can be observed if leaves are pre-treated with an inhibitor of glycollate oxidase. The most specific inhibitor seems to be 2-hydroxy-3-butynoic acid, an acetylenic compound which irreversibly inactivates the enzyme by modifying its FMN prosthetic group. Treatment of pea or wheat leaves with this compound caused glycollate accumulation but did not increase their net carbon dioxide assimilation (Jewess, Kerr, and Whitaker 1975; Kumarasinghe, Keys, and Whittingham 1977a). Hydroxysulphonates, especially pyrid-2-yl hydroxymethanesulphonate (PHMS, Fig. 7.4), are often used as inhibitors of glycollate oxidase (Corbett and Wright 1971) but they are also

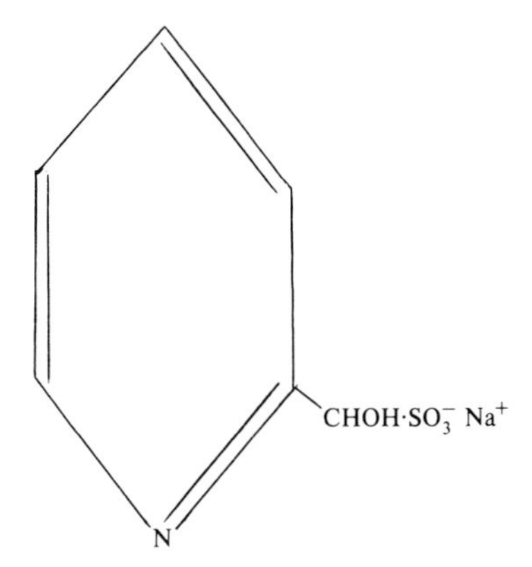

Fig. 7.4. Structure of pyrid-2-yl-hydroxymethanesulphonate (PHMS). The sodium salt is shown.

powerful inhibitors of photosynthesis. An initial claim that treatment of leaf tissues with hydroxysulphonates increased their net carbon dioxide fixation, by blocking photorespiratory carbon dioxide release from glycollate, has not been substantiated. Indeed, continued loss of carbon from the Calvin cycle to form glycollate, which cannot be further metabolized to yield carbon dioxide for reassimilation because of the presence of an inhibitor, would be expected eventually to decrease the concentration of cycle intermediates and so inhibit carbon dioxide fixation.

7.4.2 Conversion of glyoxylate to glycine, serine, and glycerate

The principal fate of the glyoxylate generated in peroxisomes is transamination to glycine, since significant amounts of glutamate-glyoxylate (eqn (7.5)) and serine-glyoxylate (eqn (7.6)) aminotransferases are present in these organelles (Rehfeld and Tolbert 1972)

$$\text{L-glutamate} + \text{glyoxylate} \longrightarrow \text{glycine} + \text{2-oxoglutarate} \quad (7.5)$$

$$\text{L-serine} + \text{glyoxylate} \longrightarrow \text{glycine} + \text{hydroxypyruvate} \quad (7.6)$$

In both cases the equilibrium position lies so far over in the direction of glycine synthesis that the reactions are essentially irreversible. Alanine is also a good amino-donor for the serine-glyoxylate aminotransferase, and asparagine and tryptophan can be transaminated, albeit with a lower V_{max} and higher K_m values (Noguchi and Hayashi 1980).

Glycine cannot be further metabolized by peroxisomes, but leaf mitochondria convert it into serine and carbon dioxide, apparently by the combined operation of a multi-enzyme complex ('glycine decarboxylase') and the enzyme serine hydroxymethyltransferase (eqn (7.8)). Most of the activity of these enzymes in leaves is located in the mitochondria (Woo 1979). The glycine decarboxylase complex of leaf mitochondria has not been studied in detail because it is difficult to isolate in a functional form, but it may well be similar to the glycine decarboxylating systems detected in animal mitochondria and some bacteria (Kikuchi 1973). These contain four different proteins, and the overall reaction catalysed may be represented by the following equation.

$$\text{glycine} + \text{tetrahydrofolate} + \text{NAD}^+ \rightleftharpoons \text{5,10 methylene-tetrahydrofolate} + CO_2 + NH_3 + \text{NADH} \quad (7.7)$$

Serine hydroxymethyltransferase can then combine the C_1-folate derivative with another glycine molecule, to yield serine

$$\text{glycine} + \text{5,10 methylene-tetrahydrofolate} \rightleftharpoons \text{L-serine} \quad (7.8)$$

Reaction (7.7) is associated with the mitochondrial electron-transport chain, the NADH being oxidized with a P/O ratio close to three (Moore, Jackson, Halliwell, Dench, and Hall 1977) and two moles of glycine being decarboxylated per mole of oxygen taken up (Arron, Spalding, and Edwards 1979a). The glycine-serine conversion in isolated mitochondrial fractions can be inhibited by adding isonicotinyl-hydrazide (Kisaki, Yoshida, and Imai 1971; Bird, Cornelius Keys, and Whittingham 1972*a, b*), hydroxylamine or glycine hydroxamate (Zelitch 1979b). Although reactions (7.8) and (7.7) are theoretically reversible the linkage of glycine decarboxylation to the electron-transport chain effectively prevents reversal of reaction (7.7) and there is little glycine formation when serine is supplied to illuminated leaves (Wang and Burris 1963). Mitochondria from etiolated or non-photosynthetic plant tissues, such as roots or leaf stalks, show little ability to decarboxylate glycine (Gardestrom, Bergman, and Ericson 1980; Arron and Edwards 1980).

Since the carbon dioxide released during conversion of glycine to serine originates from the carboxyl group of glycine, which was derived from the carboxyl group of glycollate, this reaction is a good candidate for the origin of carbon dioxide released during photorespiration and isotopic labelling studies, discussed below, indicate that it is the major contributor to photorespiratory carbon dioxide release. Indeed, [1^{14}C] glycine fed to illuminated leaves is converted to $^{14}CO_2$ as fast as or faster than [1^{14}C] glycollate (e.g. Kisaki and Tolbert 1970; Zelitch 1972a) and mutants of *Arabidopsis thaliana* deficient in glycine decarboxylase or serine hydroxymethyltransferase activities in the leaf show little photorespiration (Somerville and Ogren 1981). The rate of the glycine decarboxylation reaction in good quality leaf mitochondrial preparations is adequate to account for observed rates of photorespiration, but rates are greatly diminished if mitochondria are damaged (Woo and Osmond 1976, 1977). In leaves of nitrogen-starved plants, however, the transamination reactions (eqns 7.6 and 7.5) will be slowed down and glyoxylate decarboxylation in the peroxisomes may make a more significant contribution to carbon dioxide release (Somerville and Ogren 1981).

One problem does arise in considering glycine decarboxylation, however. The rates of photorespiration, and hence of glycine-to-serine conversion, can often be large (see below). If glycine decarboxylation has a P/O ratio of 3, considerable quantities of ATP will be made inside the mitochondria. Indeed, this could be regarded as a mechanism for the export of ATP from chloroplasts to cytoplasm (Chapter 6). The evidence discussed in Section 7.2.1 of the present Chapter indicates, however, that the rate of oxidative phosphorylation may be severely decreased in illuminated leaves. It could therefore be that some or all of the NADH generated by reaction (7.7) is dealt with without passing through the electron-transport chain. One possible alternative use for NADH is in the reductive conversion of the ammonia generated by glycine decarboxylation into L-glutamate by the action of mitochondrial glutamate dehydrogenase, although the extent to which this reaction is responsible for ammonia assimilation is debatable. An alternative fate of ammonia is conversion into glutamine by glutamine synthetase, an ATP-requiring reaction (Jackson, Dench, Morris, Lui, Hall, and Moore 1979; Keys, Bird, Cornelius, Lea, Walsgrove, and Miflin 1978). These points are further discussed in Chapter 10. NADH is also consumed in the further metabolism of serine (eqn 7.9).

Not all the C_1-tetrahydrofolate derivative generated by glycine decarboxylation (eqn (7.7)) need be used for serine synthesis. One would expect that, especially in young leaves, some of it might be diverted into such reactions as the conversion of homocysteine to methionine or the synthesis of purines (e.g. Foo and Cossins 1978).

Serine obtained from glycine might provide the substrate for the serine-glyoxylate aminotransferase in peroxisomes (eqn (7.6)). Peroxisomes also contain a D-glycerate dehydrogenase enzyme (eqn (7.9)) which catalyses the reaction

$$\text{hydroxypyruvate} + \text{NADH} + \text{H}^+ \rightleftarrows \text{D-glycerate} + \text{NAD}^+ \qquad (7.9)$$

and since the NADH/NAD^+ ratio of the cytoplasm is often high in illuminated leaves, it is logical to suppose that hydroxypyruvate obtained from serine by reaction (7.6) becomes reduced into D-glyceric acid. Leaf peroxisomes contain an isoenzyme of malate dehydrogenase, which could be involved in provision of NADH for reaction (7.9), since leaf mitochondria can export NADH by means of a malate/oxaloacetate shuttle (Woo, Jokinen, and Canvin 1980) or possibly by a malate/aspartate shuttle similar to that which operates in chloroplasts (Chapter 6). D-Glycerate can be phosphorylated to yield 3-phosphoglycerate by a kinase enzyme that has been reported to be present in both chloroplasts and cytoplasm in spinach leaves (Heber, Kirk, Gimmler, and Schafer 1974) but to be exclusively located in the chloroplasts in wheat (Usuda and Edwards 1980). Hence it is possible to represent the pathway of photorespiration as a cycle in which glycollate leaves the chloroplast, two molecules of glycollate form carbon dioxide plus one molecule of serine and the serine is converted into phosphoglycerate, which re-enters the Calvin cycle. Such a formulation is shown in Fig. 7.5, emphasizing the large amount of shuttling of metabolites between different organelles that occurs during glycollate metabolism. This scheme was originally based on enzyme localization studies (Tolbert 1971), but more direct evidence for the shuttles has now been obtained. Purified illuminated type A chloroplasts will not fix $^{14}CO_2$ into glycine, but label enters glycine if a peroxisome preparation is mixed with the chloroplasts. Little serine is detected until a mitochondrial fraction is also added (Buchholz, Reupke, Bickel, and Schultz 1979). These preliminary experiments clearly support the scheme of Fig. 7.5. Glycine appears to enter leaf mitochondria by simple diffusion (Day and Wiskich 1980).

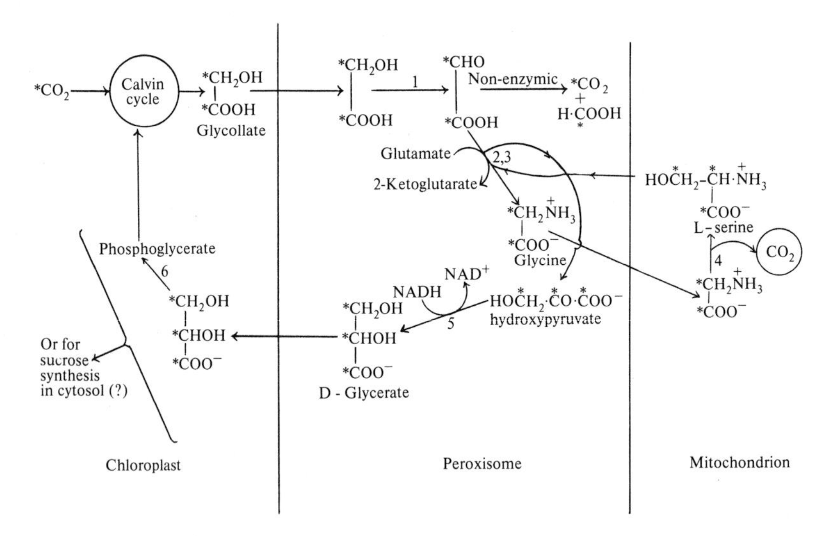

Fig. 7.5. Mechanisms of serine synthesis in green leaves in relation to photorespiration. Scheme for carbon dioxide release during photorespiration (based on Tolbert (1979)). Since the glycollate formed from $^{14}CO_2$ is uniformly labelled, then glycine and serine produced by this pathway will be labelled as shown. Enzymes: 1, glycollate oxidase; 2, glutamate-glyoxylate aminotransferase; 3, serine-glyoxylate aminotransferase; 4, glycine decarboxylase and serine hydroxymethyltransferase; 5, D-glycerate dehydrogenase; 6, phosphoglycerate kinase.

It has also been suggested that glycerate derived from glycollate might sometimes act as a precursor for the synthesis of sucrose in the cytoplasm (e.g. Bird, Cornelius, Keys, and Whittingham 1978). However, little sucrose is formed from [^{14}C] glycollate supplied to leaves in the dark (e.g. Tamas and Bidwell 1971). Light is required before significant label enters sucrose. Presumably export of triose phosphates from the chloroplasts to provide carbon and/or NADH and ATP is also required to allow sucrose synthesis to occur. Similarly, [^{14}C] serine supplied to illuminated wheat leaves does not label sucrose significantly unless carbon dioxide is also present in the atmosphere surrounding the leaves (Waidyanatha, Keys, and Whittingham 1975; Bird, Cornelius, Keys, Kumarasinghe, and Whittingham 1975). In any case, if glycerate kinase in wheat leaves is exlusively located in the chloroplast, conversion of glycerate into sucrose in the cytoplasm cannot occur.

7.4.3 Serine synthesis from sources other than glycine

Plant tissues possess two metabolic pathways for the synthesis of serine in addition to the serine hydroxymethyltransferase reaction, viz. synthesis of serine from phosphoglycerate by a 'phosphorylated route' and by a 'non-phosphorylated route'. These pathways are summarized in Fig. 7.6. Chloroplasts contain a phosphoglycerate phosphatase enzyme and some plants, including spinach, have a relatively large pool of free glycerate in the leaves. How the chloroplast phosphatase is regulated has not been studied, but its pH optimum is acidic which suggests that it would only have a small activity at the pH of the stroma in the illuminated chloroplast (Mulligan and Tolbert 1980). When $^{14}CO_2$ is fed to illuminated leaf tissues, label enters glycerate much more slowly than it enters glycollate (Tolbert 1979). Conversion of glycerate to hydroxypyruvate by glycerate dehydrogenase would be disfavoured by a high NADH/NAD^+ ratio in the cytoplasm of the illuminated leaf. Hence the rate of carbon flow to serine through the 'non-phosphorylated pathway' would probably be small in the light. On the basis of measurement of enzyme activities, Cheung, Rosenblum, and Sallach (1968) concluded that the phosphorylated pathway could only occur at low rates in a range of leaf tissues, but this may be a premature conclusion, firstly because higher activities of serine phosphatase have been detected in spinach leaves (Larsson and Albertsson 1979) and secondly because there is a rapid labelling of phosphoserine during $^{14}CO_2$ assimilation by illuminated bean (*Phaseolus vulgaris*) or maize leaves (Chapman and Leech 1976; Daley and Bidwell 1977). Further investigation of this is required. In the meantime, it must be remembered that estimates of photorespiration rates by measurement of the amount of carbon passing through 'glycine plus serine' in the illuminated leaf may be in error because of these alternative routes of serine synthesis (Table 7.1).

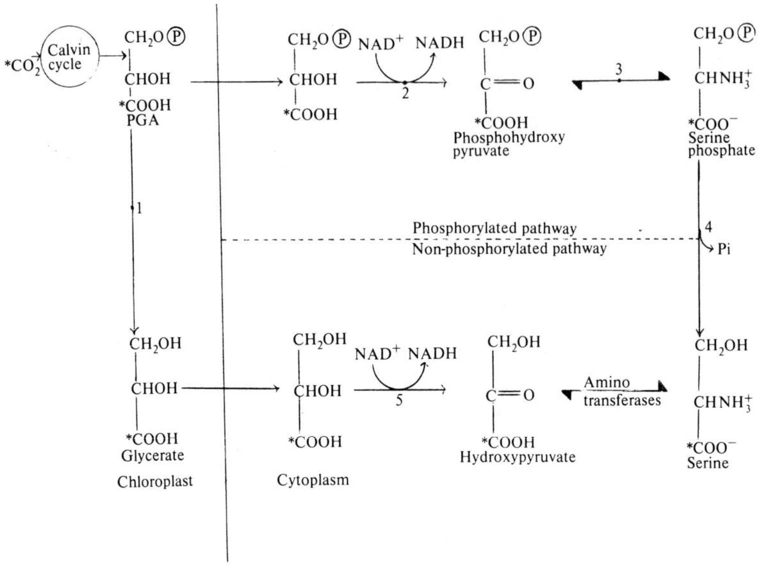

Fig. 7.6. Alternative methods of serine synthesis in leaves. The phosphoglycerate (PGA) formed at the earliest stages of photosynthesis is labelled only in the carboxyl group. The non-phosphorylated pathway of serine synthesis from PGA is probably much more important in leaves (Cheung *et al.* 1968). Enzymes: (1) PGA phosphatase; (2) PGA dehydrogenase: (3) serine phosphate aminotransferase; (4) serine phosphatase; (5) glycerate dehydrogenase. Serine supplied to illuminated leaves is not converted into glycine at significant rates.

7.4.4 The glycine-serine conversion as the source of photorespiratory carbon dioxide

It is now generally accepted that the glycine decarboxylase reaction is the principal source of the carbon dioxide released during photorespiration. The rate of formation of glycine and serine from glycollate (or from phosphoglycerate) can be studied by supplying leaf tissues with $^{14}CO_2$. Since glycollate produced from the Calvin cycle is uniformly labelled with ^{14}C the glycine and serine derived from it will be uniformly labelled also (Fig. 7.5). In the earliest stages of photosynthesis, phosphoglycerate and glycerate are labelled only in the carboxyl group, which would give a different labelling pattern in serine (Fig. 7.6). Experiments in which $^{14}CO_2$ was supplied to tobacco, tomato, peppermint, soybean, and *Coleus* leaves showed the pattern of labelling predicted by Fig. 7.5 (Rabson, Tolbert, and Kearney 1962; Hess and Tolbert 1966; Lee and Whittingham 1974). Similarly, when $^{18}O_2$ was supplied to spinach, *Atriplex*, or sunflower leaves at the carbon dioxide compensation point the isotope enrichment of glycollate reached 50–70 per cent of that of the oxygen supplied after 5-10 s. Glycine and serine achieved comparable oxygen enrichments after 40 and 180 s, respectively. $^{18}O_2$ entered 3-phosphoglycerate much more slowly, presumably via glycerate (Fig. 7.5). These results are fully consistent with a rapid flow of label from glycollate through glycine and serine (Berry *et al.* 1978). In leaf tissues or in isolated leaf cells isonicotinylhydrazide inhibits the labelling of serine by $^{14}CO_2$ or by [^{14}C] glycollate supplied to the system, and inhibitors of glycollate oxidase decrease the entry of isotope into both glycine and serine (Tolbert 1979; Servaites and Ogren 1977b). The amount of carbon flowing through glycine and serine is at least sufficient to account for observed rates of photorespiration (e.g. Bird *et al.* 1975; Canvin *et al.* 1978; Canvin 1979; Kumarasinghe, Keys, and Whittingham 1977b).

The above results also show that most, if not all, of the serine synthesized in the leaf tissues under the conditions used originated from glycine, i.e. the alternative pathways of serine synthesis (Fig. 7.6) made little contribution as would be predicted (Section 7.4.3). This need not always be the case however. For example, Servaites and Ogren (1977b) observed that treatment of soybean leaf cells with the glycollate oxidase inhibitor butyl 2-hydroxy-3-butynoate completely prevented the entry of label from $^{14}CO_2$ into glycine but did not completely eliminate the labelling of serine, suggesting that some arose from a source other than glycine. Platt, Plaut, and Bassham (1977) found that high carbon dioxide concentrations decreased the entry of carbon from $^{14}CO_2$ into glycollate and glycine in alfalfa leaves, whilst the entry of label into serine increased or remained unchanged, depending on the conditions to which the leaves were exposed. It seems that in this case both glycollate and glycerate contribute to serine synthesis. High carbon dioxide concentrations might be expected to depress the synthesis of glycollate and hence the synthesis of serine from it, but serine synthesis by the alternative routes (Fig. 7.6) might be increased at high carbon dioxide concentrations because the increased carbon dioxide fixation will allow export of more C_3 compounds from the chloroplast.

7.4.5 Alternative pathways for the release of carbon dioxide from glycollate

A part of the carbon dioxide released during photorespiration may come from the direct decarboxylation of glyoxylate by hydrogen peroxide in leaf peroxisomes, but this is a minor source in comparison with glycine decarboxylation (Section 7.4.1). Zelitch (1972c) and Elstner and Heupel (1973) have proposed another origin for carbon dioxide released during photorespiration: the non-enzymic decarboxylation of glyoxylate to formate by hydrogen peroxide or oxygen radicals (e.g. superoxide) generated by illuminated chloroplasts. As hydrogen peroxide and superoxide radicals are generated in illuminated chloroplasts *in vivo* and chloroplasts contain no catalase activity (Chapter 8) this mechanism is feasible provided that glyoxylate moves from peroxisomes to chloroplasts. However, one would expect most glyoxylate to be metabolized to glycine at its site of formation, consistent with the rapid labelling of glycine shown by studies on whole leaves using ^{14}C and $^{18}O_2$. The chloroplast decarboxylation mechanism would require light energy to generate the hydrogen peroxide and superoxide needed, but release of $^{14}CO_2$ from [1-^{14}C] glycollate supplied to leaves or leaf cells occurs rapidly in the dark and is affected by temperature in exactly the same way in both light and dark, which suggests that the same metabolic pathway is involved (Tamas and Bidwell 1971; Grodzinski and Butt 1977; Oliver 1979b). None of the reactions of the glycine/serine pathway (Fig. 7.5) requires light. Finally, chloroplasts contain significant activities of an NADPH-dependent glyoxylate reductase enzyme, which has a high affinity for glyoxylate and so might be expected to reduce any glyoxylate that does arrive at the chloroplast back into glycollate (Tolbert, Yamazaki, and Oeser 1970). It seems unlikely that the chloroplast glyoxylate decarboxylation mechanism is a significant contributor to photorespiratory carbon dioxide release, although its occurrence at a very slow rate cannot be ruled out. Isolated plant mitochondrial fractions can also slowly decarboxylate any glyoxylate that reaches them to yield formate in a reaction that requires oxygen, Mn^{2+}, and thiamin pyrophosphate (Prather and Sisler 1972).

Small amounts of formate can therefore be produced from glyoxylate in peroxisomes, mitochondria, and chloroplasts. Formate can be oxidized to carbon dioxide by an NAD^+-dependent formate dehydrogenase located in the mitochondria, or by the peroxidatic action of catalase in the peroxisomes (Fig. 7.3) (Halliwell 1974) and this might account for the low rate of carbon dioxide release from the hydroxymethyl carbon atom of glycollate supplied to illuminated leaves (Zelitch 1966). Alternatively, formate can be converted into formyltetrahydrofolate by a synthetase located in the cytosol of the leaf (Halliwell 1973) and this seems to be its major fate (Calmes and Viala 1978; Tolbert 1979; Grodzinski 1979). Thus it can enter the pool of one-carbon units ('tetrahydrofolate pool'), from which it might supply methylenetetrahydrofolate to help convert glycine into serine. These pathways are summarized in Fig. 7.7.

There have been claims that chloroplasts from greening potato peelings (Ramaswamy, Behere, and Nair 1976) and leaves of *Vicia faba* (Kent 1972) fix carbon dioxide directly into formate, but these remain to be confirmed. In any case, chloroplasts from potato peelings function to produce the alkaloid solanidine: they do not operate the Calvin cycle (Chapter 10).

7.4.5 Metabolism of carbon monoxide

Carbon monoxide is an important atmospheric pollutant. Leaves of some C_3 plants, such as bean, are capable of assimilating ^{14}CO supplied to them and a considerable percentage of the radioactivity accumulates in serine, suggesting that carbon monoxide enters the folate pool by some mechanism (Fig. 7.7). The rates of carbon monoxide uptake are low, however, (Bidwell and Fraser 1974) and both C_3 and C_4 plants can also release it in the light. An origin from

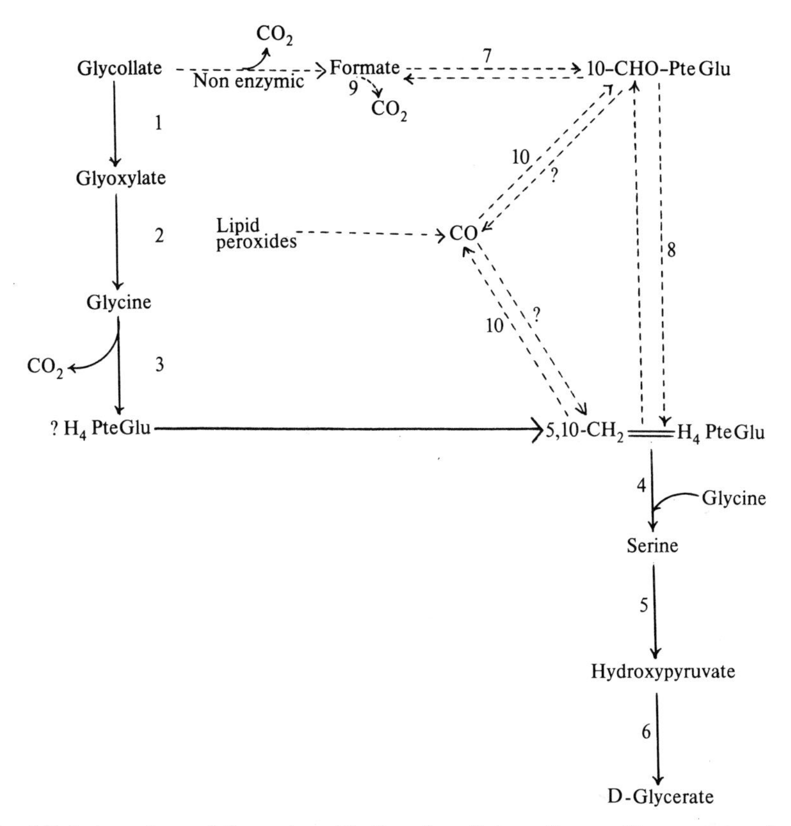

Fig. 7.7 Interaction of formate with the glycollate pathway. Enzymes involved: (1) glycollate oxidase; (2) glutamate-glyoxylate and serine-glyoxylate aminotransferases; (3) glycine decarboxylase complex; (4) serine hydroxymethyltransferase; (5) serine-glyoxylate aminotransferase; (6) hydroxypyruvate reductase: (7) formyltetrahydrofolate synthetase; (8) methenyltetrahydrofolate cyclohydrolase and NADP-linked methylenetetrahydrofolate dehydrogenase; (9) catalase and NAD^+-formate dehydrogenase; (10) carbon monoxide assimilation and release. The dotted lines indicate minor pathways of metabolism and the solid lines major pathways. The question marks indicate intermediates or pathways that have not been elucidated.

the folate pool has been suggested for the released carbon monoxide as well (Fischer and Lüttge 1978). Insufficient data are available at present to assess the net impact of plants on atmospheric carbon monoxide concentrations. Carbon monoxide is a by-product of lipid peroxidation (Chapter 8) and thus could be produced by this process in chloroplasts at high light intensity (Bauer, Conrad, and Seiler 1980).

7.5 Does photorespiration occur in C_4 and in CAM plants?

The leaves of C_4 plants do not show the gas exchange phenomena characteristic of photorespiration (Section 7.2). This could mean that photorespiratory carbon dioxide release does not occur in C_4 plants, or it might mean that any carbon dioxide released from glycollate is reassimilated by PEP carboxylase in the mesophyll and so does not appear outside the plant.

C_4 leaves do have the enzymic capacity for photorespiration. The ribulose diphosphate carboxylase enzyme of the bundle-sheath cells has significant oxygenase activity when assayed *in vitro* and the enzymes of the glycollate pathway are present in C_4 leaves, all being concentrated in the bundle-sheath cells (Rathnam 1978) with the exception of glycerate kinase, which is found in the mesophyll (Usuda and Edwards 1980). For example, mitochondria isolated from bundle-sheath cells are capable of decarboxylating glycine at significant rates, especially in the case of NAD^+-malic enzyme plants (aspartate formers of type two – see Chapter 5) which are known to have large mitochondria in these cells (Woo and Osmond 1977). In general, however, the activities of glycollate pathway enzymes are lower in C_4 than in C_3 leaves. Isolated bundle-sheath cells, as well as whole C_4 leaves, can release $^{14}CO_2$ from [^{14}C] glycine or [^{14}C] glycollate and leaves of maize and *Panicum miliaceum* have been shown to accumulate glycollate at low rates in the light after treatment with inhibitors of glycollate oxidase (Zelitch 1972b; Chollet and Ogren 1975; Servaites, Schrader, and Edwards 1978). Label from $^{14}CO_2$ supplied to C_4 leaves can appear in glycine and serine, but the flux of carbon through these compounds is low and, as already discussed, serine need not have originated from glycollate (Morot-Gaudry, Farineau, and Huet 1980).

It therefore seems that C_4 leaves are potentially capable of releasing carbon dioxide via the glycollate pathway, albeit at rates lower than those of C_3 leaves. However, the function of C_4 photosynthesis is thought to be that of increasing carbon dioxide assimilation by raising the carbon dioxide concentration in the vicinity of ribulose diphosphate carboxylase in the bundle-sheath cells (Chapter 5). Such an increased carbon dioxide concentration would decrease the oxygenase activity of this enzyme and so reduce the rate of glycollate formation (Chollet and Ogren 1975). Hence it seems likely that *in vivo* the flux through the glycollate pathway would be even less than that permitted by the low enzyme activities. In agreement with this, glycollate synthesis by isolated maize bundle sheath strands was severely depressed by including C_4 compounds in the reaction mixture (Oliver 1978). Also, photosynthesis by C_4 leaf slices treated with inhibitors of the C_4 cycle is strongly inhibited by 21 per cent oxygen, whereas oxygen concentration has little effect on C_4 metabolism normally (Rathman and Chollet 1980).

The ribulose diphosphate carboxylase activity within the leaves of CAM plants exhibits an oxygenase activity (Osmond 1978) and the leaf peroxisomes contain a complement of enzymes of glycollate metabolism similar to that shown in Fig. 7.5 (Herbert, Burkhard, and Schnarrenberger 1978). Hence part of the metabolic requirement for photorespiration is present, although it must be pointed out that Arron, Spalding, and Edwards (1979b) and Day (1980) found only low rates of glycine decarboxylation by intact mitochondria isolated from CAM leaves. Photorespiratory carbon dioxide release might conceivably occur during assimilation by ribulose diphosphate carboxylase of the carbon dioxide released from malate in the light, but one would not expect carbon dioxide to escape from the leaf because of the closed stomata. A high internal carbon dioxide concentration within the leaves due to malate decarboxylation would in any case be expected to suppress oxygenase activity and hence glycollate synthesis (Spalding, Stumpf, Ku, Burris, and Edwards 1979).

Many CAM plants grown under controlled conditions can fix external carbon dioxide in the late light phase, when the malic acid stored in the vacuoles has been used up. Fixation occurs by the combined operation of ribulose diphosphate carboxylase and PEP carboxylase (Chapter 5) and there is some evidence that photorespiratory carbon dioxide release occurs under these conditions (Osmond 1978).

7.6 The function of photorespiration

The function of photorespiration is not understood. Why should C_3 plants divert a significant proportion of the carbon dioxide that they fix into glycollate, which then has to be decarboxylated and the carbon dioxide reassimilated with expenditure of more energy? It could be argued that the glycollate pathway serves to synthesise serine and glycine in the cytoplasm, or that it is a means of export of ATP from chloroplast to cytoplasm, since glycine decarboxylation is linked to the mitochondrial electron transport chain with a P/O ratio of three. However, there are other routes by which serine is made in leaves, and by which ATP can leave chloroplasts (Chapter 6).

Perhaps a more likely suggestion was made by Lorimer and Andrews (1973, 1981). Since all ribulose diphosphate carboxylases known, even those from anaerobic bacteria, show oxygenase activity, they suggested that phosphoglycollate formation is chemically unavoidable when photosynthesis occurs in an oxygen-containing environment because an intermediate generated during the carboxylase reaction is sensitive to attack by oxygen (Fig. 7.2). As phosphoglycollate is an inhibitor of the Calvin cycle, the glycollate pathway might have evolved as a means of removing it and recycling three-quarters of the carbon back into the Calvin cycle as glycerate (Fig. 7.5).

Ribulose diphosphate carboxylase becomes largely inactivated when exposed to ribulose diphosphate in the absence of carbon dioxide (Chapter 3). Another view of photorespiration would be that it serves to ensure that even when there is very little carbon dioxide in the air surrounding C_3 leaves, the chloroplast receives carbon dioxide from glycollate decarboxylation to prevent such an inactivation. Leaves exposed to a high light intensity in the absence of carbon dioxide might have a problem in dealing with the excess energy generated within the chloroplast (Krause, Lorimer, Heber, and Kirk 1978). Reduced components closely associated with photosystem I can reduce oxygen to the superoxide radical, $O_2{\cdot}^-$, and singlet oxygen can be formed directly from excited states of chlorophyll (Chapter 8). In the absence of an acceptor for NADPH (i.e. no carbon dioxide present) both reactions will be speeded up, causing a rate of generation of toxic oxygen metabolites that is too great for the chloroplast defences (Chapter 8) to cope with, so leading to irreversible membrane damage. It might further be argued that the function of photorespiration is to ensure that this cannot happen by making carbon dioxide continually available for refixation (Bolhar-Nordenkampf 1976; Krause *et al.* 1978; Halliwell 1978). This suggestion is being actively investigated (e.g. Asami & Akazawa, 1978; Powles, Osmond, and Thorne 1979) but it is probably too early to assess its significance at the moment.

References

Andrews, T. J., Lorimer, G. H. and Tolbert, N. E. (1971). *Biochemistry* **10**, 4777–81.
Arron, G. P. and Edwards, G. E. (1980). *Plant Sci. Lett.* **18**, 229–35.
— Spalding, M. H., and Edwards, G. E. (1979*a*). *Biochem. J.* **184**, 457–60.
— — — (1979*b*). *Pl. Physiol., Lancaster* **64**, 182–6.
Asami, S. and Akazawa, T. (1977). *Biochemistry* **16**, 2202–7.
— — (1978). *Pl. Physiol., Lancaster* **62**, 981–6.
Bauer, K., Conrad, R., and Seiler, W. (1980). *Biochim. biophys. Acta* **589**, 46–55.
Beck, E. (1979) in *Encyclopaedia of plant physiology*, Vol. 6 (ed. M. Gibbs and E. Latzko) pp. 327–37. Springer. Berlin.
Berry, J. A., Osmond, C. B., and Lorimer, G. H. (1978). *Pl. Physiol., Lancaster* **62**, 954–67.
Bidwell, R. G. S. and Fraser, D. E. (1972). *Can. J. Bot.* **50**, 1435–1439.
Bird, I. F., Cornelius, M. J., Keys, A. J., and Whittingham, C. P. (1972*a*). *Phytochemistry* **11**, 1587–94.
— — — — (1972*b*). *Biochem. J.* **128**, 191–2.
— — — — (1978). *Biochem. J.* **172**, 23–7.
— — — Kumarasinghe, S., and Whittingham, C. P. (1975) in *Proceedings of the third international congress on photosynthesis* (ed. M. Avron) pp. 1291–301. Elsevier, Amsterdam.
Bjorkman, O., Hiesey, W. M., Nobs, M., Nicholson, F., and Hart, R. W. (1968). *Carnegie Inst. Wash. Year Book* **66**, 228–32.
Bolhar-Nordenkampf, H. R. (1976). *Biochem. Physiol. Pflanzen.* **169**, 121–39.
Boveris, A., Oshino, N, and Chance, B. (1972). *Biochem. J.* **128**, 617–30.
Bravdo, B. A. and Canvin, D. T. (1979). *Pl. Physiol., Lancaster* **63**, 399–401.
Buchholz, B., Reupke, B., Bickel, H., and Schultz, G. (1979). *Phytochemistry* **18**, 1109–11.

Burke, J. J. and Trelease, R. N. (1975). *Pl. Physiol., Lancaster* **56**, 710-17.
Calmes, T. and Viala, G. (1978). *C. r. hebd. Séanc. Acad. Sci. Paris* **D287**, 1039-42.
Canvin, D. T. (1979) in *Encyclopaedia of plant physiology,* Vol. 6 (eds. M. Gibbs and E. Latzko) pp. 368-96. Springer, Berlin.
— Fock, H. and Lloyd, N. D. H. (1978) in *Photosynthesis '77* (eds. D. O. Hall, J. Coombs and T. W. Goodwin) pp. 323-34. Biochemical Society, London.
Chance, B. (1952). *Science, N.Y.* **116**, 202-4.
— Sies, H. and Boveris, A. (1979). *Physiol. Rev.* **59**, 527-605.
Chapman, D. J. and Leech, R. M. (1976) *FEBS Lett.* **68**, 160-4.
Cheung, G. P., Rosenblum, I. Y., and Sallach, H. J. (1968). *Pl. Physiol., Lancaster* **43**, 1813-1820.
Chollet, R. (1978). *Pl. Physiol., Lancaster* **61**, 929-32.
— and Ogren, W. L. (1975). *Bot. Rev.* **41**, 137-179.
Christeller, J. T. and Laing, W. A. (1979). *Biochem. J.* **183**, 747-50.
— and Tolbert, N. E. (1978). *J. biol. Chem.* **253**, 1780-5.
Christen, P. and Gasser, A. (1980). *Eur. J. Biochem.* **107**, 73-77
— Anderson, T. K. and Healy, M. J. (1974). *Experientia* **30**, 603-5.
Collins, N., Brown, R. H. and Merrett, M. J. (1975). *Biochem. J.* **150**, 373-7.
Coombs, J. and Whittingham, C. P. (1966). *Proc. R. Soc.* **B164**, 511-20.
Corbett, J. R. and Wright, B. J. (1971). *Phytochemistry* **10**, 2015-24.
Creswell, C. F. and Amory, A. M. (1981) in *Proceedings of the Fifth International Congress on Photosynthesis*. In press.
Daley, L. S. and Bidwell, R. G. S. (1977). *Pl. Physiol., Lancaster* **60**, 109-14.
Day, D. A. (1980). *Plant Physiol.* **65**, 675-9.
— and Wiskich, J. T. (1980). *FEBS Lett.* **112**, 191-4.
Decker, J. P. (1955). *Pl. Physiol., Lancaster* **30**, 82-4.
— (1959). *Pl. Physiol., Lancaster* **34**, 103-6.
Eickenbusch, J. D. and Beck, E. (1973). *FEBS Lett.* **31**, 225-8.
Elstner, E. F. and Heupel, A. (1973). *Biochim. biophys. Acta* **325**, 182-8.
Fischer, K. and Lüttge, U. (1978). *Nature, Lond.* **275**, 740-1.
Fock, H., Klug., and Canvin, D. T. (1979). *Planta* **145**, 219-23.
Foo, S. S. K. and Cossins, E. A. (1978). *Phytochemistry* **17**, 1711-5.
Frederick, S. E., Gruber, P. J., and Tolbert, N. E. (1973). *Pl. Physiol., Lancaster* **52**, 318-23.
Gardestrom, P., Bergman, A. and Ericson, I. (1980). *Pl. Physiol., Lancaster* **65**, 389-91.
Goldworthy, A. (1966). *Phytochemistry* **5**, 1013-9.
— (1970). *Bot. Rev.* **36**, 321-40.
Graham, D. and Chapman, E. A. (1979) in *Encyclopaedia of plant physiology,* Vol. 6 (eds. M. Gibbs and E. Latzko) pp. 150-162. Springer, Berlin.
Grodzinski, B. (1979). *Pl. Physiol., Lancaster* **63**, 289-93.
— and Butt, V. S. (1976). *Planta* **128**, 225-31.
— — (1977). *Planta* **133**, 261-6.
— and Woodrow, L. (1981) in *Proceedings of the Fifth International Congress on Photosynthesis*. In press.
Halliwell, B. (1973). *Biochem. Soc. Trans.* **1**, 1147-50.
— (1974). *Biochem. J.* **138**, 77-85.
— (1978). *Prog. Biophys. molec. Biol.* **33**, 1-54.
— and Butt, V. S. (1974). *Biochem. J.* **138**, 217-24.
Heber, U., Kirk, M. R., Gimmler, H., and Schafer, G. (1974). *Planta* **120**, 31-46.

Heber, U., Andrews, T. J. and Boardman, N. K. (1976). *Pl. Physiol., Lancaster* **57**, 277–83.
Herbert, H., Burkhard, Ch., and Schnarrenberger, C. (1978). *Planta* **143**, 279–84.
Hess, J. L. and Tolbert, N. E. (1966). *J. biol. Chem.* **241**, 5705–11.
Jackson, C., Dench, J. E., Morris, P., Lui, S. C., Hall, D. O., and Moore, A. L. (1979). *Biochem. Soc. Trans.* **7**, 1122–4.
Jackson, W. A. and Volk, R. J. (1970). *A. Rev. Pl. Physiol.* **21**, 385–431.
Jewess, P. J., Kerr, M. W., and Whitaker, D. P. (1975). *FEBS Lett.* **53**, 292–6.
Jones, P. and Suggett, A. (1968). *Biochem. J.* **110**, 617–20.
Kent, S. S. (1972). *J. biol. Chem.* **247**, 7293–302.
Kerr, M. W. (1976). *FEBS Lett.* **64**, 266–70 (meeting report).
Keys, A. J., Sampaio, E. V. S. B., Cornelius, M. J. and Bird, I. F. (1977). *J. exp. Bot.* **28**, 525–33.
— Bird, I. F., Cornelius, M. J., Lea, P. J., Wallsgrove, R. M., and Miflin, B. J. (1978). *Nature, Lond.* **275**, 741-3.
Kikuchi, G. (1973). *Molec. Cell. Biochem.* **1**, 169–87.
Kirk, M. R. and Heber, U. (1976). *Planta* **132**, 131–41.
Kisaki, T. and Tolbert, N. E. (1970). *Plant Cell Physiol.* **11**, 247–58.
— Yoshida, N. and Imai, I. (1971). *Plant Cell Physiol.* **12**, 275–88.
Krause, G. H., Thorne, S. W., and Lorimer, G. H. (1977). *Archs Biochem. Biophys.* **183**, 471–9.
— Lorimer, G. H., Heber, U., and Kirk, M. R. (1978) in *Photosynthesis '77* (eds. D. O. Hall, J. Coombs, and T. W. Goodwin) pp. 299–310. Biochemical Society, London.
Ku, S. B. and Edwards, G. E. (1977). *Pl. Physiol., Lancaster* **59**, 986–990.
Kumarasinghe, K. S., Keys, A. J., and Whittingham, C. P. (1977*a*). *J. exp. Bot.* **28**, 1163–8.
— — — (1977*b*). *J. exp. Bot.* **28**, 1247–57.
Laing, W. A., Ogren, W. L., and Hageman, R. H. (1974). *Pl. Physiol., Lancaster* **54**, 678–85.
Larsson, C. and Albertsson, E. (1979). *Physiol. Plant.* **45**, 7–10.
Lee, R. B. and Whittingham, C. P. (1974). *J. exp. Bot.* **25**, 277–87.
Leek, A. E., Halliwell, B., and Butt, V. S. (1972). *Biochim. biophys. Acta* **286**, 299–311.
Lorimer, G. H. and Andrews, T. J. (1973). *Nature, Lond.* **243**, 359–60.
— — (1981). In *Plant biochemistry, a modern treatise*, Vol. 8. Academic Press, London.
— Krause, G. H. and Berry, T. A. (1977). *FEBS Lett.* **78**, 199–202.
— Osmond, C. B., Akazawa, T., and Asami, S. (1978). *Archs Biochem. Biophys.* **185**, 49–56.
Martin, F. A., Ozbun, J. L., and Wallace, D. H. (1972). *Pl. Physiol., Lancaster* **49**, 764–8.
McCashin, B. G. and Canvin, D. T. (1979). *Pl. Physiol., Lancaster* **64**, 354–60.
Moore, A. L., Jackson, C., Halliwell, B., Dench, J. E., and Hall, D. O. (1977). *Biochem. Biophys. Res. Commun.* **78**, 483–91.
Morot-Gaudry, J. F., Farineau, J. P. and Huet, J. C. (1980). *Pl. Physiol., Lancaster* **66**, 1079–1084.
Mulligan, R. M. and Tolbert, N. E. (1980). *Pl. Physiol., Lancaster* **66**, 1169–73.
Noguchi, T. and Hayashi, S. (1980). *J. biol. Chem.* **255**, 2267–9.
Oliver, D. J. (1978). *Pl. Physiol., Lancaster* **62**, 690–2.
— (1979*a*). *Plant Sci. Lett.* **15**, 35–40.

— (1979*b*). *Pl. Physiol., Lancaster* **64**, 1048–52.
— (1981) in *Proceedings of the Fifth International Congress on Photosynthesis*. In press.
Osmond, C. B. (1978) in *CO_2 metabolism and plant productivity* (ed. R. H. Burris) pp. 217–23. University Park Press, Baltimore, MD.
Parkinson, K. J., Penman, H. L., and Tregunna, E. G. (1974). *J. Exp. Bot.* **25**, 132–45.
Platt, S. G., Plaut, Z., and Bassham, J. A. (1977). *Pl. Physiol., Lancaster* **60**, 230–4.
Powles, S. B., Osmond, C. B., and Thorne, S. W. (1979). *Pl. Physiol., Lancaster* **64**, 982–8.
Prather, C. W. and Sisler, E. C. (1972). *Phytochemistry* **11**, 1637–47.
Quebedeaux, B. and Hardy, R. W. F. (1975). *Pl. Physiol., Lancaster* **55**, 102–7.
— Harelka, U. D., Livak, K. L., and Hardy, R. W. F. (1975). *Pl. Physiol., Lancaster* **56**, 761–4.
Rabson, R., Tolbert, N. E., and Kearney, P. C. (1962). *Archs Biochem. Biophys.* **98**, 154–63.
Ramaswamy, N. K., Behere, A. G., and Nair, P. M. (1976). *Eur. J. Biochem.* **67**, 275–82.
Rathnam, C. K. M. (1978). *Sci. Prog.* **71**, 409–35.
— and Chollet, R. (1980) in *Progress in phytochemistry*, Vol. 6 (eds. L. Reinhold, J. B. Harborne, and T. Swain) pp. 1–48. Pergamon Press, Oxford.
Raven, J. A. (1972*a*). *New Phytol.* **71**, 227–47.
— (1972*b*). *New Phytol.* **71**, 995–1014.
Rehfeld, D. W. and Tolbert, N. E. (1972). *J. biol. Chem.* **247**, 4803–11.
Richardson, K. E. and Tolbert, N. E. (1961). *J. biol. Chem.* **236**, 1280–4.
Robinson, J. M. and Gibbs. M. (1974). *Pl. Physiol., Lancaster* **53**, 790–7.
— — and Cotler, D. N. (1977). *Pl. Physiol., Lancaster* **59**, 530–4.
Scandalios, J. G., Liu, E. H., and Campeau, M. A. (1972). *Archs Biochem. Biophys.* **153**, 695–704.
Schnarrenberger, C. and Burkhard, Ch. (1977). *Planta* **134**, 109–14.
Seal, S. N. and Sen, S. P. (1970). *Plant Cell Physiol.* **11**, 119–28.
Servaites, J. C. and Ogren, W. L. (1977*a*). *Pl. Physiol., Lancaster* **60**, 693–6.
— — (1977*b*). *Pl. Physiol., Lancaster* **60**, 461–6.
— Schrader, L. E. and Edwards, G. E. (1978). *Plant Cell Physiol.* **19**, 1399–405.
Shain, Y. and Gibbs, M. (1971). *Pl. Physiol., Lancaster* **48**, 325–30.
Snyder, F. W. and Tolbert, N. E. (1974). *Pl. Physiol., Lancaster* **53**, 514–5.
Somerville, C. R. and Ogren, W. L. (1979). *Nature, Lond.* **280**, 833–5.
— — (1981) in *Proceedings of the Fifth International Congress on Photosynthesis.* In press.
Spalding, M. H., Stumpf, D. K., Ku, M. S. B., Burris, R. H. and Edwards, G. E. (1979). *Aust. J. Plant Physiol.* **6**, 557–67.
Tamas, I. A. and Bidwell, R. G. S. (1971). *Can. J. Bot.* **49**, 299–302.
Tolbert, N. E. (1971). *A. Rev. Pl. Physiol.* **22**, 45–74.
— (1973). *Curr. Topics Cell Reg.* **5**, 21–50.
— (1979) in *Encyclopaedia of plant physiology,* Vol. 6 (eds. M. Gibbs and E. Latzko) pp. 338–69. Springer, Berlin.
— Yamazaki, R. K., and Oeser, A. (1970). *J. biol. Chem.* **245**, 5129–36.
Usuda, H. and Edwards, G. E. (1980). *Pl. Physiol., Lancaster* **65**, 1017–22.
Waidyanatha, U. P. De S., Keys, A. J., and Whittingham, C. P. (1975). *J. exp.*

Bot. **26**, 15–26.
Wang, D. and Burris, R. H. (1963). *Pl. Physiol., Lancaster* **38**, 430–9.
Wolfenden, R. (1970). *Biochemistry* **9**, 3404–7.
Woo, K. C. (1979). *Pl. Physiol., Lancaster* **63**, 783–7.
— and Osmond, C. B. (1976). *Aust. J. Pl. Physiol.* **3**, 771–85.
Woo, K. C. and Osmond, C. B. (1977). *Plant Cell Physiol. Special Issue on Photosynthetic Organelles*, 315–323.
— Jokinen, M. and Canvin, D. T. (1980). *Plant Physiol.* **65**, 433–5.
Zelitch, I. (1966). *Pl. Physiol., Lancaster* **41**, 1623–31.
— (1968). *Pl. Physiol., Lancaster* **43**, 1829–37.
— (1972*a*). *Pl. Physiol., Lancaster* **60**, 109–13.
— (1972*b*). *Pl. Physiol., Lancaster* **51**, 299–305.
— (1972*c*). *Archs Biochem. Biophys.* **150**, 698–707.
— (1979*a*) in *Encyclopaedia of plant physiology* Vol. 6 (eds. M. Gibbs and E. Latzko) pp. 353–67. Springer, Berlin.
— (1979*b*). *Pl. Physiol., Lancaster* **64**, 706–11.
Zindler-Frank, E. (1976). *Z. Pflanzenphysiol.* **80**, 1–13.

8 TOXIC EFFECTS OF OXYGEN ON PLANT TISSUES

Oxygen is, by definition, essential for the life of all aerobic organisms, including plants, but it has long been known to be toxic to them at concentrations higher than those in normal air. High oxygen concentrations have been found to inhibit chloroplast development (Poskuta, Mikulska, Faltynowicz, Bielak, and Wroblewska 1974), decrease seed viability and root growth, damage the membranes of leaves and roots, stimulate leaf abscission (Marynick and Addicott 1976), and increase the incidence of growth abnormalities (Anderson and Linney 1977). Exposure to oxygen at 6 atm pressure for 15 h was lethal to a wide variety of plants, although observable damage, in the form of desiccation, shrivelling and discolouration of leaves, and the collapse of soft stems, only became apparent after the end of the treatment period (Simon 1974).

Oxygen toxicity is seen not only in whole organisms but also in cell cultures and even in isolated organelles. For example, high oxygen concentrations inhibit carbon dioxide fixation by type A chloroplast fractions (the so-called 'Warburg effect'). In part this is due to a competitive inhibition by oxygen of the carboxylase activity of ribulose diphosphate carboxylase/oxygenase (Chapter 3), but oxygen also has other inhibitory effects that will be discussed later in this chapter. Chloroplasts are especially prone to oxygen-toxicity effects because their internal oxygen concentration during photosynthesis is always somewhat greater than that in the air surrounding the leaf (Steiger, Beck, and Beck 1977).

When soybean plants were grown at low (5 per cent) oxygen concentrations, the growth of leaves, stems, and roots was actually increased (Quebedeaux and Hardy 1975), which suggests that even 21 per cent oxygen is sufficient to produce a growth-inhibiting effect. However, growth at low oxygen concentrations hindered seed development in the plants, and so some part of the reproductive process must require oxygen concentrations higher than 5 per cent. The nature of this remains to be established (Quebedeaux and Hardy 1975).

Oxygen can damage living organisms by several mechanisms, which are considered below.

8.1 Oxygen is toxic because it inactivates enzymes

This was one of the earliest proposals made to explain oxygen toxicity (for a review see Haugaard 1968). Enzyme inactivation is best illustrated by the case of certain anaerobes, which stop growing or die immediately on exposure to oxygen (Morris 1979). For example, the activity of the nitrogenase enzyme of the anaerobic bacterium *Clostridium pasteurianum* depends on maintaining some of its cofactors in a highly-reduced state. On exposure to oxygen they are irreversibly oxidized and the enzyme is inactivated (Gomez-Moreno and Ke

1979). The nitrogen-fixing enzymes of blue-green algae are also oxygen-sensitive, which poses a problem for unicellular algae such as *Gloeocapsa* that carry out nitrogen fixation and photosynthesis in the same cell. In order to prevent inactivation of the nitrogenase, the period of maximum nitrogenase activity occurs at a time in the life cycle when the rate of photosynthesis is low. The cells are also well-provided with the various protective mechanisms described later in this chapter (Dilek-Tozum and Gallon 1979).

Some enzymes present in aerobes are also inactivated by oxygen, although much less rapidly than enzymes such as nitrogenase. Inactivation is usually due to the oxidation of essential thiol (—SH) groups on the enzymes. A particularly oxygen-sensitive Calvin cycle enzyme is $NADP^+$-glyceraldehyde 3-phosphate dehydrogenase, which becomes inhibited when chloroplasts are exposed to high oxygen tensions (Ellyard and Gibbs 1969). In C_4 plants, the pyruvate-phosphate dikinase enzyme in the mesophyll chloroplasts can be inactivated at high oxygen concentrations due to oxidation of —SH groups (Rathnam and Chollet 1980).

Aerobic cells have evolved a mechanism to prevent inactivation of enzymes by oxidation of thiol groups; the cells contain high concentrations of a low-molecular-weight thiol compound, the simple tripeptide glutathione (Fig. 8.1). Since reduced glutathione (GSH) is more easily 'available' to oxygen than are enzyme thiol groups, it should be preferentially oxidized and will therefore protect the enzymes. GSH can also reactivate some enzymes that have been inhibited by previous exposure to high oxygen concentrations, presumably by reducing their oxidized —SH groups.

Reaction of GSH with oxygen gives the oxidized form (GSSG), which contains a disulphide bridge (Fig. 8.1).

$$2GSH + \tfrac{1}{2}O_2 \rightarrow GSSG + H_2O \qquad (8.1)$$

The oxidation of pure GSH by oxygen is very slow, but it is catalysed by traces of metal ions and other cofactors and proceeds rapidly in extracts of animal and plant tissues, especially at alkaline pH values. The chloroplast stroma contains glutathione at millimolar concentrations (Foyer and Halliwell 1976) and, of course, becomes alkaline during photosynthesis, reaching a pH of 8.0 (Chapter 4).

GSSG may be converted back to GSH by the activity of glutathione reductase enzymes, which are usually NADPH-dependent (eqn (8.2)). Their equilibrium position greatly favours

$$GSSG + NADPH + H^+ \rightarrow 2GSH + NADP^+ \qquad (8.2)$$

reduction of GSSG. The chloroplast stroma contains an NADPH-dependent glutathione reductase, which is equally active under both light and dark conditions. The purified spinach chloroplast enzyme has a molecular weight of

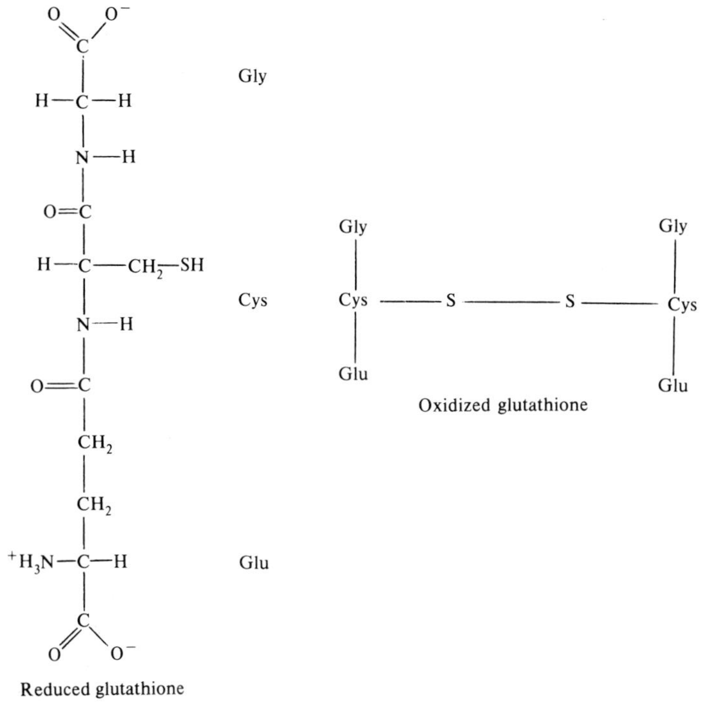

O O⁻
C
H—C—H Gly
N—H
O=C
H—C—CH_2—SH Cys
N—H
O=C
CH_2
CH_2
^+H_3N—C—H Glu
C
O O⁻

Reduced glutathione

Gly Gly
Cys —— S —— S —— Cys
Glu Glu

Oxidized glutathione

Fig. 8.1. Structures of glutathione (γ-L-glutamyl-L cysteinylglycine).

about 145 000 and consists of two apparently-identical subunits. It has a broad pH optimum (7.8–9.3). Its K_m for GSSG is approximately 200 μM and that for NADPH is approximately 3 μM. The concentration of NADPH provided in chloroplasts by the operation of the oxidative pentose phosphate pathway in the dark (Chapter 4) has been given by Lendzian and Bassham (1975) as 15 nmol mg chlorophyll^{-1}, and this concentration is higher in the light. If we assume an 'available internal volume' of 21 μl mg chlorophyll^{-1} (Hall 1976) for spinach chloroplasts, their internal concentration of NADPH in the dark may be calculated as 710 μM, which is amply sufficient to saturate the glutathione reductase. The enzyme will therefore be half-maximally active at a GSSG concentration of 200 μM, which is small in relation to the total glutathione content of chloroplasts (1.0–3.5 mM). Since substantial amounts of glutathione reductase activity are present in chloroplasts, one would expect stromal GSH/GSSG ratios to be normally kept high under both light and dark conditions, a prediction which has been confirmed experimentally (Halliwell and Foyer 1978).

As plant cells or chloroplasts are exposed to increasing oxygen concentrations, however, a point might be reached at which reaction (8.1) proceeds more rapidly than reaction (8.2) and GSH/GSSG ratios begin to fall, which would allow inactivation of enzymes such as $NADP^+$-glyceraldehyde 3-phosphate dehydrogenase. GSSG itself can inactivate enzymes by the formation of mixed disulphides with them, as shown in eqn (8.3)

$$\underset{\text{active enzyme}}{\text{enzyme} - \text{SH}} + \begin{array}{c}\text{GS} - \text{S} \\ | \\ \text{GS} - \text{S}\end{array} \rightleftharpoons \underset{\text{inactive enzyme}}{\text{enzyme} - \text{S} - \text{SG}} + \text{GSH} \qquad (8.3)$$

Mixed disulphides have been found *in vivo* in both bacterial and in animal systems (e.g. Haugaard 1968; Ernst, Levin, and London 1978) and should presumably be formed in plant cells as well. The occurrence of reaction (8.3) provides a good reason for cells and chloroplasts to keep their GSSG concentrations low under normal conditions. Indeed, the glutathione reductase activity of cotton-plant leaves is increased by exposing them to 75 per cent oxygen for 48h (Foster and Hess 1980).

GSH also plays a role in removing hydrogen peroxide in chloroplasts (Section 8.2). In some plant tissues it is involved in the metabolism of phenolic compounds (Diesperger and Sandermann 1979) and in the detoxification of herbicides such as atrazine, fluorodifen, propachlor, barban, and CDAA (*N,N*-diallyl, 2-chloroacetamide). For example, corn leaves contain an enzyme which detoxifies atrazine by conjugating it with GSH. The higher the activity of this enzyme, the greater is the resistance of the plant to the herbicide (Chasseaud 1973; Shimabukuro, Lamoureux, and Frear 1978). Some herbicide 'antidotes' appear to act by increasing intracellular GSH concentrations in plants (Lay and Casida 1978). The role played by chloroplasts in such conjugation reactions, and in the biosynthesis of GSH itself, remains to be investigated. Glutathione also appears to be a major form by which 'organic' sulphur is transported from the leaves to the roots of plants (Chapter 10).

8.2 Oxygen is toxic because it causes formation of hydrogen peroxide

Most oxygen taken up by aerobic cells is reduced to water by the addition of four electrons to each molecule (eqn (8.4))

$$O_2 + 4e^- + 4H^+ \rightarrow 2H_2O \qquad (8.4)$$

This reaction is catalysed by the cytochrome oxidase complex of the inner mitochondrial membrane (or plasma membrane in the case of bacteria). Some enzymes, however, are oxidases which transfer two electrons on to each oxygen molecule that they use, making hydrogen peroxide (eqn (8.5))

$$O_2 + 2H^+ + 2e^- \rightarrow H_2O_2 \qquad (8.5)$$

Important enzymes of this type in plant tissues are glycollate oxidase (Chapter 7), urate oxidase, and amino acid oxidases. Egneus, Heber, Matthiesen, and Kirk (1975), using isotopic methods, detected a significant uptake of oxygen during carbon dioxide fixation by type A chloroplast preparations. Carbon dioxide fixation by their preparations was stimulated by addition of catalase, indicating production of hydrogen peroxide during photosynthesis to an extent that could inhibit carbon dioxide fixation. Since catalase cannot cross the chloroplast envelope, this hydrogen peroxide must have been available to the catalase outside the chloroplast, i.e. it had passed through the envelopes. One problem in experiments of this type is the possibility of production of hydrogen peroxide by the small number of broken chloroplasts always present in a type A preparation. This hydrogen peroxide could then damage the intact chloroplasts. However, Egneus *et al.* (1975) were able to rule out this explanation because the effect of catalase did not depend on the percentage of damaged chloroplasts present in the preparation. When hydrogen peroxide is added to type A chloroplasts, carbon dioxide fixation is inhibited by 50 per cent at concentrations as low as 10^{-5}M (Kaiser 1976). The site of inhibition seems to be the fructose and sedoheptulose diphosphatase enzymes (Heldt, Chon, Lilley, and Portis 1978; Kaiser 1979), which are oxidized by hydrogen peroxide to forms that cannot participate in the Calvin cycle (Charles and Halliwell 1980).

The results of Egneus *et al.* (1975) have been supported by many other workers. For example, Hind, Mills, and Slovacek (1978) found it necessary to include catalase in the reaction mixture to obtain maximal rates of carbon dioxide fixation by type A spinach chloroplasts. Jennings and Forti (1975) observed a rapid oxygen uptake during the induction period when chloroplasts are first illuminated. They suggested that during this period, when carbon dioxide fixation is low (see Chapter 4) and the $NADPH/NADP^+$ ratio is high, the electron-transport chain reduces mainly oxygen instead of $NADP^+$. Radmer and Kok (1976) obtained similar results during studies on a range of algae. Illuminated cells of the blue-green alga *Anacystis nidulans* excrete hydrogen peroxide into the surrounding medium; there is a 'burst' of hydrogen peroxide production when the light is switched on, followed by a slower, steady rate of release from the cells (Patterson and Myers 1973). These experiments are discussed further in Chapter 2.

Some authors have been unable to detect hydrogen peroxide production by type A chloroplasts *in vitro*, however (Allen, 1978*a*,*b*; Allen and Whatley 1978). If hydrogen peroxide is formed *in vivo*, then protective mechanisms should be present to prevent or limit the inhibition of carbon dioxide fixation that would result. If these mechanisms were of variable efficiency in the chloroplast preparations studied by different workers, this could explain the different results obtained.

8.2.1 Methods for the removal of hydrogen peroxide in chloroplasts and other organelles

Leaf tissues contain high activities of the haem-containing enzyme catalase. The purified *Lens culinaris* and spinach-leaf enzymes are tetramers of identical subunits (Gregory 1968; Schiefer, Teifel, and Kindl 1976). Catalase breaks down high concentrations of hydrogen peroxide very rapidly, but it is much less effective at removing hydrogen peroxide present in low concentrations because of its low affinity (high K_m) for this substrate (Halliwell 1974). Most, if not all, of the catalase in leaf cells is located in peroxisomes, where it serves to remove hydrogen peroxide generated during the action of glycollate oxidase (Chapter 7). Variable amounts of catalase activity can be detected in isolated chloroplast preparations, but this is due to cytoplasmic contamination and little, if any, catalase is present in the stroma (Chapter 1). Small amounts of catalase activity have been reported to be associated with isolated photosystem I particles (Van Ginkel and Brown 1978) but this can be attributed to the fact that several other haem proteins present in the preparation can non-specifically degrade hydrogen peroxide at low rates (Brown, Collins, and Merrett 1975). Hence catalase cannot be involved in the removal of hydrogen peroxide generated by illuminated chloroplasts *in vivo*.

Cells can also dispose of hydrogen peroxide by using peroxidase enzymes, which catalyse hydrogen peroxide-dependent oxidation of substrates (SH_2) according to the general equation

$$SH_2 + H_2O_2 \rightarrow S + 2H_2O \tag{8.6}$$

Animal cells contain the enzyme glutathione peroxidase, a selenium-containing protein which uses GSH as substrate (eqn (8.7))

$$2GSH + H_2O_2 \rightarrow GSSG + 2H_2O \tag{8.7}$$

There is no evidence for the existence of selenium-containing glutathione peroxidases in plant tissues, but leaf extracts contain an unstable enzyme activity which brings about removal of hydrogen peroxide in the presence of GSH (Flohe and Menzel 1971). A similar activity was detected in spinach chloroplast fractions by Wolosiuk and Buchanan (1977), although it might have been due to contamination of the preparation by other subcellular fractions (Chapter 1). Elucidation of the nature of this enzyme must await further studies.

However, plants do contain a wide range of peroxidases with a very broad specificity for substrate (SH_2). Such enzymes are usually detected in tissue extracts by using artificial substrates such as guaiacol or *o*-dianisidine. Since the identity of the natural substrates of these enzymes is usually unknown, it is difficult to assess their contribution to removal of hydrogen peroxide *in vivo*. Chemiluminescence (light emission) by root and stem tissues from a wide range of plants has been detected by use of liquid scintillation counting, however, and it has been suggested to be due to the interaction of hydrogen peroxide with peroxidases *in vivo* (Abeles, Leather, and Forrence 1978), although other sources are possible (Boveris, Sanchez, Varsavsky, and Cadenas 1980). Some peroxidases play an important role in the lignification of plant cell walls and must therefore be provided with hydrogen peroxide *in vivo* (Gross, Janse, and Elstner 1977; Sagisaka 1976).

Isolated chloroplasts contain little, if any, 'nonspecific' peroxidase activity (Parish 1972). However, Groden and Beck (1979) have detected an ascorbate peroxidase in spinach chloroplasts, which catalyses the reaction

$$2H^+ + \text{ascorbate} + H_2O_2 \rightarrow \text{dehydroascorbate} + 2H_2O \qquad (8.8)$$

Since the chloroplast stroma often contains ascorbate at up to 50 mM concentrations, ascorbate peroxidase would seem a likely candidate for the role of hydrogen peroxide removal. At pH 8, dehydroascorbate is rapidly reduced by GSH in a non-enzymic reaction (Foyer and Halliwell 1977). It is therefore possible that a cycle of reactions could operate in the stroma of illuminated chloroplasts to remove hydrogen peroxide at the eventual expense of oxidizing NADPH from the electron-transport chain (Fig. 8.2). Addition of hydrogen peroxide to isolated chloroplast fractions causes an activation of the enzyme glucose 6-phosphate dehydrogenase, possibly by means of a decrease in $NADPH/NADP^+$ ratios as the cycle shown in Fig. 8.2 becomes operative (Kaiser 1979). The green alga *Euglena*, which has no catalase activity, possesses high activities of an ascorbate-specific peroxidase (Shigeoka, Nakano, and Kitaoka 1980a, b).

Both the ascorbate content of chloroplasts and their ascorbate peroxidase activity vary over a wide range, depending on leaf age and time of year (Walker

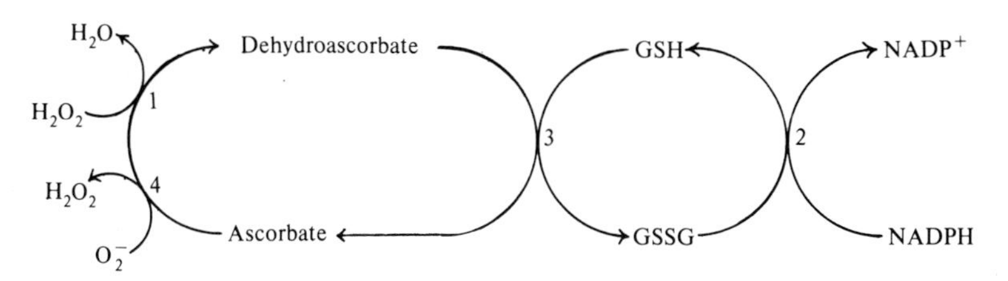

Fig. 8.2. An ascorbate-glutathione cycle for removal of hydrogen peroxide in illuminated spinach chloroplasts. Enzymes involved: 1, Ascorbate peroxidase; 2, glutathione reductase; 3, 4, are non-enzymic reactions. Reaction 3 proceeds rapidly at pH 8, the pH of the stroma in illuminated chloroplasts. Leaf tissues contain a dehydroascorbate reductase enzyme, which catalyses this reaction at acidic and neutral pH values, but it is not located within chloroplasts (Foyer and Halliwell 1976, 1977). H_2O_2 added to spinach chloroplasts is gradually consumed in the light, but not in the dark (Nakano and Asada 1980).

1971; Groden and Beck 1979). Variations in glutathione concentration might also occur. In chloroplasts with low ascorbate and/or glutathione content the cycle in Fig. 8.2 would be impaired, hydrogen peroxide could accumulate during photosynthesis and it would be necessary to add catalase to the reaction medium to obtain maximum rates of carbon dioxide fixation. Indeed, Egneus *et al.* (1975) found that added ascorbate was able to replace catalase in stimulating carbon dioxide fixation by their chloroplast preparations. In preparations with high stromal ascorbate and GSH concentrations, hydrogen peroxide should not accumulate and the effects of catalase reported by Egneus *et al.* (1975) and others would not be expected to occur (e.g. Allen 1978a, b; Allen and Whatley 1978).

8.3 Oxygen is toxic because it causes lipid peroxidation and formation of singlet oxygen

Lipids make up about 35 per cent of the dry weight of chloroplasts. The major lipids present are glycolipids together with some sulphoquinovosyl diglyceride, often called 'the plant sulpholipid'. The lipid compositions of envelope and thylakoid membranes are different (see Chapter 1). Most of the fatty acids present in an esterified form in chloroplast lipids are C_{18} acids containing one or more double bonds. α-Linolenic acid ($C_{18:3}$) can sometimes comprise over 90 per cent of the esterified fatty acids. An unsaturated fatty acid unique to photosynthetic tissues is *trans*-3-hexadecenoic acid ($C_{16:1}$), which occurs acylated to phosphatidylglycerol (Chapter 1).

Polyunsaturated fatty acid side-chains are prone to undergo a process known as lipid peroxidation, which involves the reaction of oxygen with conjugated dienes to produce lipid peroxides (Fig. 8.3). Lipid peroxides decompose to give aldehydes (e.g. malondialdehyde) and other products, including volatile hydrocarbons such as ethane and pentane (Konze and Elstner 1978; Pryor 1978). Their decomposition is accelerated in the presence of transition metal ions or haem compounds. The peroxides and some of their degradation products cause extensive damage to enzymes and to membranes, producing a decrease in electrical resistance and membrane fluidity and eventual loss of membrane integrity (Hicks and Gebicki 1978; Putvinsky, Sokolov, Roshcupkin, and Vladimirov 1979; Pauls and Thompson 1980). Disruption of, for example, the tonoplast membrane of a plant cell by lipid peroxidation might release organic acids and hydrolytic enzymes into the cytoplasm, so potentiating the damage. The loss of viability of soybean seeds that occurs during storage in warm, damp conditions has been attributed to accumulation of lipid peroxides (Stewart and Bewley 1980) and many of the damaging effects of elevated oxygen concentrations on plant tissues seem to be due to increased leakiness of membranes (Simon 1974). In contrast little, if any, lipid peroxide can be detected in extracts from plant tissues kept under normal conditions, which suggests the

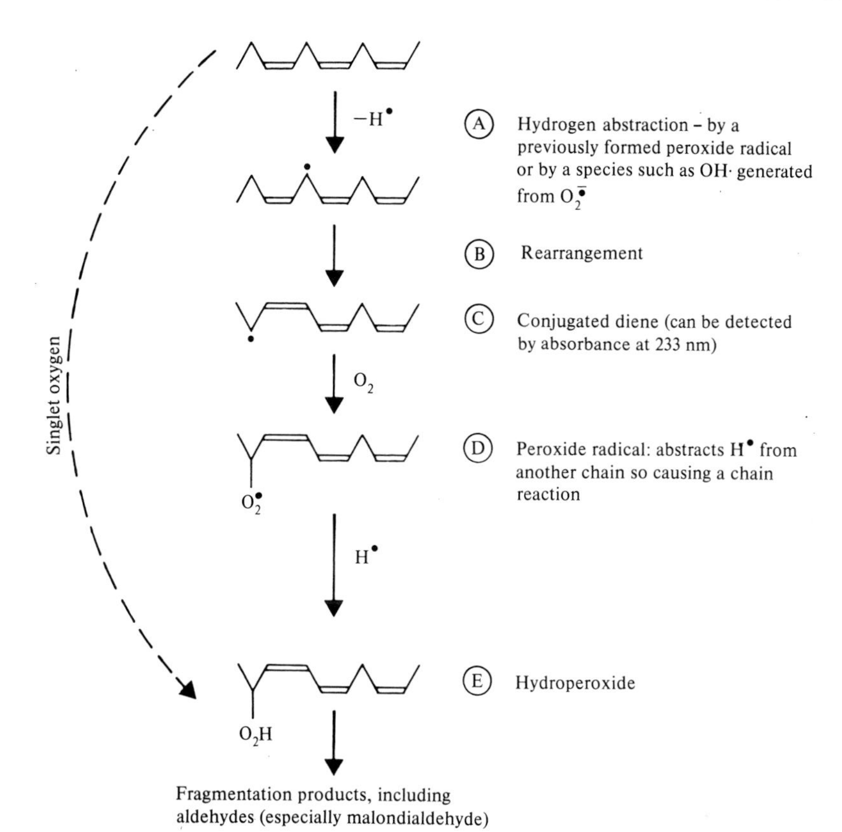

Fig. 8.3. Mechanism of peroxidation of polyunsaturated fatty acids. The chain reaction of lipid peroxidation can be effectively inhibited by abstraction of .H from another donor such as α-tocopherol (α-TH) and so such compounds inhibit lipid peroxidation.

$$\text{Lipid} - O_2^{\bullet} + \alpha\text{TH} \rightarrow \text{Lipid} - \text{OOH} + \alpha\text{-T}.$$

Singlet oxygen reacts directly with polyunsaturated fatty acids to give hydroperoxides. It has been suggested that *breakdown* of lipid peroxides also releases singlet oxygen.

occurrence of protective mechanisms that become overwhelmed at high oxygen concentrations.

The rate of peroxidation of pure lipids or fatty acids is very slow indeed, but it is greatly accelerated by the presence of transition metal ions (especially iron or copper, e.g. see Sandmann and Boger 1980), haem compounds or illuminated chlorophyll. Once initiated, peroxidation is autocatalytic (Fig. 8.3). Lipid peroxidation may be controlled by the presence of 'scavengers' which trap the peroxide radicals and so interrupt the chain reaction (Fig. 8.3). Chloroplasts contain large amounts of the scavenger α-tocopherol (Fig. 8.4), known to animal nutritionists as vitamin E (Janiszowska and Korczak 1980). Its lipophilic nature enables it to deal effectively with peroxides formed inside membranes. α-Tocopherol appears to be particularly concentrated in Photosystem I (Baszynski 1974). The radical form of this compound (αTH. in Fig. 8.3) can be reduced back to α-tocopherol by ascorbic acid (Packer, Slater, and Willson 1979) which is present in the stroma at high concentrations and could therefore interact with tocopheryl radicals at the membrane surface, so as to regenerate tocopherol for further use in preventing lipid peroxidation.

Prolonged illumination of thylakoids *in vitro* causes marked lipid peroxidation. Indeed, when polyunsaturated fatty acids are mixed with chlorophyll and illuminated, they are rapidly peroxidized (Heath and Packer 1968*a*, *b*; Rawls and van Santen 1970). The higher electronic excitation states formed on illumination of the chlorophyll molecule are capable of transferring energy

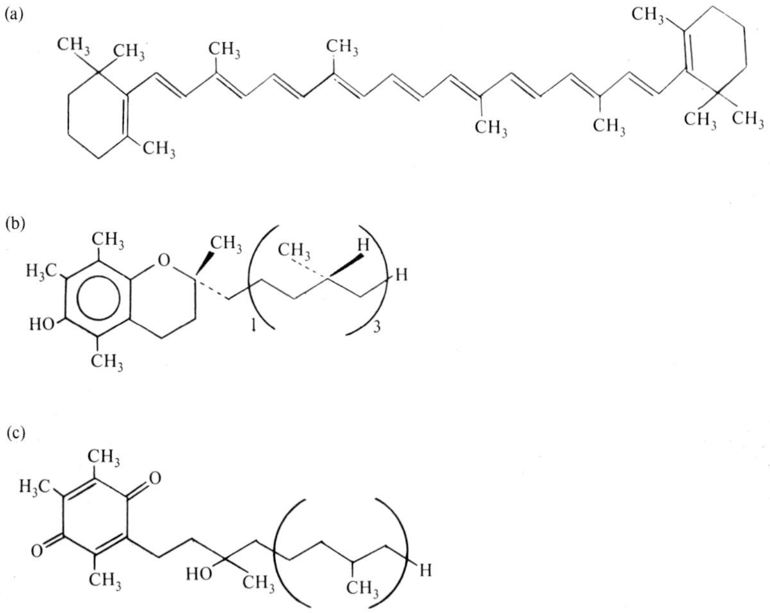

Fig. 8.4. Structure of compounds involved in controlling lipid peroxidation in chloroplasts. (a) β-Carotene. A scavenger of singlet oxygen and a quencher of excess chlorophyll excitation energy. (b) α-Tocopherol. A scavenger of singlet oxygen and a 'free radical trap' for lipid peroxides. About one molecule is present in spinach chloroplasts for every 10 molecules of chlorophyll a. (c) α-Tocopherylquinone. About one molecule is present in spinach chloroplasts for every 40 molecules of chlorophyll a.

on to the oxygen molecule, raising it from its ground state to a more-reactive excited state known as singlet oxygen $^1\Delta g$ (Fig. 8.5). The higher excited state of oxygen ($^1\Sigma g^+$) may also be formed, but it usually decays to the lower state $^1\Delta g$ before it has time to react with anything.

Singlet oxygen $^1\Delta g$ has a very short lifetime in aqueous solution, but it survives for much longer in hydrophobic environments, such as the interior of membranes. Unlike ground-state oxygen, singlet oxygen can react directly with polyunsaturated fatty acid side chains to form lipid peroxides (Fig. 8.3). Since the trapping of light by photosystems I and II causes formation of excited states of the reaction centre chlorophyll molecules and chloroplasts have a high internal oxygen concentration, production of singlet oxygen might be expected whenever there is 'excess' excitation. This accounts for the increase in the rate of lipid peroxidation observed in isolated illuminated chloroplasts or algal cells when electron transport is blocked by DCMU or monuron (CMU) and the decrease induced by adding electron acceptors such as ferricyanide (Heath and Packer 1968*a*,*b*; Pallet and Dodge 1979; Elstner and Osswald 1980). Singlet oxygen can also damage the chlorophyll molecule itself, causing loss of the characteristic green colour ('bleaching'). Indeed, the chlorophyll bleaching

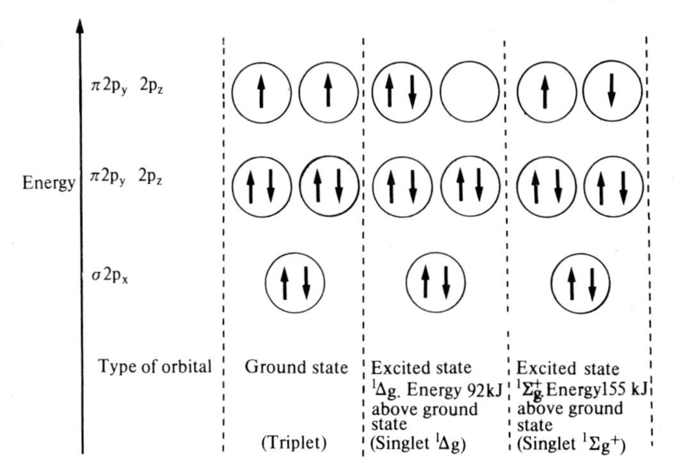

Fig. 8.5. Molecular orbital scheme for oxygen. The oxygen atom has eight electrons. In the oxygen molecule, O_2, it has been shown that only one electron pair is shared and there are two unpaired electrons, i.e. O_2 is not O=O but .O–O. and is itself a radical. The two unpaired electrons occupy molecular orbitals which are referred to as π^* orbitals (antibonding orbitals). Usually, these two electrons occupy different π^* orbitals but they have the same spin quantum number (a quantity which has only two values). This form of oxygen is referred to as being in the *triplet state* and it is the state of least energy for oxygen, i.e. the *ground state*. Two other states, of higher energy, are known in which the two electrons have opposite spin (*singlet states*). These are illustrated above. Because an input of energy is needed to generate them, they are *excited states* and oxygen in these states is far more reactive than ground state oxygen.

observed in flax cotyledons treated with CMU has been shown to require both light and oxygen (Pallett and Dodge 1979).

Since chloroplasts are likely to form singlet oxygen *in vivo*, one would expect them to be protected against it. As well as being an inhibitor of lipid peroxidation, α-tocopherol is a powerful scavenger of singlet oxygen. The products of reaction of tocopherol with singlet oxygen include tocopherylquinone (Clough, Yee, and Foote 1979), which may account for the presence of this compound in chloroplasts (Fig. 8.4). Consistent with this, illumination of isolated spinach chloroplasts under conditions favouring lipid peroxidation causes conversion of tocopherol to tocopherylquinone in the thylakoids (Yamauchi and Matsushita 1979). It must not be assumed, however, that α-tocopherol in plant tissues serves *only* as a protection against membrane damage. Other roles are perhaps suggested by the fact that a tocopherol oxidase enzyme, whose activity is under hormonal control, has been detected in many plant tissues (Barlow and Gaunt 1972; Gaunt, Matthews, and Plumpton 1980) although this enzyme has so far resisted purification and its subcellular location is unknown.

Carotenoids, which are important constituents of chloroplast membranes, react with singlet oxygen at a diffusion controlled rate and can therefore help to protect chlorophyll and membranes against damage (Foote 1970). Carotenoids are quickly destroyed on illumination of thylakoids *in vitro*, presumably as they absorb any singlet oxygen formed (Takahama and Nishimura 1975; Ridley 1977). Their protective function *in vivo* is well illustrated by maize mutants which lack them: illumination of such plants under aerobic conditions causes rapid bleaching of chlorophylls and destruction of chloroplast membranes. Illumination under anaerobic conditions causes much less damage, since singlet oxygen cannot then be generated (Anderson and Robertson 1960). Similar destructive effects of illumination in the presence of oxygen are seen in normal plants in which carotenoid biosynthesis has been inhibited by herbicides (Frosch, Jabben, Bergfeld, Kleinig, and Mohr 1979). The carotenoids present in chloroplasts are two main types, the carotenes (e.g. β-carotene, whose structure is shown in Fig. 8.4) and xanthophylls, which are oxygen-containing derivatives of carotenes. Figure 8.6 shows some typical xanthophylls. Carotenoids are also able to *directly* quench those excited states of chlorophyll that lead to singlet oxygen formation. Hence they have a dual role; preventing formation of singlet oxygen *in vivo* and helping to remove any that does happen to be formed (Koka and Song 1978; Krinsky 1979).

In addition to α-tocopherol and carotenoids, ascorbate and glutathione are also rapidly oxidized by singlet oxygen (Bodannes and Chan 1979) and might help to remove any that is formed in the stroma.

The deterioration of isolated, illuminated chloroplast thylakoids is in fact an extremely complicated process. In addition to lipid peroxidation, aldehyde formation, and carotenoid destruction, there is actual hydrolysis of glycolipids and phospholipids to release fatty acids. Both esterified and released fatty

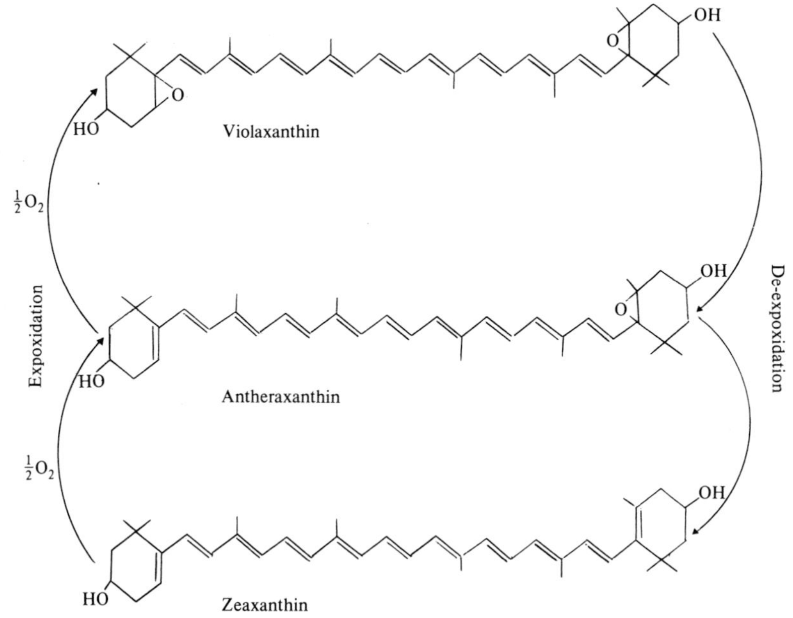

Violaxanthin

$\frac{1}{2}O_2$

Expoxidation

De-expoxidation

Antheraxanthin

$\frac{1}{2}O_2$

Zeaxanthin

Fig. 8.6. The xanthophyll cycle in the chloroplast thylakoids.

acids can undergo lipid peroxidation (Harnischfeger 1972; Hoshina, Kaji, and Nishida 1975). Galactolipid hydrolysis seems to be due to the action of galactolipase enzymes, which normally show little activity in isolated chloroplasts but seem to be 'unmasked' during membrane deterioration (Wintermans, Helmsing, Polman, Van Gisbergen, and Collard 1969; Anderson, McCarty, and Zimmer 1974). The free fatty acids produced by lipase action themselves cause severe membrane damage (Siegenthaler 1972) and inhibit photosynthesis (Mue Akamba and Siegenthaler 1979; Golbeck, Martin, and Fowler 1980).

Some leaves contain the enzyme lipoxygenase, an iron-containing protein which catalyses a direct reaction of polyunsaturated fatty acids with oxygen to give 13- and 9-hydroperoxides (Boldingh 1976). The subcellular fractionation techniques employed to date have given no clear evidence for the presence of lipoxygenase within the chloroplast, although it is often present in chloroplast fractions and could therefore facilitate their deterioration (Holden 1970; Wardale, Lambert, and Galliard 1978). The role of lipoxygenase has been elucidated more clearly in non-green plant tissues, such as tubers, fruits, and seeds. Such plant tissues contain enzymes which, on wounding of the tissue, initiate a series of reactions that result in lipid peroxidation and subsequent degradation of the peroxides (Galliard 1978). The enzymes involved are hydrolases, which release free-fatty acids from membrane-bound lipids, lipoxygenase to produce fatty acid peroxides, and enzymes for the further metabolism and cleavage of these compounds (Fig. 8.7). Some of the aldehydes so produced have characteristic smells, which are responsible for the aroma of certain plant tissues, such as sliced cucumbers and tea. The odour of crushed green leaves is caused by a process similar to that shown in Fig. 8.7 (Matthews and Galliard 1978). Chloroplast fractions from leaves of tea and Japanese silver (*Farfugium japonicum*) plants convert C_{18} unsaturated fatty acids into C_6-aldehydes, but the role of lipoxygenase in the reactions has not been established, nor is it rigorously proven that the activities are actually located within the chloroplast and not due to contamination with other organelles (Sekiya *et al.* 1978; Hatanaka, Sekiya, and Kajiwara 1977a; Hatanaka, Kajiwara, Sekiya, and Kido 1977b). The formation of lipid peroxides and aldehydes when plant tissues are damaged may play an important role in killing fungi and bacteria attempting to enter the wound (e.g. Harman, Mattick, Nash, and Nedrow 1980). Indeed, crushed leaves of

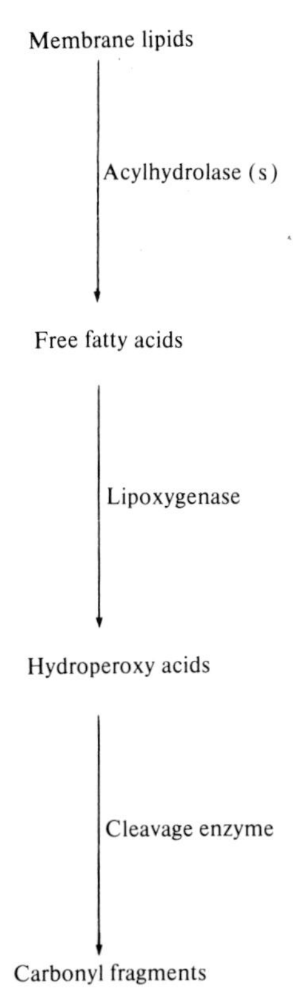

Fig. 8.7. Pathway for the breakdown of membrane lipids induced on wounding plant tissues

Japanese silver are a traditional remedy for wound infections. It has also been speculated that some of the compounds formed may be insect attractants or repellants (Hatanaka and Harada 1973).

Damage to plant tissues can result in formation of 'traumatic acid' (wound hormone), a compound which induces cell proliferation. Traumatic acid is formed by oxidative degradation of polyunsaturated fatty acids by processes related to those discussed above (Galliard 1978).

8.3.1 The xanthophyll cycle

Higher plants and algae operate a series of reactions involving carotenoids, known as the 'xanthophyll cycle' (Fig. 8.6), in the thylakoids. Epoxidation of zeaxanthin to violaxanthin, via antheraxanthin, is an oxygen-dependent reaction which occurs in the thylakoids under both light and dark conditions. De-epoxidation is an ascorbate-dependent process (Sokolove and Marsho 1976) that occurs only in the light, apparently because the de-epoxidase enzyme is located on the inner side of the thylakoid membrane and only functions at the low pH generated there because of proton translocation on illumination (see Chapter 2). The accessibility of violaxanthin to the de-epoxidase also appears to be increased by light-dependent conformational changes in the membrane (Sieferman and Yamamoto 1975*a*,*b*). Despite the work that has been done to elucidate the reactions of the xanthophyll cycle, its function *in vivo* remains obscure. In particular, it does not seem to be involved in the protective action of carotenoids against singlet oxygen (Ridley 1977).

8.4 Oxygen is toxic because it causes formation of oxygen-free-radicals

Some cellular systems catalyse oxidation reactions in which a single electron is transferred from the substrate onto each molecule of oxygen used, so producing an oxygen-free-radical known as superoxide, O_2^- (eqn (8.9)). The term 'radical' can be applied to any species that has an unpaired electron.

$$O_2 + e^- \rightarrow O_2^- \qquad (8.9)$$

Examples of superoxide-producing enzymes present in some plant tissues are nitropropane dioxygenase, galactose oxidase, and xanthine oxidase. Indeed, a mixture of xanthine and xanthine oxidase is often used as a source of O_2^- for experiments *in vitro*. 'Non-specific' peroxidases can generate O_2^- during oxidation of certain substrates, e.g. NADH (Halliwell 1978).

The autoxidation of some reduced compounds also proceeds with single electron transfer to oxygen to give O_2^-. Such compounds include reduced flavins, pteridines, diphenols, and ferredoxin (eqn (8.10))

$$\text{Ferredoxin}_{\text{reduced}} + O_2 \rightarrow \text{Ferredoxin}_{\text{oxidized}} + O_2^- \qquad (8.10)$$

Allen (1975) has proposed that reduced ferredoxin has a sufficiently low redox potential to further reduce O_2^- to hydrogen peroxide (eqn (8.11)), but this reaction has not yet been rigorously proved to occur.

$$\text{Ferredoxin}_{\text{reduced}} + O_2^- + 2H^+ \rightarrow H_2O_2 + \text{Ferredoxin}_{\text{oxidized}} \quad (8.11)$$

Isolated illuminated chloroplast thylakoids slowly take up oxygen in the absence of added electron acceptors. This was first observed by Mehler (1951) and is hence often referred to as the 'Mehler reaction'. It appears to result from the reduction of O_2 to O_2^- by electron acceptors associated with photosystem I (Asada, Kiso, and Yoshikawa 1974; Harbour and Bolton 1975; Miller and McDowall 1975; Jursinic 1980; Lien and San Pietro 1979). Addition of ferredoxin increases the amount of oxygen uptake, since it is reduced by PSI much more quickly than is oxygen and the reduced ferredoxin can then undergo reaction (8.10) and possibly reaction (8.11). *In vivo*, however, reduced ferredoxin also passes electrons on to $NADP^+$ via ferredoxin-$NADP^+$ reductase and possibly into a cyclic electron-transport pathway (Chapter 2). Thus electrons from PSI can pass through at least three routes, as shown in Fig. 8.8, of which route C is preferred (Lien and San Pietro 1979). If the supply of $NADP^+$ were limited, however, the rate of electron flow along pathway C would be expected to be decreased and more O_2^- should be made by route B and, to a lesser extent, by route A (Fig. 8.8). In the induction period observed when chloroplasts are illuminated after a dark period (see Chapter 4), the $NADP^+$/NADPH ratio is very low, and O_2^- formation by routes B and C would account for the observations, discussed previously, of a substantial oxygen uptake during this period (Jennings and Forti 1974; Radmer and Kok 1976). However, in the experiments of Egneus *et al.* (1975) carbon dioxide was not in limiting amounts, and yet substantial oxygen reduction was occurring.

Elstner, Stoffer, and Heupel (1975) and Elstner, Wildner, and Heupel (1976) and others (for a review see Trebst 1974) have suggested that another compound which they have isolated from chloroplast fractions is required to mediate ferredoxin-dependent oxygen uptake. It has been called 'oxygen reducing factor' and seems to be a phenolic compound, possibly a derivative of *p*-coumaric acid (see Chapter 9). Chloroplasts contain enzymes that hydroxylate monophenols to give diphenols, which can then be oxidized into quinones (Chapter 9). Both diphenols and quinones react rapidly with O_2^- to produce semiquinones, and so it is not surprising that effects on oxygen uptake can be demonstrated when such compounds are added to illuminated chloroplasts *in vitro*. Whether these effects have physiological significance remains to be determined. Pteridines have been detected in chloroplast preparations (Fuller and Nugent 1969; Iwai, Bunno, Kobayashi, and Suzuki 1976). Since reduced pteridines autoxidize to yield O_2^-, they could also be involved in oxygen uptake by chloroplasts, although direct evidence for this is lacking.

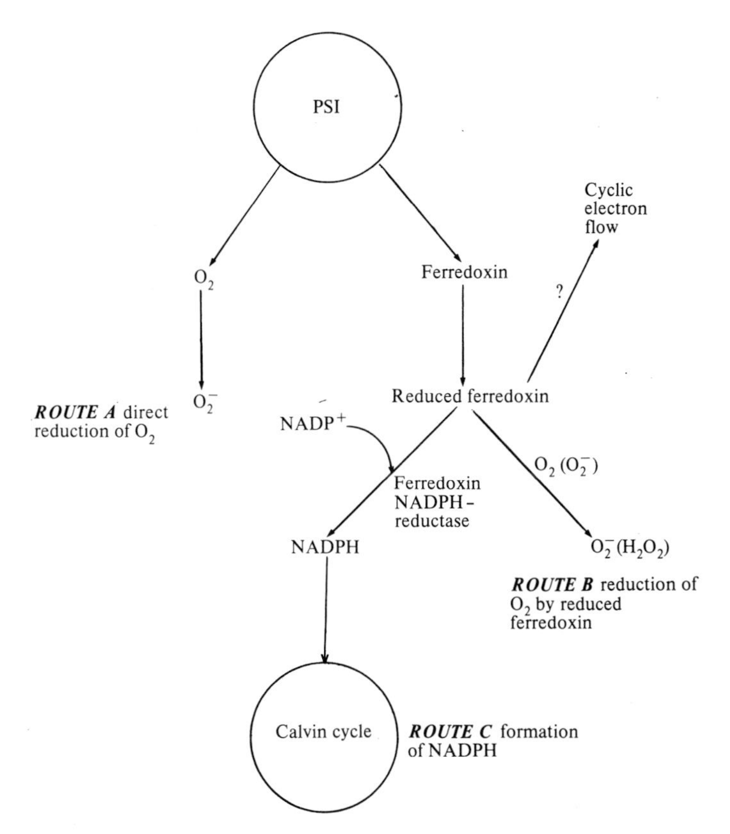

Fig. 8.8. Possible pathways of electron flow from reduced photosystem I. The occurrence of cyclic electron transport *in vivo* has been questioned (see Chapter 2).

8.4.1 Properties of superoxide

O_2^- in aqueous solution can act either as a reducing agent, giving up its extra electron, or as a weak oxidizing agent, becoming reduced to hydrogen peroxide. For example it can reduce cytochrome f (eqn (8.12)) and plastocyanin (eqn. (8.13)) but it oxidizes ascorbic acid (eqn (8.14)) and possibly ferredoxin (eqn (8.11) above).

$$\text{cytochrome f}\,(Fe^{3+}) + O_2^- \rightarrow \text{cytochrome f}\,(Fe^{2+}) + O_2 \qquad (8.12)$$

$$\text{plastocyanin}\,(Cu^{2+}) + O_2^- \rightarrow \text{plastocyanin}\,(Cu^{+}) + O_2 \qquad (8.13)$$

$$2H^+ + 2O_2^- + \text{ascorbate} \rightarrow 2H_2O_2 + \text{dehydroascorbate} \qquad (8.14)$$

Superoxide oxidizes thiol compounds such as glutathione, but these reactions are slow.

Superoxide also appears to convert the metal ion Mn^{2+} into a more reactive species. For example, O_2^- is insufficiently reactive to oxidise NADH but if Mn^{2+} is added to the system there is a rapid oxidation of the coenzyme. Apparently the Mn^{2+} reacts with O_2^- to form a complex (probably MnO_2^+ – Bielski and Chan 1978), that oxidizes the NADH. Perhaps this or similar reactions might be involved in the water-splitting reactions associated with PSII (Chapter 2) (Sayre and Homann 1979).

By comparison with other oxygen radicals, however, O_2^- is rather unreactive. For example, it cannot react directly with membrane lipids to cause peroxidation. Most O_2^- formed in biochemical systems reacts with itself as shown in eqn (8.15).

$$O_2^- + O_2^- + 2H^+ \rightarrow H_2O_2 + O_2 \tag{8.15}$$

This non-enzymic 'dismutation' of O_2^- produces oxygen and hydrogen peroxide. The results of Asada *et al.* (1974) indicate that the hydrogen peroxide generated by illuminated chloroplasts arises from O_2^- (generated as in Fig. 8.8) by reaction (8.15), and that the oxygen uptake demonstrated in the experiments of Egneus *et al.* (1975), Radmer and Kok (1976) and Jennings and Forti (1975) was due to formation of O_2^- and then of H_2O_2.

There is considerable evidence to show that hydrogen peroxide and O_2^- can react together in biochemical systems to form the hydroxyl radical, ·OH. Pure hydrogen peroxide and O_2^- will not react together at significant rates *in vitro* unless traces of iron salts are added, when the reactions shown in eqns (8.16) and (8.17) occur (Halliwell 1979). Of course, iron salts are present in all biochemical reagents and in extracts of plant tissues (Brown 1978), so that these reactions can readily occur *in vivo*.

$$Fe^{3+} - \text{salt} + O_2^- \rightarrow Fe^{2+} - \text{salt} + O_2 \tag{8.16}$$

$$Fe^{2+} - \text{salt} + H_2O_2 \rightarrow \cdot OH + OH^- + Fe^{3+} - \text{salt} \tag{8.17}$$

$$\text{Net } H_2O_2 + O_2^- \xrightarrow[\text{catalyst}]{\text{Fe-salt}} O_2 + \cdot OH + OH^- \tag{8.18}$$

Reduced ferredoxin may be able to bring about generation of ·OH from hydrogen peroxide by a related mechanism (Elstner, Saran, Bors and Lengfelder 1978) and it has been suggested that illuminated chlorophyll might be able to generate ·OH radicals as well as singlet oxygen (Harbour and Bolton 1978).

Hydroxyl radicals are the most reactive species known to chemistry: they will attack and damage almost every molecule found in living cells. For example, they can hydroxylate the purine and pyrimidine bases present in DNA, so giving rise to mutations, and they can abstract hydrogen radicals from membrane lipids and so trigger peroxidation (Fig. 8.3).

There is some evidence to suggest that part of the oxygen formed during the breakdown of O_2^- radicals is singlet oxygen $^1\Delta g$. It is not at present clear whether singlet oxygen arises during reaction (8.15) or during reactions (8.16) and (8.17). Experimental results are very confused, largely because of the lack of specificity of many so-called 'specific' scavengers of singlet oxygen (for a review see Halliwell 1979). The damaging effects of $^1\Delta gO_2$ have already been discussed.

Although O_2^- is scarcely pleasant for the cells, its real danger therefore seems to lie in its ability to form ·OH and singlet oxygen. Protective mechanisms against these species are therefore essential. Mechanisms of protection against singlet oxygen have already been discussed.

8.4.2 Superoxide dismutase

Since the pioneering work of McCord and Fridovich (1969) it has been clearly established that all aerobic organisms contain enzymes, known as 'Superoxide Dismutases', which greatly accelerate reaction (8.15). O_2^- is thus rapidly removed, which should prevent its conversion into ·OH radicals. Further, the oxygen produced during the enzyme-catalysed reaction (8.15) is ground-state oxygen, not singlet oxygen. The superoxide dismutase (SOD) content of tomato fruits has been reported to vary inversely with their susceptibility to the 'sunscald' damage caused by exposure to strong sunlight (Rabinowitch and Sklan 1980) which does suggest that SOD plays a role in protection against species generated in illuminated pigment systems.

Superoxide dismutase was originally isolated from mammalian red blood cells and found to contain copper and zinc at its active site. All superoxide dismutase since discovered have also been found to be metalloproteins. The copper–zinc enzyme is found in eukaryotic cells, and is often accompanied by a manganese-containing dismutase. Mangano-enzymes are also found in bacteria. Some bacteria and blue-green algae contain superoxide dismutases with iron as the prosthetic group. Whatever the metal ion present, the enzymes catalyse the same reaction (Fridovich 1978). There has recently been a report of the presence of an iron-containing dismutase in mustard (*Brassica campestris*) leaves (Salin and Bridges 1980).

Most of the superoxide dismutase activity of green leaves is located in the chloroplasts, where it is present as a copper–zinc enzyme (Jackson, Dench, Moore, Halliwell, Foyer, and Hall 1978). Some of it is bound to the thylakoids, from which it can be released by washing in hypotonic solutions, and the rest is apparently free in the stroma. The purified superoxide dismutase protein from spinach chloroplasts is a dimer of identical subunits, each of which has an active site containing one atom of copper and one of zinc (Asada, Urano, and Takahashi 1973). Although several reports of a manganese-containing dismutase in chloroplast fractions have appeared, this activity appears to be due to cytoplasmic contamination. However, disruption and subfractionation of photosystem II results in the release of a manganese-containing peptide that can react with O_2^- (Foyer and Hall 1979). Whether this peptide and/or superoxide play any role in the water splitting process has yet to be determined, but the stepwise nature of this process suggests that oxygen radicals in some form must be involved. In C_4 leaves superoxide dismutase is present in both mesophyll and bundle-sheath cells (Foster and Edwards 1980).

Chloroplast copper-zinc superoxide dismutase is slowly but irreversibly inhibited by its product, hydrogen peroxide (Asada, Yoshikawa, Takahashi, Maedi, and Enmanji 1975) and so it cannot remain active unless efficient methods for disposal of hydrogen peroxide are operative (see Section 2). Ascorbic acid present in the chloroplast stroma can also help to remove O_2^- (eqn (8.14) and Fig. 8.2).

It is possible that the amounts of superoxide dismutase and ascorbate present in chloroplasts are only sufficient to cope with the normal rate of O_2^- generation *in vivo*, so that an increased rate would swamp them. This helps to account for the toxic effects of elevated oxygen concentrations, which presumably increase O_2^- generation in chloroplasts because of increased competition of oxygen for electrons from photosystem I (Fig. 8.8). Indeed, a mutant of the green alga *Chlorella* with abnormally high superoxide dismutase activity is much more resistant to elevated oxygen concentrations than the wild-type strains (Pulich 1974). Hence any compound which increased O_2^- generation *in vivo* at normal oxygen concentrations would be expected to be toxic.

Such a mechanism may well be involved in the damaging effects of the bipyridyl herbicides paraquat (also known as 'methyl viologen') and diquat (Fig. 8.9). Their severe toxic effects on plants are only seen in the presence

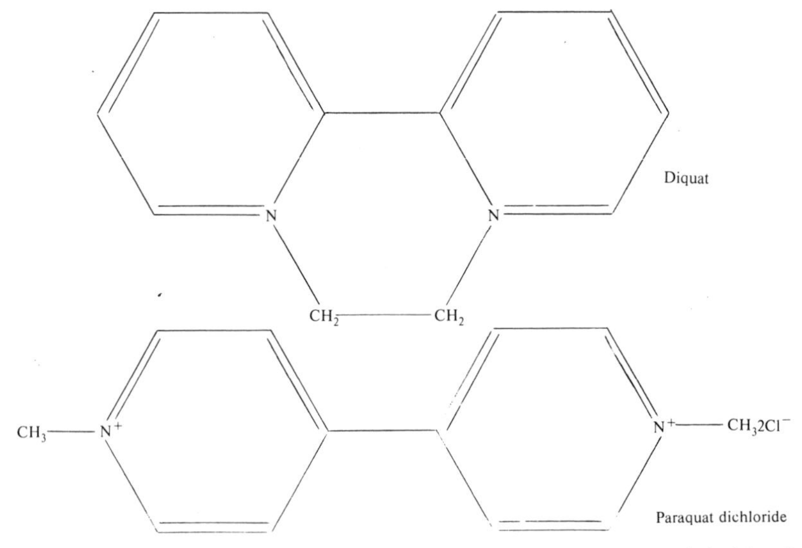

Fig. 8.9. Structures of the bipyridyl herbicides: 'Methylviologen' is identical with 'paraquat dichloride'.

of both light and oxygen, and increase in extent if the oxygen concentration around the plant is raised (Mees 1960). Bipyridyl herbicides are taken up rapidly by the roots and enter the leaves. The chloroplast envelope is freely permeable to paraquat and diquat, which can be reduced directly and rapidly by photosystem I in the light to give bipyridyl radical ions, $BP^{+}\cdot$ (eqn (8.19)). This cannot, of course, occur in the dark.

$$BP^{2+} + H_2O \xrightarrow[\text{electron transport}]{h\nu} \tfrac{1}{2}O_2 + BP^{+}\cdot \quad (8.19)$$

Such radicals rapidly autoxidize by the transfer of single electrons to oxygen, making O_2^- (Farrington, Ebert, Land, and Fletcher 1973).

$$BP^{+}\cdot + O_2 \rightarrow BP^{2+} + O_2^- \quad (8.20)$$

Addition of paraquat to chloroplasts therefore induces a rapid light-dependent oxygen uptake (Whitehouse, Ludwig, and Walker 1971). This can be used as a measure of the rate of electron transport in disrupted chloroplasts under certain conditions (Chapter 2), but in intact chloroplasts the enormously increased generation of O_2^-, and hence of ·OH, hydrogen peroxide and singlet oxygen derived from it, inhibits carbon dioxide fixation. This is followed by lipid peroxidation and disruption of the chloroplast membranes, which is easily observed by electron microscopy of sections of paraquat-treated leaves (Dodge 1975). Figure 8.10 shows an example of such damage. In agreement with the above mechanism, addition of the related compound benzyl viologen was found to increase the rate of peroxidation of isolated illuminated thylakoids, and this increase was prevented by adding an excess of superoxide dismutase (Takahama and Nishimura 1976). A low-molecular-weight copper chelate that scavenges O_2^- was found to decrease the toxicity of paraquat to flax cotyledons (Youngman, Dodge, Lengfelder, and Elstner 1979). The leaves of some paraquat-tolerant strains of ryegrass are found to contain increased activities of superoxide dismutase and catalase as compared with the leaves of sensitive plants (Harper and Harvey 1978). The significance of the increase in catalase activity is not easy to evaluate, however, since this enzyme is not present within the chloroplasts. Increased leaf superoxide dismutase activity has also been observed in paraquat-resistant strains of other plants (R. J. Youngman, personal communication).

The toxicity of sulphur dioxide to leaves may also in part be due to increased superoxide production, since aqueous solutions of this gas contain the sulphite ion, which autoxidizes to produce O_2^- (Tanaka and Sugahara 1980).

8.5 Overall significance of oxygen effects in chloroplasts

Superoxide, hydrogen peroxide, singlet oxygen, the hydroxyl radical (·OH), and lipid peroxides are species whose formation is damaging to the chloroplast and must therefore be carefully controlled. Chloroplasts are especially affected

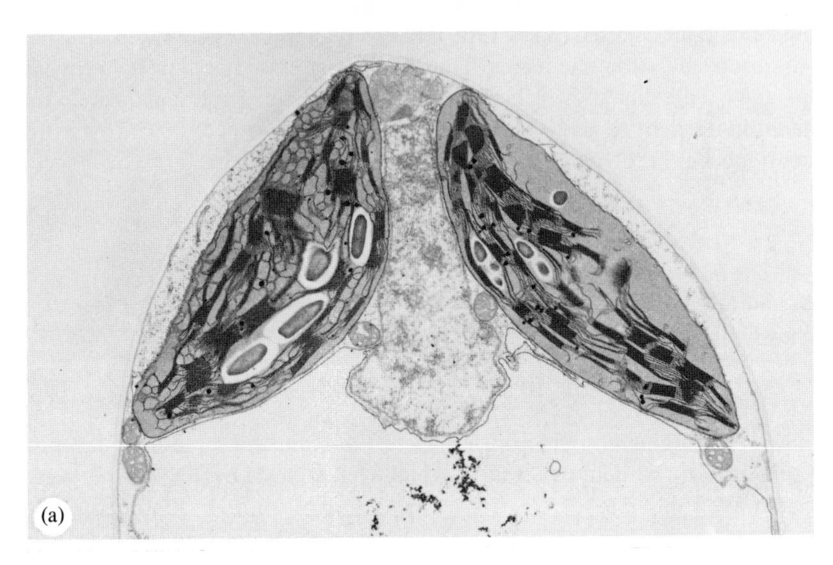

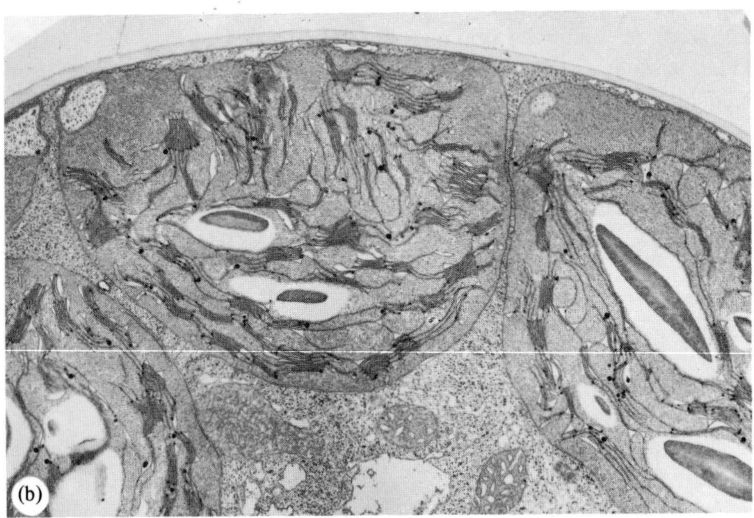

Fig. 8.10. Effects of paraquat treatment of leaves on chloroplast structure. (a). shows the appearance of chloroplasts from leaves of ryegrass (*Lolium perenne* cv. Kent Indigenous) under the electron microscope (X9000). (b). Two hours after paraquat treatment of the leaves the chloroplasts are swollen and the thylakoid membrane stacking extensively disrupted. By contrast, the mitochondria are normal (X18 000).

by them because of a high internal oxygen concentration, the presence of molecules which can reduce O_2 to O_2^- (e.g. reduced ferredoxin) and the presence of pigments which can sensitise the formation of singlet oxygen and possibly of the hydroxyl radical. To allow their continued functioning, chloroplasts have multiple protective mechanisms against these species. The main mechanisms are superoxide dismutase (removes O_2^- and hence prevents formation of ·OH and singlet O_2 from it), ascorbic acid (reacts rapidly with O_2^-, ·OH and singlet O_2 and also helps remove H_2O_2 by the ascorbate peroxidase reaction), reduced glutathione (reacts rapidly with ·OH and singlet oxygen, protects enzyme –SH groups and helps to regenerate ascorbate from dehydroascorbate), α-tocopherol (inhibits the chain reaction of lipid peroxidation and scavenges singlet oxygen), and carotenoids (decrease formation of singlet oxygen by absorbing excess excitation energy from chlorophyll by direct transfer, and can also quench singlet oxygen). Not until these systems are overloaded should toxic effects appear.

If spinach chloroplasts or leaf cells are incubated in carbon dioxide-free air, their ability to fix carbon dioxide supplied to them subsequently is impaired (Krause, Kirk, Heber, and Osmond 1978). It may be that in the absence of carbon dioxide no $NADP^+$ is available to accept electrons from photosystem I, so that they are 'shunted off' to oxygen (Fig. 8.8). Any excess excitation of chlorophyll would cause more singlet oxygen to be formed. It has been suggested that the function of photorespiration (Chapter 7) is to prevent the damage that would result from excessive oxygen-radical production by continually recycling carbon dioxide, so that chloroplasts *in vivo* are never completely depleted of it.

Formation of lipid peroxides and aldehydes on wounding of plant tissues (Section 3) may be an example of a situation where peroxidative reactions are put to a useful purpose. The ripening and senescence of fruits also seems to be a controlled oxidative process. As fruits ripen, there is a decrease in free –SH groups present and an increased formation of lipid peroxides and of hydrogen peroxide. It seems that the progress of ripening is correlated with an increase in peroxide content of the fruit, which presumably leads to the breakdown of membranes and cell-walls that is seen during ripening (Brennan and Frenkel 1977; Maguire and Haard 1975). It is interesting to speculate that these changes could be brought about by the progressive withdrawal of the protective systems described above.

References

Abeles, F. B., Leather, G. R., and Forrence, L. E. (1978). *Pl. Physiol., Lancaster* **62**, 696–8.
Allen, J. F. (1975). *Biochem. Biophys. Res. Commun.* **66**, 36–43.
— (1978*a*). *Plant Sci. Lett.* **12**, 151–9.
— (1978*b*). *Plant Sci. Lett.* **12**, 161–7.

Allen, J. F. and Whatley, F. R. (1978). *Pl. Physiol., Lancaster* **61**, 957–60.
Anderson, I. C. and Robertson, D. S. (1960). *Pl. Physiol., Lancaster* **35**, 531–4.
Anderson, R. A. and Linney, T. L. (1977). *Chem. biol. Interact.* **19**, 317–25.
Anderson, M. M., McCarty, R. E., and Zimmer, E. A. (1974). *Pl. Physiol., Lancaster* **53**, 699–701.
Asada, K., Urano, M., and Takahashi, M. (1973). *Eur. J. Biochem.* **35**, 257–66.
— Kiso, K. and Yoshikawa, K. (1974). *J. biol. Chem.* **249**, 2175–81.
— Yoshikawa, K., Takahashi, M., Maeda, Y., and Enmanji, K. (1975). *J. biol. Chem.* **250**, 2801–7.
Barlow, S. M. and Gaunt, J. K. (1972). *Phytochemistry* **11**, 2161–70.
Baszynski, T. (1974). *Biochim. biophys. Acta* **347**, 31–5.
Bielski, B. H. J. and Chan, P. C. (1978). *J. Am. Chem. Soc.* **100**, 1920–1.
Bodannes, R. S. and Chan, P. C. (1979). *FEBS Lett.* **105**, 195–6.
Boldingh, J. (1976). In *The Lipids*, (eds. R. Paoletti and G. Porcellati). Vol. 1. Raven Press, New York.
Boveris, A., Sanchez, R. A., Varsavsky, A. I., and Cadenas, E. (1980). *FEBS Lett*; **113**, 29–32.
Brennan, R. and Frenkel, C. (1977). *Pl. Physiol., Lancaster* **59**, 411–6.
Brown, J. C. (1978). *Plant Cell Envir.* **1**, 249–57.
Brown, R. H., Collins, N., and Merrett, M. J. (1975). *Pl. Physiol., Lancaster* **55**, 1123–4.
Charles, S. A. and Halliwell, B. (1980). *Biochem. J.* **189**, 373–6.
Chasseaud, F. (1973). *Drug Metab. Rev.* **2**, 185–220.
Clough, R. L., Yee, B. G., and Foote, C. S. (1979). *J. Am. Chem. Soc.* **101**, 683–6.
Diesperger, H. and Sandermann, H. (1979). *Planta* **146**, 643–8.
Dilek Tozum, S. and Gallon, J. (1979). *J. gen. Microbiol.* **111**, 313–26.
Dodge, A. D. (1975). *Sci. Prog.* **62**, 447–66.
Egneus, H., Heber, U., Matthiesen, U., and Kirk, M. (1975). *Biochim. biophys. Acta* **408**, 252–68.
Ellyard, P. W. and Gibbs, M. (1969). *Pl. Physiol., Lancaster* **44**, 1115–21.
Elstner, E. F., Stoffer, C., and Heupel, A. (1975). *Z. Naturforsch.* **30C**, 53–6.
— Saran, H., Bors, W., and Lengfelder, E. (1978). *Eur. J. Biochem.* **89**, 61–6.
— Wildner, G. F. and Heupel, A. (1976). *Archs Biochem. Biophys.* **173**, 623–30.
— and Osswald, W. (1980). *Z. Naturforsch.* **35C**, 129–35.
Ernst, V., Levin, D. H., and London, I. M. (1978). *Proc. natn. Acad. Sci. U.S.A.* **75**, 4110–4.
Farrington, J. A., Ebert, M., Land, E. J., and Fletcher, K. (1973). *Biochim. biophys. Acta* **314**, 372–81.
Flohe, L. and Menzel, H. (1971). *Plant Cell Physiol.* **12**, 325–33.
Foster, J. G. and Edwards, G. E. (1980). *Plant Cell Physiol.* **21**, 895–906.
— and Hess, J. L. (1980). *Pl. Physiol., Lancaster* **66**, 482–7.
Foote, C. S. (1970). *Ann. N. Y. Acad. Sci.* **171**, 139–48.
Foyer, C. H. and Halliwell, B. (1976). *Planta* **133**, 21–5.
— — (1977). *Phytochemistry* **16**, 1347–50.
— and Hall, D. O. (1979). *FEBS Lett*, **101**, 324–8.
Fridovich, I. (1978). *Science, N.Y.* **201**, 875–80.
Frosch, S., Jabben, M., Bergfeld, R., Kleinig, H., and Mohr, H. (1979). *Planta* **145**, 497–505.
Fuller, R. C. and Nugent, N. A. (1969). *Proc. natn. Acad. Sci. U.S.A.* **63**, 1311–8.

Galliard, T. (1978). In *Biochemistry of wounded plant tissues* (ed. G. Kahl) pp. 155-201. de Gruyter, Berlin.
Gaunt, J. K., Matthews, G. M., and Plumpton, E. S. (1980). *Biochem. Soc. Trans.* **8**, 186-7.
Golbeck, J. H., Martin, I. F. and Fowler, C. F. (1980). *Pl. Physiol., Lancaster* **65**, 707-13.
Gomez-Moreno, C. and Ke, B. (1979). *Molec. Cell Biochem.* **26**, 111-8.
Gregory, R. P. F. (1968). *Biochim. biophys. Acta* **159**, 429-39.
Groden, D. and Beck, E. (1979). *Biochim. biophys. Acta* **546**, 426-35.
Gross, G. G., Janse, C. and Elstner, E. F. (1977). *Planta* **136**, 271-6.
Hall, D. O. (1976). In *The intact chloroplast, Topics in photosynthesis* Vol. 1 (ed. J. Barber) pp. 135-170. Elsevier, Amsterdam.
Halliwell, B. (1974). *New Phytol.* **73**, 1075-86.
— (1978). *Planta* **140**, 81-8.
— (1979). In *Strategies of microbial life in extreme environments* (ed. M. Shilo) pp. 195-221. Verlag Chemie, Weinheim & New York.
— and Foyer, C. H. (1978). *Planta* **139**, 9-17.
Harbour, J. R. and Bolton, J. R. (1975). *Biochem. Biophys. Res. Commun.* **64**, 803-7.
— — (1978). *Photochem. Photobiol.* **28**, 231-4.
Harman, G. E., Mattick, L. R., Nash, G., and Nedrow, B. L. (1980). *Can. J. Bot.* **58**, 1541-7.
Harnischfeger, G. (1972). *Planta* **104**, 316-28.
Harper, D. B. and Harvey, B. M. R. (1978). *Plant Cell Env.* **1**, 211-15.
Hatanaka, A. and Harada, T. (1973). *Phytochemistry* **12**, 2341-6.
— Sekiya, J. and Kajiwara, T. (1977a). *Phytochemistry* **18**, 107-16.
— Kajiwara, T., Sekiya, J., and Kido, Y. (1977b). *Phytochemistry* **16**, 1828-9.
Haugaard, N. (1968). *Physiol. Rev.* **48**, 311-45.
Heath, R. L. and Packer, L. (1968*a*). *Archs Biochem. Biophys.* **125**, 189-98.
— — (1968*b*). *Archs Biochem. Biophys.* **125**, 850-7.
Heldt, H. W., Chon, C. J., Lilley, R. McC., and Portis, A. (1978). In *Photosynthesis '77* (eds. D. O. Hall, J. Coombs, and T. W. Goodwin) pp. 469-78. Biochemical Society, London.
Hicks, M. and Gebicki, J. M. (1978). *Biochem. Biophys. Res. Commun.* **80**, 704-8.
Hind, G., Mills, J. D., and Slovacek, R. E. (1978). In *Photosynthesis '77* (eds. D. O. Hall, J. Coombs, and T. W. Goodwin). pp. 591-600. Biochemical Society, London.
Holden, M. (1970). *Phytochemistry* **9**, 507-12.
Hoshina, S., Kaji, T., and Nishida, K. (1975). *Plant Cell Physiol.* **16**, 465-74.
Iwai, K., Bunno, M., Kobashi, M., and Suzuki, T. (1976). *Biochim. biophys. Acta* **444**, 618-22.
Jackson, C., Dench, J., Moore, A. L., Halliwell, B., Foyer, C. H. and Hall, D. O. (1978). *Eur. J. Biochem.* **91**, 339-44.
Janiszowska, W. and Korczak, G. (1980). *Phytochemistry* **19**, 1391-2.
Jennings, R. C. and Forti, G. (1975). In *Proceedings of the third international congress on photosynthesis* (ed. M. Avron) pp. 735-43. Elsevier, Amsterdam.
Jursinic, P. A. (1980). *Photochem. Photobiol.* **32**, 61-5.
Kaiser, W. (1976). *Biochim. biophys. Acta* **440**, 476-82.
— (1979). *Planta* **145**, 377-82.
Koka, P. and Song, P. S. (1978). *Photochem. Photobiol.* **28**, 509-15.

Konze, J. R. and Elstner, E. F. (1978). *Biochim. biophys. Acta* **528**, 213-21.
Krause, G. H., Kirk, M., Heber, U., and Osmond, C. B. (1978). *Planta* **142**, 229-33.
Krinsky, N. H. (1979). In *Strategies of microbial life in extreme environments* (ed. M. Shilo) pp. 163-77. Verlag Chemie, Weinheim & New York.
Lay, M. M. and Casida, J. E. (1978). In *Chemistry and action of herbicide antidotes* (eds. F. M. Pallos and J. E. Casida) pp. 151-60. Academic Press, New York.
Lendzian, K. and Bassham, J. A. (1975). *Biochim. biophys. Acta* **396**, 260-75.
Lien, S. and San Pietro, A. (1979). *FEBS Lett.* **99**, 189-93.
Maguire, Y. P. and Haard, N. F. (1975). *Nature, Lond.* **258**, 599-600.
Marynick, M. C. and Addicott, F. T. (1976). *Nature, Lond.* **264**, 668-9.
Matthews, J. A. and Galliard, T. (1978). *Phytochemistry* **17**, 1043-4.
McCord, J. M. and Fridovich, I. (1969). *J. biol. Chem.* **244**, 6049-55.
Mees, G. (1960). *Ann. Appl. Biol.* **48**, 601-12.
Mehler, A. H. (1951). *Archs Biochem. Biophys.* **33**, 65-77.
Miller, R. W. and MacDowall, F. D. H. (1975). *Biochim. biophys. Acta* **387**, 176-87.
Morris, J. G. (1979). In *Strategies of microbial life in extreme environments* (ed. M. Shilo) pp. 149-62. Verlag Chemie, Weinheim & New York.
Mue Akamba, L. and Siegenthaler, P. A. (1979). *FEBS Lett.* **99**, 6-10.
Nakano, Y. and Asada, K. (1980). *Plant Cell Physiol.* **21**, 1295-307.
Packer, J. E., Slater, T. F., and Willson, R. L. (1979). *Nature, Lond.* **278**, 737-8.
Pallett, K. E. and Dodge, A. D. (1979). *Z. Naturforsch.* **34C**, 1058-61.
Parish, R. W. (1972). *Eur. J. Biochem.* **31**, 446-55.
Patterson, C. O. P. and Myers, J. (1973). *Pl. Physiol., Lancaster* **51**, 104-9.
Pauls, K. P. and Thompson, J. E. (1980). *Nature, Lond.* **283**, 504-6.
Poskuta, J., Mikulska, M., Faltynowicz, M., Bielak, B., and Wroblewska, B. (1974). *Z. Pflanzenphysiol.* **73**, 387-93.
Pryor, W. A. (1978). *Photochem. Photobiol.* **28**, 787-801.
Pulich, W. M. (1974). *J. Cell. Biol.* **62**, 904-7.
Putvinsky, A. V., Sokolov, A. I., Roshcupkin, D. I., and Vladimirov, Y. A. (1979). *FEBS Lett.* **106**, 53-5.
Quebedeaux, B. and Hardy, R. W. F. (1975). *Pl. Physiol., Lancaster* **55**, 102-7.
Rabinowitch, H. D. and Sklan, D. (1980). *Planta* **148**, 162-7.
Radmer, R. J. and Kok, B. (1976). *Pl. Physiol., Lancaster* **58**, 336-40.
Rathnam, C. K. M. and Chollet, R. (1980). In *Progress in phytochemistry*, Vol. 6 (eds. L. Reinhold, J. B. Harborne, and T. Swain) pp. 1-48. Pergamon Press, Oxford.
Rawls, H. R. and Van Santen, P. J. (1970). *Ann. N.Y. Acad. Sci.* **171**, 135-7.
Ridley, S. M. (1977). *Pl. Physiol., Lancaster* **59**, 724-32.
Robinson, J. M., Smith, M. G., and Gibbs, M. (1978). *Pl. Physiol., Lancaster* **61**, P100 (proceedings).
Sagisaka, S. (1976). *Pl. Physiol., Lancaster* **57**, 308-9.
Salin, M. L., and Bridges, S. M. (1980). *Archs Biochem. Biophys.* **201**, 369-74.
Sandman, G. and Boger, P. (1980). *Pl. Physiol., Lancaster* **66**, 797-800.
Sayre, R. T. and Homann, P. H. (1979). *Archs Biochem. Biophys.* **196**, 525-33.
Schiefer, S., Teifel, W., and Kindl, H. (1976). *Hoppe-Seylers Z. Physiol. Chem.* **357**, 163-75.
Sekiya, J., Kajiwara, T., and Hatanaka, A. (1978). *Plant Cell Physiol.* **19**, 553-9.
Shigeoka, S., Nakano, Y., and Kitaoka, S. (1980*a*). *Biochem. J.* **186**, 377-80.
— — — (1980*b*). *Archs Biochem. Biophys.* **201**, 121-7.

Shimabukuro, R. H., Lamoureux, G. L., and Frear, D. S. (1978). In *Chemistry and action of herbicide antidotes* (ed. F. M. Pallos and J. G. Casida) pp. 133–49. Academic Press, New York.
Siefermann, D. and Yamamoto, H. Y. (1975*a*). *Biochim. biophys. Acta* **387**, 149–58.
— — (1975*b*). *Archs Biochem. Biophys.* **171**, 70–7.
Siegenthaler, P. A. (1972). *Biochim. biophys. Acta* **275**, 182–91.
Simon, E. W. (1974). *New Phytol.* **73**, 377–420.
Sokolove, P. M. and Marsho, T. V. (1976). *Biochim. biophys. Acta* **430**, 321–46.
Steiger, H. M., Beck, E. and Beck, R. (1977). *Pl. Physiol., Lancaster* **60**, 903–6.
Stewart, R. R. C. and Bewley, J. D. (1980). *Pl. Physiol., Lancaster* **65**, 245–8.
Takahama, U. and Nishimura, M. (1975). *Plant Cell Physiol.* **16**, 737–48.
— — (1976). *Plant Cell Physiol.* **17**, 111–8.
Tanaka, Y. and Sugahara, K. (1980). *Plant Cell Physiol.* **21**, 601–11.
Trebst, A. (1974). *A. Rev. Pl. Physiol.* **25**, 423–58.
Van Ginkel, G. and Brown, J. S. (1978). *FEBS Lett.* **94**, 284–6.
Walker, D. A. (1971). *Meth. Enzymol.* **23A**, 211–20.
Wardale, D. A., Lambert, E. A., and Galliard, T. (1978). *Phytochemistry* **17**, 205–12.
Whitehouse, D. G., Ludwig, L. J., and Walker, D. A. (1971). *J. exp. Bot.* **22**, 772–91.
Wintermans, J. F. G. M., Helmsing, P. J., Polman, B. J. J., Van Gisbergen, J., and Collard, J. (1969). *Biochim. biophys. Acta* **189**, 95–105.
Wolosiuk, R. A. and Buchanan, B. B. (1977). *Nature, Lond.* **266**, 565–7.
Yamauchi, R. and Matsushita, S. (1979). *Agr. Biol. Chem.* **43**, 2157–61.
Youngman, R. J., Dodge, A. D., Lengfelder, E., and Elstner, E. F. (1979). *Experientia* **35**, 1295–6.

9 SYNTHESIS OF PHENOLIC COMPOUNDS, MEMBRANE LIPIDS, CHLOROPHYLLS, AND CAROTENOIDS IN CHLOROPLASTS

9.1 Biosynthesis of phenolic compounds

The biosynthetic pathways that lead to lignin and flavonoids are major metabolic routes in plant tissues (Fig. 9.1), although the subcellular location of many of the enzymes involved has not been accurately determined. The pathways begin with the non-oxidative deamination of L-phenylalanine to yield cinnamic acid and ammonia, catalysed by phenylalanine ammonia lyase. This enzyme appears to have multiple localizations in plant tissues, having been detected in soluble (e.g. Poulton, McRee, and Conn 1980), microsomal, mitochondrial (Towers 1974), and even in chloroplast (Loffelhardt, Ludwig, and Kindl 1973; Czichi and Kindl 1975; Saunders and McClure 1975; Ranjeva, Alibert, and Boudet 1977 *a,b*; Nishizawa, Wolosiuk, and Buchanan 1979) fractions. Unfortunately, most of the studies performed have not been carried out rigorously enough to *prove* that chloroplasts contain this enzyme, since the possibilities of contamination of the chloroplasts by cytoplasm or adsorption of enzymes on to the envelope (Chapter 1) were not ruled out. However, the lyase detected in spinach chloroplast fractions by Nishizawa *et al.* (1979) was reported to be activated by the thioredoxin system (Chapter 3), which does suggest that it is localized within the chloroplast. If some lyase activity is localized in chloroplasts *in vivo*, then the cinnamic acid it produces would have to leave the chloroplast for further metabolism, since cinnamate hydroxylase is located in the endoplasmic reticulum, being a cytochrome-P_{450}-linked enzyme (Ranjeva *et al.* 1977*a, b*; Gross 1978; Mayer and Harel 1979). Tyrosine ammonia lyase, which forms *p*-coumaric acid directly, is found only in some grasses.

The *p*-coumarate hydroxylase activity of leaf tissues has been shown by several well-designed biochemical and cytochemical experiments to be largely, if not completely, localized in the chloroplasts, at least in tobacco (*Nicotiana tabacum*), spinach (*Spinacia oleracea*), and spinach-beet (*Beta vulgaris*) leaves (Parish 1972; Bartlett, Poulton, and Butt 1972; Henry 1975, 1976; Tolbert 1973), although *Petunia* chloroplasts were reported not to contain this enzyme (Ranjeva *et al.* 1977*a,b*). The purified spinach-beet hydroxylase requires molecular oxygen and a reducing agent (red:H_2) to function, as shown in eqn (9.1). It contains copper and has a molecular weight of 40 000 (Vaughan, Eason, Paton, and Ritchie 1975).

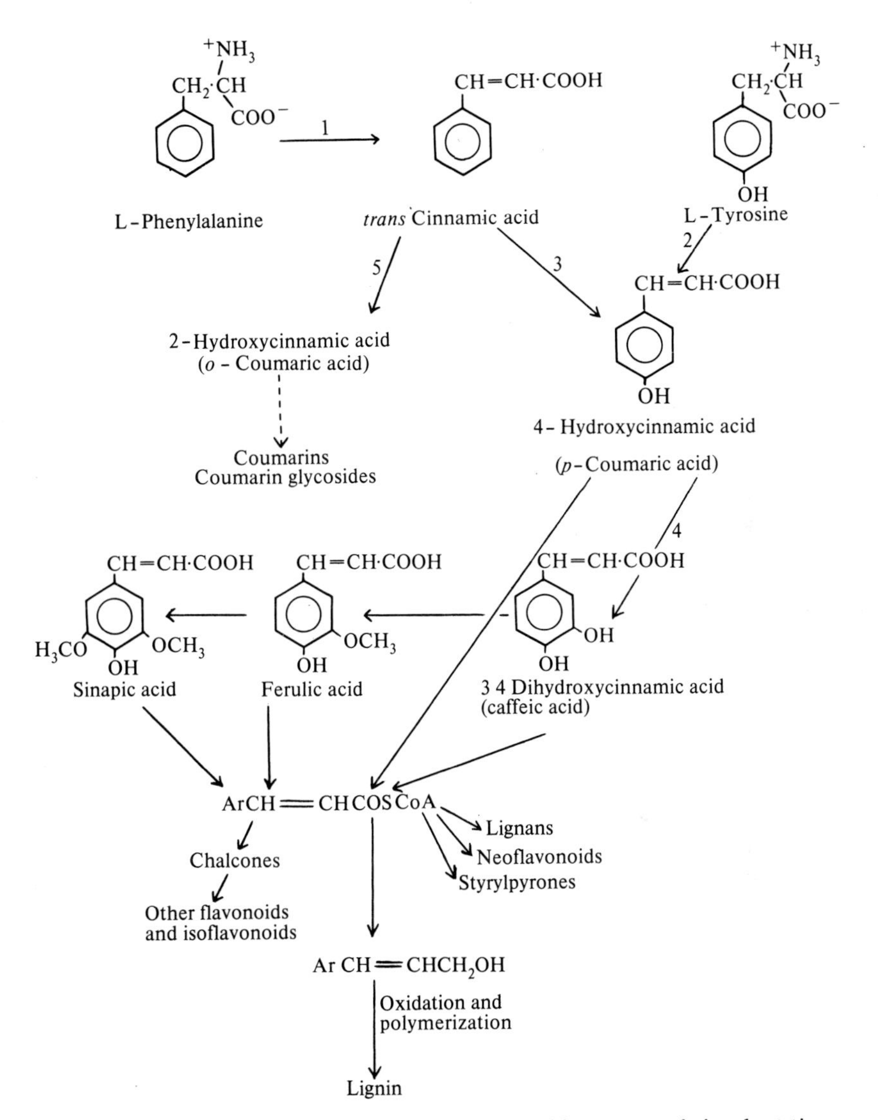

Fig. 9.1. Biosynthetic pathways of phenylpropanoid compounds in plant tissues. Enzymes involved: 1, phenylalanine ammonia lyase (found in a wide range of plant tissues); 2, tyrosine ammonia lyase (found only in some grasses); 3, cinnamate hydroxylase; 4, *p*-coumarate hydroxylase; 5, *o*-coumarate hydroxylase. Enzymes catalysing the later stages of the reactions are not shown in detail. Further details may be found in Harborne (1980).

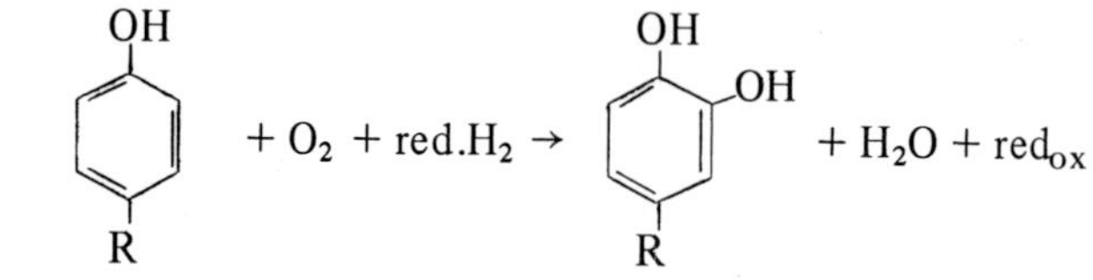

$$p\text{-coumarate} + O_2 + \text{red.}H_2 \rightarrow \text{caffeate} + H_2O + \text{red}_{ox} \qquad (9.1)$$

OH, R — *p*-coumarate; OH, OH, R — caffeate

p-Coumarate hydroxylase is also capable of oxidising diphenols into quinones, as shown in eqn (9.2) for caffeic acid

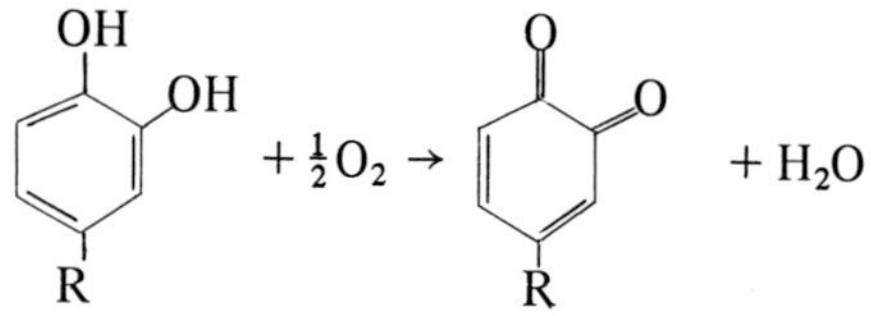

$$\text{caffeate} + \tfrac{1}{2}O_2 \rightarrow \text{caffeoyl-}o\text{-quinone} + H_2O \qquad (9.2)$$

OH, OH, R — caffeate; O, O, R — caffeoyl-*o*-quinone

Indeed, the reaction catalysed by spinach-beet *p*-coumarate hydroxylase probably occurs in two parts. Firstly, the *p*-coumarate is hydroxylated to caffeate. Caffeate is then oxidized to caffeoyl-*o*-quinone (eqn (9.2)), which is reduced back to caffeic acid by the reducing agent (Vaughan and Butt 1969). Since the reducing agent does not appear to interact directly with the active site but simply reduces quinones generated by reaction (9.2) the requirement is not very specific: for example ascorbate, NADH, NADPH, and tetrahydropteridines are all effective with the spinach-beet enzyme (McIntyre and Vaughan 1975). Once the added reducing agent is exhausted, the accumulated caffeic acid is rapidly oxidized by reaction (9.2). If *p*-coumarate hydroxylase is supplied with a monophenol *in vitro*, a lag period is often seen before hydroxylation begins. The lag can be eliminated by addition of small amounts of a diphenol, and it has therefore been suggested that the 'diphenol oxidase' activity of the enzyme is required to reduce the copper at the active site to give a form that can bring about hydroxylation (Mason 1955; Vanneste and Zuberbuhler 1974; McIntyre and Vaughan 1975; Butt 1979). A possible reaction mechanism is shown in Fig. 9.2. Hence the diphenol oxidase activity of the enzyme is essential for its hydroxylase activity and the enzyme is often referred to as 'phenolase' or as 'the phenolase complex'. It is very similar to the enzyme in mushrooms, tyrosinase, which converts L-tyrosine into dihydroxyphenylalanine (DOPA) and then oxidizes this product to DOPA-quinone (Sato 1969). The ratio of the hydroxylase to the diphenol oxidase activities of plant phenolases is extremely variable and is affected by the assay and storage conditions (Butt 1979; Mayer and Harel 1979). Although evidence has been presented that *p*-coumarate hydroxylases containing little diphenol oxidase activity are present in *Sorghum* (Stafford 1974) and in parsley (*Petroselinum hortense*) cell cultures (Schill and Grisebach 1973),

Fig. 9.2. A likely reaction sequence for the action of spinach-beet *p*-coumarate hydroxylase. Based on Vanneste and Zuberbuhler (1974), McIntyre and Vaughan (1975) and Butt (1979). It is assumed that hydroxylase and diphenol oxidase activities share the same active sites on the enzyme, although this has not been rigorously proved (see Mayer and Harel 1979). The enzyme is a copper-protein. Red H_2 is the reducing agent added to the reaction mixture.

neither enzyme has been rigorously proved not to be a phenolase. Similarly, reports of diphenol oxidase enzymes without hydroxylase activity might have been in error because of the prolonged lag period often seen in hydroxylation when catalyic quantities of diphenols are not added, and because of the ease with which hydroxylase activity is lost during enzyme purification (Mayer and Harel 1979; Kahn and Pomerantz 1980).

Chloroplast phenolase activity is tightly bound to the thylakoids, from which it can often be extracted by detergents, washing with solutions of high ionic strength, or exposure to proteolytic enzymes, although Tolbert (1973) found that inactive trypsin was just as effective in releasing phenolase as the native enzyme! Phenolase extraction is usually accompanied by an increase in enzyme activity. Removal of the coupling factor CF_1 (Chapter 2) from chloroplasts by washing with EDTA enhances the *bound* phenolase activity, suggesting that CF_1 may 'cover up' some phenolase in the thylakoid membrane. Tolbert (1973) reported very high activities of diphenol oxidase in chloroplast fractions isolated from many plants: in some cases (e.g. alfalfa) activity was low in the dark and increased on illumination of the organelles. The existence of such high activities of phenolase as an integral component of the chloroplast membranes suggests that this enzyme plays a major role in chloroplast metabolism. Hydroxylation of *p*-coumaric acid to caffeic acid is a key reaction in plant metabolism (Fig. 9.1) and is presumably carried out by the chloroplasts *in vivo*. If so, then reducing agents must be made available to keep caffeic acid in its reduced form until it has left the chloroplast, since subsequent stages of its metabolism are catalysed by cytoplasmic enzymes (Hrazdina, Alscher-Herman, and Kish 1980; Ranjeva *et al.* 1977*a,b*). NADPH or ascorbate would be likely candidates for reductants *in vivo*. Although pteridines have been found in isolated chloroplasts and their reduced forms act as electron donors to phenolase, the physiological significance of this is debatable (Fuller and Nugent 1969). Continous oxidation and reduction of diphenols such as caffeic acid by phenolase and reductants would be expected to contribute to oxygen uptake by chloroplasts, which might confuse estimates of the rate of pseudocyclic electron flow by this method (Chapter 2). Caffeic acid will be present within chloroplasts *in vivo*, since it is synthesized there, and several other diphenols have been detected in chloroplast fractions (Saunders and McClure 1976). Unfortunately in most cases the possibility of cytoplasmic contamination has not been ruled out and a rigorous re-investigation of the occurrence of phenols in *type A* (Chapter 1) chloroplasts is required, along the lines of the preliminary studies by Weissenbock and Schneider (1974). Concentrations of phenolic compounds in chloroplasts must in any case be carefully controlled since they are powerful inhibitors of carbon dioxide fixation (Tissut, Chevallier, and Douce 1980).

The *ortho*-hydroxylation of cinnamic acid to give *o*-coumaric acid, a precursor of coumarins (Fig. 9.1) has been reported to be catalysed by chloroplast fractions from sweetclover (*Melitotus alba*), *Hydrangea*, and *Petunia* (Gestetner and Conn 1974; Kindl 1971; Ranjeva *et al.* 1977*a,b*). Chloroplast fractions from some plants have been reported to convert L-phenylalanine into benzoic acid, a precursor of salicylate (Loffelhardt and Kindl 1975) and synthesis of methylsalicyclic acid from acetate has been claimed to occur in barley chloroplast fractions (Kannangara, Henningsen, Stumpf, and Wettstein 1971). *Petunia* chloroplast fractions contain enzymes for the synthesis of chlorogenic acid and of naringenin (Ranjeva *et al.* 1977*a,b*). In none of the above experiments were rigorous subcellular fractionation and latency experiments (Chapter 1) carried out to *prove* a localization within the chloroplast, however.

9.2 Synthesis of chloroplast lipids

The major lipids present in chloroplasts, other than the photosynthetic pigments, which are considered in Section 9.3, comprise galactolipids (monogalactosyl and digalactosyldiglycerides), sulphoquinovosyl diglyceride, and phospholipids. Most of the fatty acids they contain are C_{18} acids with one or more double bonds, in particular α-linolenic acid (Chapter 1).

9.2.1 Synthesis of fatty acids

The first stage in lipid biosynthesis is the assembly of long-chain fatty acids, a process which has been demonstrated to occur in isolated chloroplast fractions by several groups (e.g. Nakamura and Yamada 1975; Yamada and Nakamura 1975; Leech and Murphy 1976; Givan and Harwood 1976; Roughan, Holland, Slack, and Mudd 1979a). Of course, such fractions can be contaminated with other organelles and it is possible that these play some part in the biosynthesis, just as the light-dependent synthesis of glycine by isolated type A chloroplast fractions actually involves a cooperation between chloroplasts and peroxisomes (Chapters 1 and 7).

The pathway of fatty acid biosynthesis in chloroplast fractions begins with the conversion of acetyl-coenzyme A (acetyl SCoA) to malonyl SCoA by a set of proteins known collectively as *acetyl SCoA carboxylase*. The overall reaction may be represented by the following equation (Mohan and Kekwick 1980)

$$\text{acetyl SCoA} + HCO_3^- + H^+ + \text{ATP} \rightarrow \text{malonyl SCoA} + \text{ADP} + \text{Pi} \quad (9.3)$$

The spinach chloroplast carboxylase complex consists of three proteins (Kannangara and Stumpf 1972, 1973). One of these is bound to the thylakoids and contains the prosthetic group biotin (Fig. 9.3) attached to a lysine residue at the active site. It is therefore called the *biotin carboxyl carrier protein* (BCCP). A second protein, *BCCP carboxylase*, located in the stroma, catalyses the ATP-dependent carboxylation of BCCP,

$$\text{BCCP} - \text{biotin} + HCO_3^- + H^+ + \text{ATP} \longrightarrow \text{BCCP} - \text{biotin} - CO_2 + \text{ADP} + \text{Pi} \quad (9.4)$$

HCO_3^- rather than carbon dioxide is the true substrate of this enzyme. Finally a stromal *transcarboxylase* enzyme transfers the 'activated carbon dioxide' from the biotin of BCCP on to acetyl SCoA

$$\text{BCCP-biotin-}CO_2 + CH_3COSCoA \rightarrow \underset{\text{malonyl SCoA}}{HOOC{\cdot}CH_2{\cdot}COSCoA} + \text{BCCP-biotin} \quad (9.5)$$

Subsequent stages of fatty acid biosynthsis in chloroplast fractions appear to resemble those used by bacteria such as *E. coli*, in that they are brought about by a multi-enzyme complex which contains an *acyl carrier protein* (ACP). All the ACP of spinach leaves is located in the chloroplasts (Ohlrogge, Kuhn, and Stumpf 1979). It has a prosthetic group, 4′-phosphopantetheine, which is

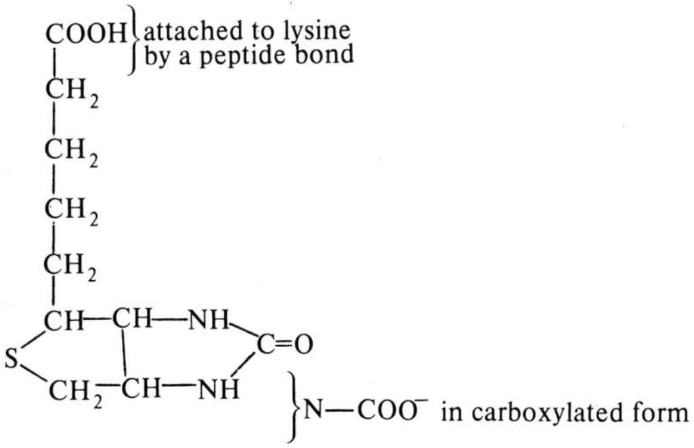

```
COOH } attached to lysine
 |   } by a peptide bond
CH2
 |
CH2
 |
CH2
 |
CH2
 |
 CH—CH—NH
/   |     \
S   |      C=O
\   |     /
 CH2—CH—NH
             } N—COO⁻ in carboxylated form
```

Fig. 9.3. Biotin, a prosthetic group involved in conversion of acetyl-coenzyme A to malonyl-coenzyme A in chloroplasts.

linked to serine at its active site and possesses a free —SH group (Fig. 9.4). The exact details of fatty acid biosynthesis in chloroplasts have not been established, but they may well resemble those used by the *E. coli* multi-enzyme complex (Fig. 9.5), especially as the antibiotic cerulenin, which inhibits β-ketoacyl-ACP synthetase in *E. coli*, also inhibits fatty acid synthesis in preparations from barley and spinach leaves. The overall action of the multi-enzyme complex is to convert acetyl SCoA into palmitoyl-ACP. The NADPH and ATP it requires are presumably provided from the chloroplast electron-transport chain, which in part explains why the rate of fatty acid biosynthesis in the light is much greater than that in the dark, although light may also directly influence the activity of one or more of the enzymes of the biosynthetic pathway (Roughan, Kagawa, and Beevers 1980a). However, fatty acids can be made to some extent in the dark, presumably by the use of NADPH from the oxidative pentose phosphate pathway and ATP from chloroplast glycolysis (Chapter 4) (Nakamura and Yamada 1979).

The chloroplast envelope membranes are impermeable to acetyl SCoA (Chapter 6) and so most studies of fatty acid biosynthesis *in vitro* have used acetate as a precursor. Acetate crosses the envelope rapidly, probably by simple diffusion of the unionised form (Jacobson and Stumpf 1972), and it is then converted into acetyl SCoA by a thiokinase enzyme. How acetate could be provided in the quantities required *in vivo*, however, is not clear although some acetate is made during cysteine synthesis in chloroplasts (Chapter 10). Pyruvate will also act as a precursor of fatty acids in isolated illuminated chloroplast fractions (Yamada and Nakamura 1975) although it is less efficient than is acetate (Roughan *et al.* 1979*a*). Consistent with this, subcellular fractionation experiments have shown that pea (*Pisum sativum*) chloroplasts contain pyruvate dehydrogenase, a multi-enzyme complex which catalyses the reaction

$$\text{pyruvate} + NAD^+ \rightarrow \text{acetyl SCoA} + CO_2 + NADH + H^+ \qquad (9.6)$$

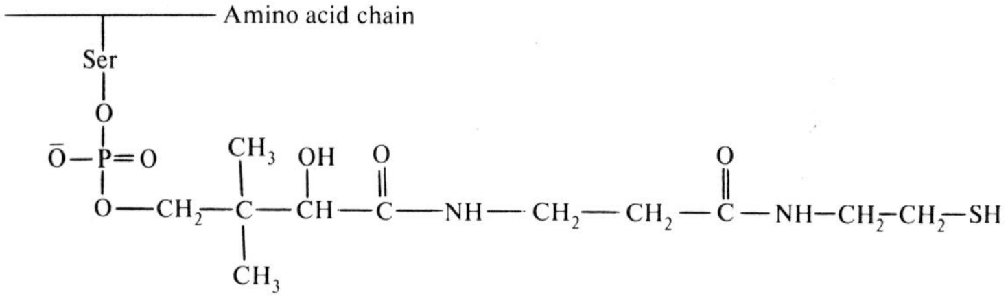

Amino acid chain
Ser
O
$\bar{O}$—P=O
O—CH_2—C(CH_3)(CH_3)—CH(OH)—C(=O)—NH—CH_2—CH_2—C(=O)—NH—CH_2—CH_2—SH

Fig. 9.4. 4′-phosphopantetheine at the active site of acyl carrier protein. ACP has been purified from spinach chloroplast fractions and is similar to, but not identical with, the *E. coli* protein. The region of the primary structure around the prosthetic group is identical in both these proteins.

1. AcetylSCoA reacts with —SH on the acyl carrier protein catalysed by a transferase enzyme in the multi-enzyme complex (*ACP acyltransferase*)
   acetylSCoA + ACP – SH ⟶ ACP – S – acetyl + HSCoA
2. The acetyl group is transferred from the ACP to a cysteine residue on another enzyme of the complex (*β-ketoacyl-ACP synthetase*)
3. MalonylSCoA reacts with ACP (*ACP malonyltransferase*)
   malonylSCoA + ACP – SH ⟶ ACP – S – malonyl + HSCoA
4. β-ketoacyl-ACP synthetase joins together the malonyl residue on the ACP and the acetyl residue at its own active site. The CO_2 that was originally introduced by acetylSCoA carboxylase is released.
   acetyl – S – enzyme + malonyl – S – ACP ⟶ acetoacetyl – S – ACP + CO_2 + enzyme —SH
5. Acetoacetyl – S – ACP is reduced by NADPH from the electron-transport chain (*β-ketoacyl-ACP reductase*)
   acetoacetyl – S – ACP + NADPH + H^+ ⟶ hydroxybutyryl – S – ACP + $NADP^+$
6. Water is eliminated from hydroxybutyryl – S – ACP (*dehydratase*)
   $CH_3CHOHCH_2CO$ – S – ACP ⟶ $CH_3{\cdot}CH = CH.COS$ – ACP + H_2O
   crotonyl – S – ACP
7. Crotonyl – S – ACP is reduced by NADPH from the electron-transport chain (*enoyl-ACP reductase*)
   crotonyl – S – ACP + NADPH + H^+ ⟶ $CH_3CH_2CH_2COS$ – ACP + $NADP^+$
   butyryl-S-ACP
8. The butyryl group is transferred from ACP to the —SH group at the active site of *β-ketoacyl – ACP synthetase*, so permitting another malonyl group to bind (reaction (3)). The cycle then repeats. After seven complete cycles palmitoyl–S–ACP is produced, which can be hydrolysed to free palmitic acid (*thioesterase*).

Overall equation:

$$8\ \text{acetylSCoA} + 14\text{NADPH} + 14\text{H}^+ + 7\text{ATP} + \text{H}_2\text{O} \longrightarrow \text{palmitic acid} + 8\ \text{HSCoA} + 14\text{NADP}^+ + 7\text{ADP} + 7\text{Pi}$$

Fig. 9.5. Probable sequence of reactions involved in fatty acid biosynthesis in the chloroplast multi-enzyme complex.

The chloroplast enzyme requires Mg^{2+} and thiamin pyrophosphate for activity, like its counterparts in animal tissues (Williams and Randall 1979; Elias and Givan 1979). The pyruvate it requires is presumably imported from the cytoplasm, crossing the envelope by simple diffusion of the unionized form (Chapter 6), since the later stages of the glycolytic pathway do not seem to occur in chloroplasts (Chapter 4).

The major products of fatty acid biosynthesis by isolated illuminated chloroplast fractions are palmitic acid ($C_{16:0}$), stearic acid ($C_{18:0}$), and oleic acid ($C_{18:1}$). Palmitoyl-ACP produced by the multi-enzyme complex can be converted by an elongase system in chloroplast fractions to give stearyl-ACP, malonyl SCoA apparently donating the extra two carbon atoms (Givan and Harwood

1976). The elongase requires NADPH and can be distinguished from the multi-enzyme complex in that it is not inhibited by cerulenin at concentrations that prevent palmitate production from acetyl SCoA. Desaturation of stearyl-ACP to oleyl-ACP is catalysed by a stromal enyme complex (Jacobson, Jaworski, and Stumpf 1974), which requires oxygen and reduced ferredoxin. The acyl-ACP compounds are then hydrolysed to give the free acids. Free oleic acid can be converted to oleyl SCoA by a thiokinase enzyme that has been reported to be present in the chloroplast envelope (Roughan and Slack 1977). Perhaps this enzyme is so arranged in the envelope that oleic acid from the stroma gives rise to oleyl SCoA outside the chloroplast, releasing fatty acyl SCoA derivatives for use by other subcellular fractions.

Most of the fatty acids found in chloroplast lipids are polyunsaturated, especially linolenic acid (Chapter 1), yet these are not formed in significant quantities when isolated illuminated chloroplasts are incubated with acetate or pyruvate. Isotopic labelling studies on whole leaves support the sequence of reactions

$$\underset{(9\text{-}C_{18:1})}{\text{oleate}} \longrightarrow \underset{(9,12\text{-}C_{18:2})}{\text{linoleate}} \longrightarrow \underset{(9,12,15\text{-}C_{18:3})}{\alpha\text{-linolenate}} \qquad (9.7)$$

(Murphy and Stumpf 1979), yet these conversions have not been demonstrated in chloroplasts. This may be because they occur in other subcellular organelles (e.g. the endoplasmic reticulum; Tremolieres, Drapier, Dubacq, and Mazliak 1980a) using oleyl SCoA from the chloroplast, or because desaturation occurs *after* the fatty acids have been esterified into lipids. For example, oleate incorporated into diacylgalactosylglycerol can be converted into linoleate and α-linolenate by isolated spinach chloroplast fractions (Roughan, Mudd, McManus, and Slack 1979b).

9.2.2 Conversion of fatty acids into chloroplast lipids

Our understanding of lipid biosynthesis in chloroplasts is complicated by the fact that the microsomal fraction of leaves also has substantial lipid-synthesizing activities, and many of the chloroplast preparations used to study lipid synthesis may well have been contaminated with microsomes.

The first stage in the biosynthesis of lipids in chloroplasts is the esterification of fatty acyl SCoA residues with glycerol 3-phosphate, as shown in Fig. 9.6. The enzyme catalysing the first esterification (reaction A) is found in the chloroplast stroma, whereas the second esterification (reaction B) and the phosphatidate phosphatidohydrolase enzyme (reaction C) are both located in the chloroplast envelope (Joyard and Douce 1979; Douce and Joyard 1979). However, chloroplasts are unable to synthesize the glycerol 3-phosphate needed for these reactions (Leech and Murphy 1976; McKee and Hawke 1979), so they must import it from the cytoplasm. The leaf cytoplasm synthesizes glycerol

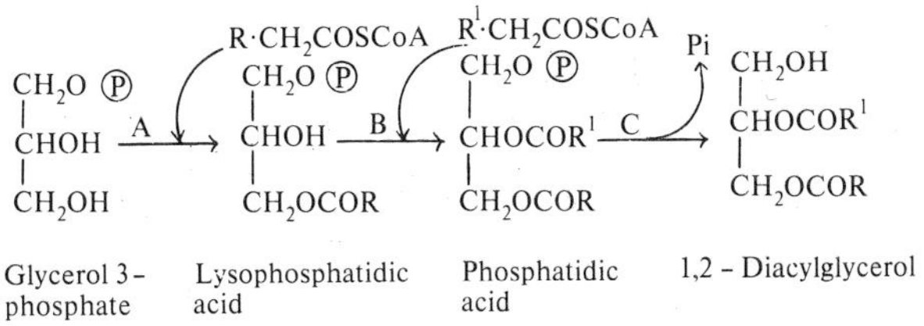

$$\begin{array}{c} CH_2O\,(P) \\ | \\ CHOH \\ | \\ CH_2OH \end{array} \xrightarrow[A]{R\cdot CH_2COSCoA} \begin{array}{c} CH_2O\,(P) \\ | \\ CHOH \\ | \\ CH_2OCOR \end{array} \xrightarrow[B]{R^1\cdot CH_2COSCoA} \begin{array}{c} CH_2O\,(P) \\ | \\ CHOCOR^1 \\ | \\ CH_2OCOR \end{array} \xrightarrow[C]{\uparrow Pi} \begin{array}{c} CH_2OH \\ | \\ CHOCOR^1 \\ | \\ CH_2OCOR \end{array}$$

Glycerol 3-phosphate — Lysophosphatidic acid — Phosphatidic acid — 1,2 - Diacylglycerol

Fig. 9.6. Conversion of glycerol 3-phosphate to 1,2-diacylglycerol in chloroplasts R and R′ represent long fatty-acid side-chains.

3-phosphate from dihydroxyacetone phosphate using a dehydrogenase enzyme (eqn (9.8))

$$\text{dihydroxyacetone phosphate} + \text{NADH} + \text{H}^+ \rightleftarrows \text{NAD}^+ + \text{glycerol 3-phosphate} \quad (9.8)$$

and glycerol 3-phosphate enters in exchange for dihydroxyacetone phosphate exported from the illuminated chloroplast *in vivo* (Chapter 6) (Santora, Gee, and Tolbert 1979).

1,2-Diacylglycerol can be further metabolized by enzymes located in the chloroplast envelope to give monogalactosyl and di-galactosyldiglycerides, which are major components of the chloroplast lipids (Chapter 1). Three enzyme-catalysed reactions have been demonstrated in isolated envelope fractions (Givan and Harwood 1976; Douce and Joyard 1979; Van Besouw and Wintermans 1979), namely

$$\text{diacylglycerol} + \text{UDP-galactose} \rightarrow \text{monogalactosyldiglyceride} + \text{UDP} \quad (9.9)$$

$$\text{monogalactosyldiglyceride} + \text{UDP-galactose} \rightarrow \text{digalactosyldiglyceride} + \text{UDP} \quad (9.10)$$

$$2\ \text{monogalactosyldiglyceride} \rightarrow \text{digalactosyldiglyceride} + \text{diacylglycerol} \quad (9.11)$$

The relative importance of the latter two reactions in the synthesis of digalactosyl diglycerides has not yet been established.

Since the enzymes catalysing reactions (9.9-9.11) are apparently exclusively located in the chloroplast envelope, there must be a large flow of galactolipids from envelope to thylakoids. How this is achieved is unknown, but it might involve some kind of protein carrier molecule, perhaps of the type known to mediate exchange of lipid molecules between different organelles in extracts of potato tubers (Kader 1975; Williams, Simpson, and Chapman 1979). Furthermore, enzymes involved in the synthesis of UDP-galactose have not been detected in chloroplasts, so this compound must be imported from the cytoplasm (Leech and Murphy 1976; McKee and Hawke 1979; Roughan, Holland and Slack 1980b) where it is presumably formed by the reaction sequence

$$\text{glucose 1-phosphate} \xrightarrow[\text{pyrophosphorylase}]{\text{UTP} \quad \text{PPi}} \text{UDP-glucose} \underset{\text{epimerase}}{\rightleftharpoons} \text{UDP-galactose} \qquad (9.12)$$

Studies of sucrose biosynthesis have also shown the inability of chloroplasts to synthesize UDP-glucose (Chapter 4). How UDP-galactose crosses the chloroplast envelope has not yet been established, although it is possible that the envelope-bound enzymes that utilize it are so arranged in the membrane that they can utilize external UDP-galactose (Fig. 9.7). Oleate that has been incorporated into position 1 of monogalactosyldiglyceride can be further desaturated to linoleate and α-linolenate by chloroplast fractions (Roughan *et al.* 1979b).

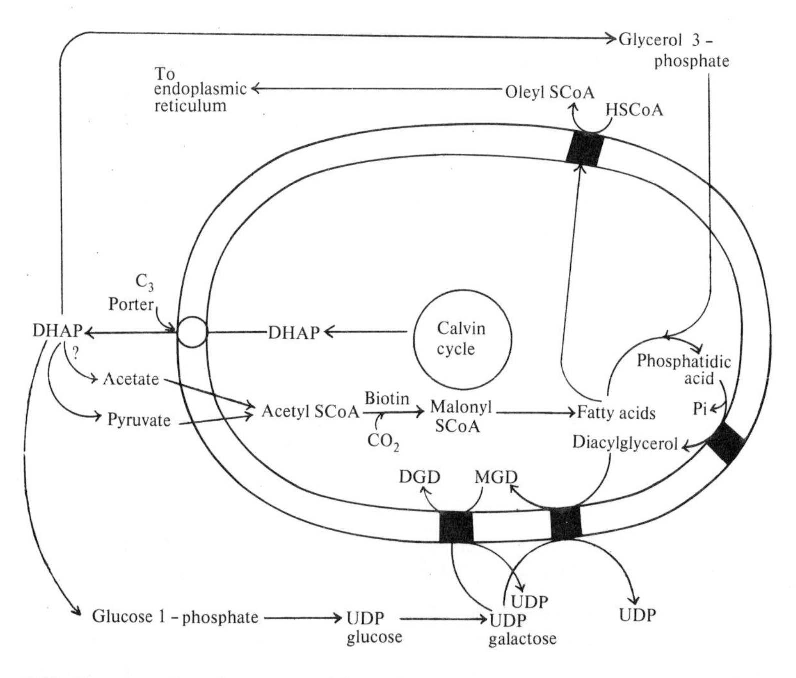

Fig. 9.7. Co-operation between chloroplast and cytoplasm in the synthesis of chloroplast lipids. Abbreviations: DHAP, dihydroxyacetone phosphate; MGD-monogalactosyldiglyceride; DGD-digalactosyldiglyceride. The physiological significance of acetate as a precursor of lipids is unclear, although it is readily converted into fatty acids by illuminated chloroplasts *in vitro* (see text).

1,2-Diacylglycerol is probably also the precursor of sulphoquinovosyl diglyceride, the 'plant sulpholipid' uniquely associated with the chloroplasts of leaf tissues (Chapter 1). How the sulphoquinovose moiety is synthesized and incorporated into the molecule has not been established, although it is possible that phosphoadenosine phosphosulphate (PAPS), which is formed in chloroplasts (Chapter 10), is the actual sulphur donor, since type A spinach chloroplasts purified on a Percoll gradient (Chapter 1) incorporate label from $^{35}SO_4^{2-}$ into sulpholipid (Haas, Siebertz, Wrage, and Heinz 1980).

Diacylglycerol and phosphatidic acid are also precursors of phosphatidylcholine and phosphatidylglycerol, both of which are found in chloroplasts. Phosphatidylcholine is present in much greater amounts in the envelope than in the thylakoids (Chapter 1). When $^{14}CO_2$ is supplied to leaves, label enters phosphatidylcholine much more rapidly than it enters the galactolipids, and transfer of fatty acids from phosphatidylcholine to mono- and di-galactosyldiglycerides has been suggested to occur *in vivo* (Leech and Murphy 1976; Simpson and Williams 1979).

The metabolic routes by which phosphatidylcholine and phosphatidylglycerol can be synthesized in leaf tissues are summarized in Fig. 9.8. However, chloroplasts do *not* appear to contain the enzymes required to catalyse these reactions or to synthesize CDP-choline. These enzymes are located in the endoplasmic reticulum of the leaf cell and therefore appear in the microsomal fraction on subcellular fractionation (Quinn and Williams 1978). Indeed, isolated chloroplast fractions are incapable of synthesizing these lipids unless substantial microsomal contamination is present (Roughan *et al.* 1980*a,b*; Tremolieres, Dubacq, Drapier, Muller, and Mazliak 1980b). There must then be some mechanism of transferring phospholipids from endoplasmic reticulum to chloroplasts, perhaps involving transfer proteins, although these have not yet been found in spinach leaves (Murphy and Kuhn 1981).

It must be concluded that biosynthesis of lipids in chloroplasts involves an extensive co-operation between chloroplast and cytoplasm, some aspects of which are summarized in Fig. 9.7.

9.3 Chlorophylls

Chlorophyll consists of a porphyrin 'head' part to which is attached a phytol 'tail' (Chapter 1). Addition of the 'tail' is one of the last steps in biosynthesis of the chlorophyll molecule, indicating that the porphyrin and phytol skeletons are synthesized separately. Synthesis of the porphyrin can be considered in three stages: synthesis of δ-aminolaevulinic acid, its conversion to protoporphyrin IX, and finally the conversion of this molecule to protochlorophyllide, which becomes reduced to chlorophyllide when chloroplasts are illuminated (Givan and Harwood 1976).

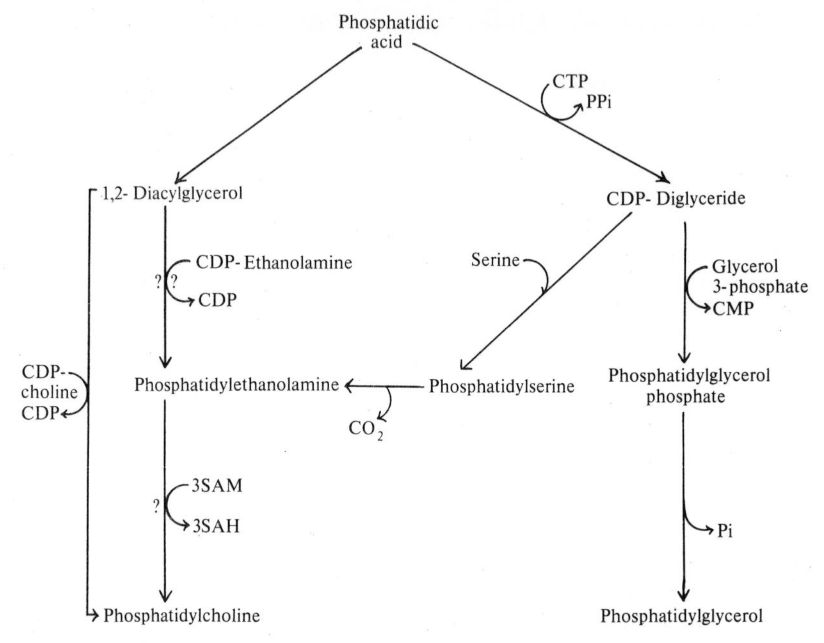

Fig. 9.8. Metabolic pathways to and from phosphatidylcholine and phosphatidylglycerol in leaf tissues. Based on Marshall and Kates (1974), Macher and Mudd (1974) and Quinn and Williams (1978). The evidence for some of the routes given (marked ?) is not yet rigorous. The enzymes involved in these syntheses are located in the endoplasmic reticulum of the leaf. Phosphatidylethanolamine is included for completeness, although this lipid is not found in chloroplasts. Abbreviations: SAM, *S*-adenosylmethionine; SAH, *S*-adenosylhomocysteine.

9.3.1 Synthesis of δ-amino laevulinic acid (ALA)

Animal cells, yeasts, and bacteria contain the enzyme ALA synthetase, which catalyses a reaction between glycine and succinyl-coenzyme A to give carbon dioxide and ALA. However, in plants ALA synthetase has not been detected in chloroplasts, nor can these organelles synthesize glycine (Chapter 7) or succinyl SCoA (Givan and Harwood 1976). Indeed, chloroplasts appear to form ALA by another route that starts from glutamate, 2-oxoglutarate or a related compound with five carbon atoms (Harel 1978; Beale 1978). In this reaction, the five-carbon skeleton is incorporated intact into ALA, C_1 of the precursor becoming C_5 of ALA as shown in eqn (9.13); some possible intermediates are shown in brackets. The rate of ALA synthesis may be the site of regulation of the whole pathway leading to chlorophyll (Harel 1978).

The enzymic pathway by which reaction (9.13) occurs has not yet been fully characterised (Beale 1978), although it has been reported to occur in

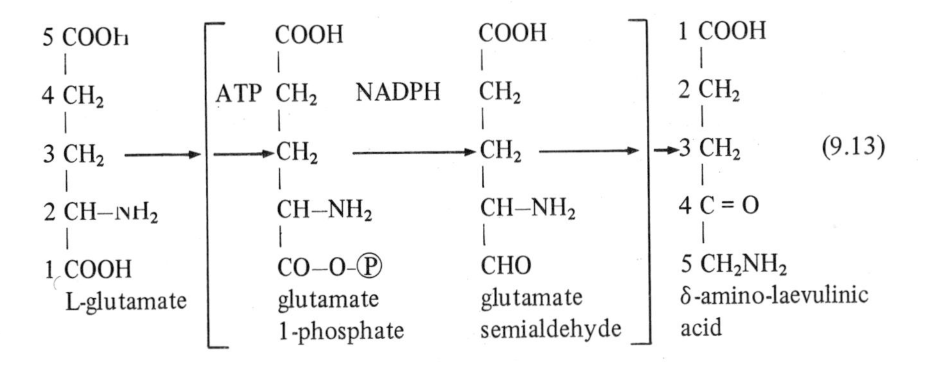

isolated chloroplast fractions. This does not, of course, mean that all the enzymes involved are actually located within the chloroplast (e.g. see Fig. 9.7).

9.3.2 *Conversion of δ-amino laevulinic acid to protoporphyrin IX*

The enzymic steps involved are outlined in Fig. 9.9. All of them can be catalysed by isolated chloroplast fractions, and rigorous fractionation studies have shown that some of the enzymes required are located within the chloroplast stroma (Givan and Harwood 1976). It therefore seems likely that all the enzymes involved are present within chloroplasts *in vivo*, although this has not been completely proven. Protoporphyrin IX is presumably also a precursor of the prosthetic haem groups of the chloroplast cytochromes (Givan and Harwood 1976), i.e. it is the site of a metabolic branch-point.

9.3.3 *Synthesis of chlorophyll from protoporphyrin IX*

The next stage in biosynthesis of the chlorophyll molecule is the incorporation of the magnesium ion into protoporphyrin IX. How this is achieved is not completely understood, but an intact plastid fraction from greening cucumber cotyledons has been demonstrated to catalyse incorporation of Mg^{2+} in the presence of ATP and glutamic acid (Castelfranco, Weinstein, Schwarcz, Pardo, and Wezelman 1979).

Magnesium protoporphyrin IX is then methylated by chloroplast enzymes which use *S*-adenosylmethionine (SAM) as methyl donor, and the resulting magnesium protoporphyrin IX monomethyl ester is converted by a complex series of reactions into protochlorophyllide a. These reactions appear to be catalysed by chloroplast enzymes (Givan and Harwood 1976) but the details remain to be elucidated; it is also not clear how the required SAM is provided. Protochlorophyllide a is then reduced to chlorophyllide a in a light-dependent process which requires the participation of a protein molecule ('the holochrome protein') to which the protochlorophyllide becomes attached. The reducing equivalents required are probably derived from NADPH (Harel 1978). Finally,

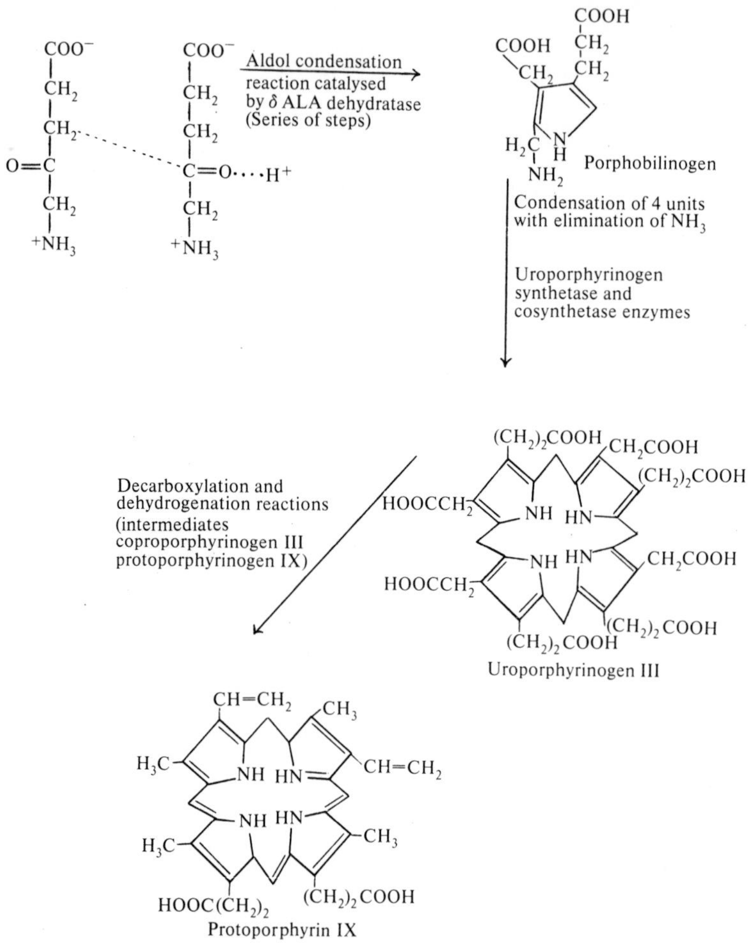

Aldol condensation reaction catalysed by δ ALA dehydratase (Series of steps)

Porphobilinogen

Condensation of 4 units with elimination of NH_3

Uroporphyrinogen synthetase and cosynthetase enzymes

Uroporphyrinogen III

Decarboxylation and dehydrogenation reactions (intermediates coproporphyrinogen III protoporphyrinogen IX)

Protoporphyrin IX

Fig. 9.9. Synthesis of protoporphyrin IX from δ-ALA. Several stages have been omitted for simplicity. Further details may be found in the reviews by Harel (1978), Granick and Beale (1978) and Battersby, Fookes, Matcham, and McDonald (1980). The enzymes required are probably located in the chloroplast stroma.

chlorophyll a is formed by esterification of chlorophyllide a with a precursor of the hydrophobic long-chain alcohol phytol, which is then converted into phytol itself. Chlorophyll b, which differs from chlorophyll a in that a —CHO group replaces $-CH_3$ at one position (Chapter 1), is probably formed from chlorophyll a *in vivo*, although this has not been rigorously proven (Harel 1978). Figure 9.10

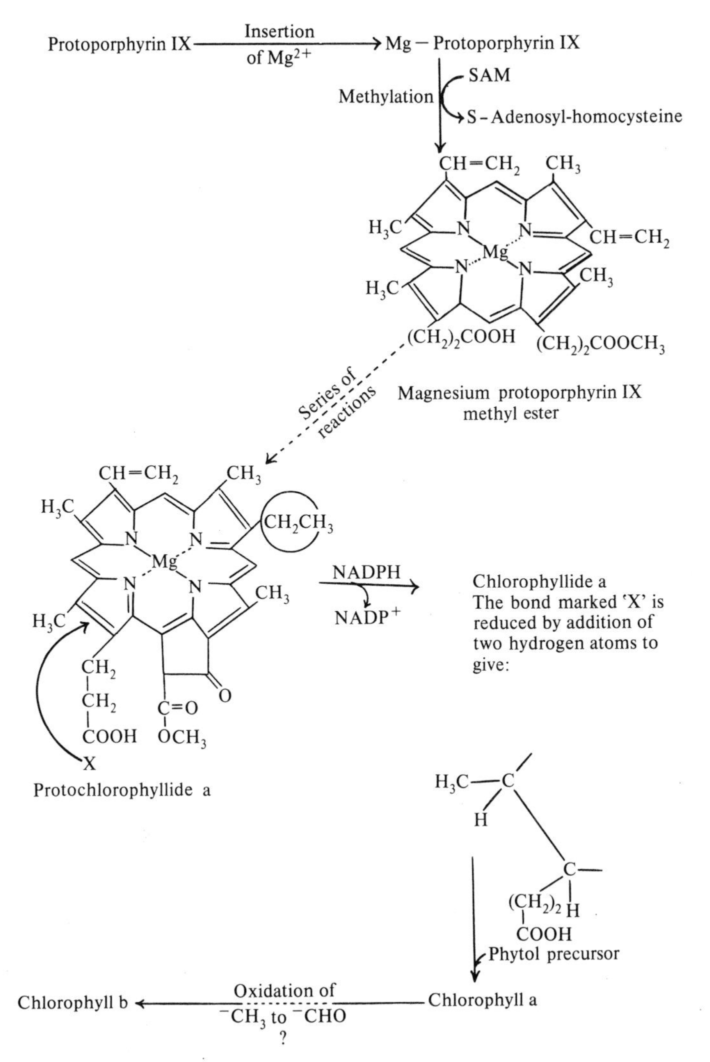

Fig. 9.10. Synthesis of chlorophylls a and b from protoporphyrin IX. Several stages have been omitted for simplicity. Further details may be found in the reviews by Harel (1978), Granick and Beale (1978), and Battersby *et al.* (1980). The enzymes required may be bound to the chloroplast thylakoids. A form of protochlorophyllide a in which the ethyl group circled is replaced by a vinyl group has also been detected in etiolated plant tissues (Belanger and Rebeiz 1980). This can give rise to modified chlorophyll a and b molecules (Rebeiz, Belanger, Freyssinet, and Saab 1980), which may be differently distributed in the various chlorophyll–protein complexes of the thylakoids (Freyssinet, Rebeiz, Fentom, Khanna, and Govindjee 1980). Other modified forms might also occur.

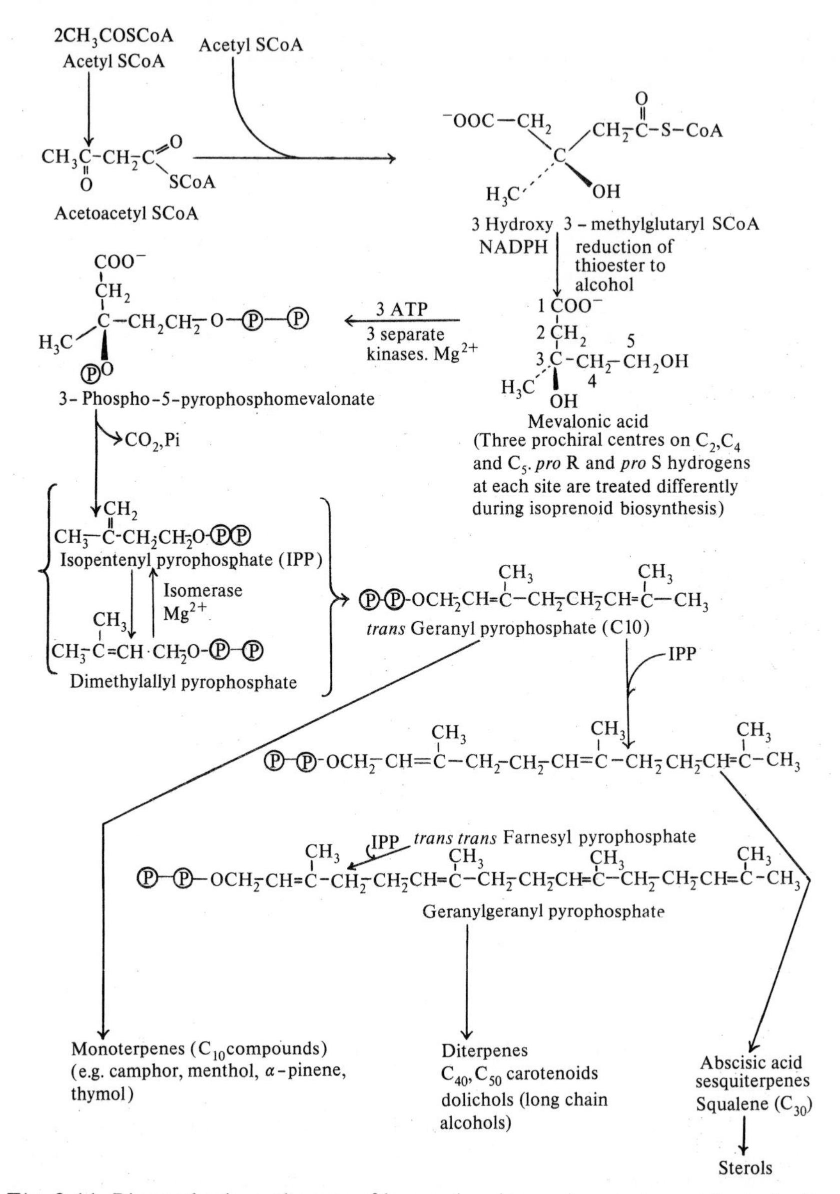

Fig. 9.11. Biosynthetic pathways of isopentenylpyrophosphate and its derivatives in plant tissues. Terpenes, carotenoids, and steroids all arise from the C_5 compound isopentenyl pyrophosphate. Many steps have been omitted for simplicity. For further details see Porter and Anderson (1976) and Goodwin (1979, 1980). Goodwin (1979) gives an excellent discussion of the stereochemistry of these complex reactions.

summarizes the above reactions. Slightly modified forms of chloropylls a and b have also been detected in leaf tissues (Fig. 9.10).

9.4 Carotenoids

Biosynthesis of carotenoids and terpenes in plant tissues begins with the formation of the basic 'isoprenoid' building block isopentenyl pyrophosphate, which is synthesized from 3-hydroxyl-3-methylglutaryl SCoA as shown in Fig. 9.11. Although most of these biosynthetic steps have been demonstrated in extracts from higher plants, only hydroxymethylglutaryl SCoA reductase, mevalonate kinase, and mevalonate 5-phosphokinase activities have been reported to be present in isolated chloroplast fractions (Brooker and Russell 1975; Givan and Harwood 1976). The subcellular location of the enzymes converting isopentenyl pyrophosphate into geranyl, farnesyl, and geranylgeranyl pyrophosphates and their derivatives has not been rigorously studied, although it seems likely that at least the terminal stages of carotenoid synthesis occur within the chloroplast. Synthesis of α-tocopherol, which plays an important role in the protection of chloroplast membranes against light-induced damage (Chapter 8), has been reported to occur in isolated spinach chloroplast fractions supplied with homogentisic acid (Soll, Douce, and Schultz 1980). This compound is also a precursor of plastoquinone in chloroplast fractions (Soll, Kemmerling, and Schultz 1980; Hutson and Threlfall 1980). Phytol originates from a geranylgeranyl unit.

References

Bartlett, D. J., Poulton, J. E., and Butt, V. S. (1972). *FEBS Lett.* **23**, 265–7.

Battersby, A. R., Fookes, C. J. R., Matcham, G. W., and McDonald, E. (1980). *Nature, Lond.* **285**, 17–21.

Beale, S. J. (1978). In *Photosynthesis '77* (eds. D. O. Hall, J. Coombs and T. W. Goodwin) pp. 507–16. Biochemical Society, London.

Belanger, F. C. and Rebeiz, C. A. (1980). *J. biol Chem.* **255**, 1266–72.

Brooker, J. D. and Russell, D. W. (1975). *Archs Biochem. Biophys.* **167**, 730–7.

Butt, V. S. (1979). In *Biochemistry of plant phenolics* (eds. T. Swain, J. B. Harborne and C. F. Van Sumere) pp. 433–56. Plenum Press, New York.

Castelfranco, P. A., Weinstein, J. D., Schwarcz, S., Pardo, A. D., and Weselman, B. C. (1979). *Archs Biochem. Biophys.* **192**, 592–8.

Czichi, U. and Kindl, H. (1975). *Hoppe-Seylers Z. Physiol. Chem* **356**, 475–85.

Douce, R. and Joyard, J. (1979). In *Plant organelles, Methodological surveys in biochemistry,* Vol. 9 (ed. E. Reid) pp. 47–59. Ellis Horwood, Chichester.

Elias, B. A. and Givan, C. V. (1979). *Plant Sci. Lett.* **17**, 115–22.

Freyssiner, G., Rebeiz, C. A., Fenton, J. M., Khanna, R., and Govindjee (1980). *Photobiochem. Photobiophys.* **1**, 203–12.

Fuller, R. C. and Nugent, W. A. (1969). *Proc. natn. Acad. Sci., U.S.A.* **63**, 1311–18.

Gestetner, B. and Conn, E. E. (1974). *Archs Biochem. Biophys.* **163**, 617–24.

Givan, C. V. and Harwood, J. L. (1976). *Biol. Rev.* **51**, 365–406.

Goodwin, T. W. (1979). *A. Rev. Pl. Physiol.* **30**, 369–404.

— (1980). In *Secondary plant products* (eds. E. A. Bell and B. V. Charlwood) pp. 257–87. Springer, Berlin.

Granick, S. and Beale, S. I. (1978). *Adv. Enzymol.* **46**, 33-203.
Gross, G. G. (1978). *Rec. Adv. Phytochem.* **11**, 141-84.
Haas, R., Siebertz, H. P., Wrage, K., and Heinz, E. (1980). *Planta* **148**, 238-44.
Harborne, J. B. (1980) In *Secondary plant products* (eds. E. A. Bell and B. V. Charlwood) pp. 329-402. Springer, Berlin.
Harel, E. (1978). *Prog. Phytochem.* **5**, 127-80.
Henry, E. W. (1975). *J. Microsc. Biol. Cell.* **22**, 109-16.
— (1976). *Z. Pflanzenphysiol.* **78**, 446-52.
Hrazdina, G., Alscher-Herman, R., and Kish, V. M. (1980). *Phytochemistry* **19**, 1355-9.
Hutson, K. G. and Threlfall, D. R. (1980). *Biochim. biophys. Acta* **632**, 630-48.
Jacobson, B. S. and Stumpf, P. K. (1972). *Archs Biochem. Biophys.* **153**, 656-63.
— Jaworski, J. G. and Stumpf, P. K. (1974). *Pl. Physiol., Lancaster* **54**, 484-6.
Joyard, J. and Douce, R. (1979). *FEBS Lett.* **102**, 147-50.
Kader, J. C. (1975). *Biochim. biophys. Acta* **380**, 31-44.
Kahn, V. and Pomerantz, S. H. (1980). *Phytochemistry* **19**, 379-85.
Kannangara, C. G. and Stumpf, P. K. (1972). *Archs Biochem. Biophys.* **152**, 83-91.
— — (1973). *Archs Biochem. Biophys.* **155**, 391-9.
— Henningsen, K. W., Stumpf, P. K., and Wettstein, D. V. (1971). *Eur. J. Biochem.* **21**, 334-8.
Kindl, H. (1971). *Hoppe-Seyler's Z. Physiol. Chem.* **352**, 78-84.
Leech, R. M. and Murphy, D. J. (1976). In *The intact chloroplast, Topics in photosynthesis,* Vol. 1 (ed. J. Barber) pp. 365-401. Elsevier, Amsterdam.
Loffelhardt, W. and Kindl, H. (1975). *Hoppe-Seyler's Z. Physiol. Chem.* **356**, 487-93.
— Ludwig, B. and Kindl, H. (1973). *Hoppe-Seyler's Z. Physiol. Chem.* **354**, 1006-12.
Macher, B. A. and Mudd, J. B. (1974). *Pl. Physiol., Lancaster* **53**, 171-5.
Marshall, M. D. and Kates, M. (1974). *Can. J. Biochem.* **52**, 469-82.
Mason, H. S. (1955). *Adv. Enzymol.* **16**, 105-73.
Mayer, A. M. and Harel, E. (1979). *Phytochemistry* **18**, 193-215.
McIntyre, R. J. and Vaughan, P. F. T. (1975). *Biochem. J.* **149**, 447-61.
McKee, J. W. A. and Hawke, J. C. (1979). *Archs Biochem. Biophys.* **197**, 322-32.
Mohan, S. B. and Kekwick, R. G. O. (1980). *Biochem. J.* **187**, 667-76.
Murphy, D. J. and Stumpf, P. K. (1979). *Pl. Physiol., Lancaster* **64**, 428-30.
— and Kuhn, D. N. (1981). *Biochem. J.* **194**, 257-64.
Nakamura, Y. and Yamada, M. (1975). *Plant Cell Physiol.* **16**, 139-49.
— — (1979). *Plant Sci. Lett.* **14**, 291-5.
Nishizawa, A. N., Wolosiuk, R. A., and Buchanan, B. B. (1979). *Planta* **145**, 7-12.
Ohlrogge, J. B., Kuhn, D. N., and Stumpf, P. K. (1979). *Proc. natn. Acad. Sci. U.S.A.* **76**, 1194-8.
Parish, R. W. (1972). *Eur. J. Biochem.* **31**, 446-55.
Porter, J. W. and Anderson, D. G. (1967). *A. Rev. Pl. Physiol.* **18**, 197-228.
Poulton, J. E., McRee, D. E. and Conn, E. E. (1980). *Pl. Physiol., Lancaster* **65**, 171-5.
Quinn, P. J. and Williams, W. P. (1978). *Prog. Biophys. Molec. Biol.* **34**, 109-73.
Ranjeva, R., Alibert, G., and Boudet, A. M. (1977a). *Plant Sci. Lett.* **10**, 225-34.
— — — (1977b). *Plant Sci. Lett.* **10**, 235-42.
Rebeiz, C. A., Belanger, F. C., Freyssinet, G., and Saab, D. G. (1980). *Biochim.*

biophys. Acta **590**, 234–47.
Roughan, P. G. and Slack, C. R. (1977). *Biochem. J.* **162**, 457–9.
— Holland, R., Slack., C. R., and Mudd, J. B. (1979a). *Biochem. J.* **184**, 565–9.
— Mudd, J. B., McManus, T. T., and Slack, C. R. (1979b). *Biochem. J.* **184**, 571–4.
— Kagawa, T. and Beevers, H. (1980a). *Plant Sci. Lett.* **18**, 221–8.
— Holland, R. and Slack, C. B. (1980b). *Biochem. J.* **188**, 17–24.
Santora, G. T., Gee, R., and Tolbert, N. E. (1979). *Archs Biochem. Biophys.* **196**, 403–11.
Sato, M. (1969). *Phytochemistry* **8**, 353–62.
Saunders, J. A. and McClure, J. W. (1975). *Phytochemistry* **14**, 1285–9.
— — (1976). *Phytochemistry* **15**, 809–10.
Schil, L. and Grisebach, H. (1973). *Hoppe-Seylers Z. Physiol. Chem.* **354**, 1555–62.
Simpson, E. E. and Williams, J. P. (1979). *Pl. Physiol., Lancaster* **63**, 674–6.
Soll, J., Douce, R., and Schultz, G. (1980). *FEBS Lett.* **112**, 243–6.
— Kemmerling, M., and Schultz, G. (1980). *Archs Biochem. Biophys.* **204**, 544–50.
Stafford, H. A. (1974). *A. Rev. Pl. Physiol.* **25**, 459–86.
Tissut, M., Chevallier, D., and Douce, R. (1980). *Phytochemistry* **19**, 495–500.
Tolbert, N. E. (1973). *Pl. Physiol., Lancaster* **51**, 234–44.
Towers, G. H. N. (1974). In *MTP international review of science*, Vol. II (ed. D. H. Northcote), pp. 247–76. Butterworths, London.
Tremolieres, A., Drapier, D., Dubacq, J. P., and Mazliak, P. (1980a). *Plant Sci. Lett.* **918**, 257–269.
— Dubacq, J. P., Drapier, D., Muller, M., and Mazliak, P. (1980b). *FEBS Lett.* **114**, 135–8.
Van Besouw, A. and Wintermans, J. F. G. M. (1979). *FEBS Lett.* **102**, 33–7.
Vanneste, W. H. and Zuberbuhler, A. (1974). In *Molecular mechanism of oxygen activation* (ed. O. Hayaishi) pp. 371–94. Academic Press, New York.
Vaughan, P. F. T. and Butt, V. S. (1969). *Biochem. J.* **113**, 109–15.
— Eason, R., Paton, J. Y., and Ritchie, G. A. (1975). *Phytochemistry* **14**, 2382–6.
Weissenbock, G. and Schneider, V. (1974). *Z. Pflanzenphysiol.* **72**, 23–35.
Williams, J. P., Simpson, E. E., and Chapman, D. J. (1979). *Pl. Physiol., Lancaster* **63**, 669–73.
Williams, M. and Randall, D. D. (1979). *Pl. Physiol., Lancaster* **64**, 1099–103.
Yamada, M. and Nakamura, Y. (1975). *Plant Cell Physiol.* **16**, 151–62.

10 NITROGEN AND SULPHUR METABOLISM IN PHOTOSYNTHETIC TISSUES

10.1 Nitrogen assimilation by nitrate reductase

The nitrate ion, NO_3^-, is the most abundant form of inorganic nitrogen in the soil and, except in the case of those plant species that are symbiotically associated with nitrogen-fixing bacteria, it is the major source of nitrogen available to the plant for the synthesis of amino acids, purines, pyrimidines, and other nitrogenous compounds.

The first stage in NO_3^- assimilation is a reduction to the nitrite ion, NO_2^-, the oxidation number of the nitrogen decreasing from +5 to +3. This is catalysed by an enzyme, nitrate reductase, whose overall reaction may be represented by the equation

$$NO_3^- + H^+ + NAD(P)H \rightarrow NAD(P)^+ + NO_2^- + H_2O \qquad (10.1)$$

Nitrate reductase is a complex enzyme that contains FAD, molybdenum, and a cytochrome (b_{557}) as components essential for the enzyme activity (Hewitt 1974; Vennesland and Guerrero 1979). The sequence of electron flow during operation of the enzyme has been suggested to be

$$NADH \rightarrow FAD \rightarrow cytb_{557} \rightarrow molybdenum \rightarrow NO_3^- \qquad (10.2)$$

NADH seems to be the preferred reducing agent both *in vitro* and *in vivo* for the enzymes from most higher plant sources (Wells and Hageman 1974; Hewitt 1974; Lee 1980), although nitrate reductase from the Nigerian plant *Erythina senegalensis* has been reported to be equally active with both NADH and NADPH (Stewart and Orejambo 1979) and ferredoxin-dependent enzymes have been found in some bacteria and blue-green algae. Nitrate reductases are powerfully inhibited by hydrogen cyanide and they are also capable of reducing chlorate ions (ClO_3^-) to the highly toxic chlorite, which can inactivate this and other enzymes. Such a reaction probably accounts for the damaging effects of chlorate on plant tissues, hence its use as a weedkiller (Vennesland and Guerrero 1979). Plants can incorporate tungsten into nitrate reductase in place of molybdenum, producing an inactive enzyme. As a result, tungsten is toxic to plant tissues.

Nitrate reductase can be detected in both the roots and leaves of most plants. In woody species, such as apple, and in some other plants such as cranberry and rhododendron, most nitrate reduction occurs in the roots (Haynes and Goh 1978) and little NO_3^- is supplied to the leaves. In other plants, however, the leaves make a major contribution (Pate 1980; Haynes and Goh 1978) and we shall be largely concerned here with the processes operating in leaf tissues. A minor source of NO_3^- to leaf tissues is uptake of the atmospheric pollutant gas nitrogen dioxide, which dissolves in leaf water to form a mixture of nitric (HNO_3) and nitrous (HNO_2) acids (Rogers, Campbell, and Volk 1979).

The nitrate reductase activity of both leaves and roots is exclusively located in the cytosol of the cell (Lee 1980). Reports of its association with peroxisomes, proplastids, chloroplasts, or chloroplast envelopes have been ruled out by the use of improved subcellular fractionation and enzyme assay techniques (Dalling, Tolbert, and Hageman 1972a,b; Wallsgrove, Lea, and Miflin 1979). Isolated chloroplast *fractions* may show nitrate reductase activity because of cytoplasmic contamination, but it is easily removed by washing (see Chapter 1).

Nitrate reductase in the cytosol cannot derive reducing power directly from the chloroplast electron-transport system, since the envelope is impermeable to NADH and NADPH (Chapter 6). However, illumination greatly increases the rate of nitrate reduction by leaf tissues, and some leaves are incapable of reducing nitrate in the dark. There is a slow decrease in the nitrate reductase activity of leaves kept in the dark and an increase on re-illumination, possibly owing to a circadian rhythm, but these changes are not fast enough to account for the rapid light-induced increase in nitrate assimilation. It seems likely that carbon dioxide fixation by the Calvin cycle allows export of C_3 compounds from the chloroplast to generate NADH in the cytosol by using glycolytic enzymes (Chapter 6). Indeed, removal of carbon dioxide from the atmosphere surrounding illuminated leaves often inhibits nitrate reduction (e.g. Kessler 1964; Rathnam 1978). However, Canvin and Atkins (1974) showed that barley or corn leaves illuminated in the absence of carbon dioxide could still assimilate NO_3^- rapidly. In this case, NADH might have been provided by the mobilization of stored carbohydrate reserves or by the export of reducing power from the chloroplast using the malate/oxaloacetate shuttle described in Chapter 6 (Neyra and Hageman 1976; Aslam, Huffaker, Rains, and Rao 1979; Rathnam 1978). Nitrate reduction in the cytosol might also use up some of the NADH generated by the oxidation of glycine in leaf mitochondria during photorespiration (see Fig. 10.1 and Sawhney, Naik, and Nicholas 1978), since plant mitochondria can export NADH using a malate/oxaloacetate shuttle system (Woo, Jokinen, and Canvin 1980). Indeed, one might expect there to be some relationship between nitrogen metabolism and photorespiration because the rapid synthesis of glycine from glycollate requires glutamate as an amino donor in the peroxisomes (see Fig. 10.1 and Chapter 7). Fair, Tew, and Cresswell (1973, 1974) have claimed that the carbon dioxide compensation point of barley leaves, used by them as an index of photorespiration rates, is closely correlated with the ratio of nitrate reductase to ribulose diphosphate carboxylase activities within the leaf. Increasing the concentration of nitrate supplied to C_4 plants has also been reported to raise the compensation point from almost zero to a measurable value (Cresswell, Tew, and Baxter 1975).

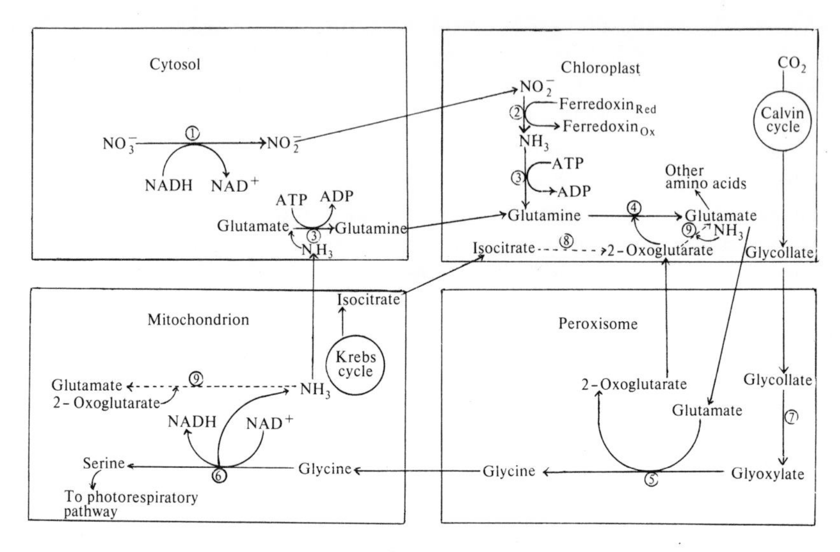

Fig. 10.1. Pathways of nitrate (NO_3^-) assimilation in green leaves in relation to photorespiration. The further metabolism of serine produced from glycine has been omitted for simplicity: details may be found in Chapter 7. Solid lines represent major pathways of metabolism and dotted lines minor pathways. Enzymes involved: (1) nitrate reductase; (2) nitrite reductase; (3) glutamine synthetase; (4) glutamate synthetase; (5) glutamate-glyoxylate aminotransferase; (6) glycine decarboxylase; (7) glycollate oxidase; (8) isocitrate dehydrogenase; (9) glutamate dehydrogenase.

10.2 Reduction of NO_2^- to ammonia by nitrite reductase

Nitrite produced by the action of nitrate reductase in leaves and roots can be further reduced to ammonia (oxidation number of nitrogen equal to −3) by the action of the enzyme nitrite reductase, which catalyses the overall reaction

$$NO_2^- + 6e^- + 8H^+ \rightarrow NH_4^+ + 2H_2O \qquad (10.3)$$

Nitrite or ammonia rarely accumulate in plant tissues whereas NO_3^- often does, so it would seem that the rate of the nitrite reductase enzyme is controlled by the supply of its substrate, i.e. nitrate reductase is the rate-limiting step for nitrogen assimilation.

Nitrite reductase is a complex enzyme that contains an iron-porphyrin prosthetic group, known as sirohaem, as well as a non-haem-iron-sulphur centre (Hewitt 1974; Vennesland and Guerrero 1979; Hucklesby, Cammack, and Hewitt 1979). A powerful reducing agent is required to supply the six electrons needed to reduce each NO_2^- ion: reduced ferredoxin is especially active *in vitro*. Subcellular fractionation experiments have shown that the nitrite reductase activity of leaf tissues is exclusively located in the chloroplasts and absolutely dependent on light, so it presumably uses reduced ferredoxin derived from the electron-transport chain *in vivo* as well (Miflin and Lea 1976, 1977; Givan 1979). The NO_2^- required must enter the chloroplast from the cytoplasm (Fig. 10.1). The nitrite reductase activity of root tissues is also located in plastids. Its source of reductant is unknown, although NADPH generated by the oxidative pentose phosphate pathway is a likely candidate (Butt and Beevers 1961; Washitani and Sato 1977; Lee 1980).

10.3 Assimilation of ammonia

Ammonia is highly toxic to plant tissues, the roots being especially sensitive (Haynes and Goh 1978). For example, it uncouples photophosphorylation and mitochondrial oxidative phosphorylation. Thus any ammonia formed by nitrite reductase must be rapidly assimilated.

Until fairly recently, it was thought that ammonia assimilation takes place by a reductive amination of 2-oxoglutarate, catalysed by glutamate dehydrogenase (eqn (10.4))

$$\text{2-oxoglutarate} + NH_3 + H^+ + NAD(P)H \rightleftarrows \text{L-glutamate} + H_2O + NAD(P)^+ \quad (10.4)$$

Purified glutamate dehydrogenases are effective with both NADPH and NADH as electron donor. The enzyme has been detected in both the mitochondria and the chloroplasts of green leaf tissues. For example, the chloroplast enzyme from lettuce leaves was found to be equally effective with both NADH and NADPH, although it would presumably use NADPH *in vivo*, whereas the mitochondrial enzyme is seven times more active with NADH (Lea and Thurman 1972). However, the K_m of the chloroplast glutamate dehydrogenase for NH_4^+ ions was found to be 5.8 mM and its maximum velocity to be too low to account for observed rates of ammonia assimilation. Since 2 mM NH_4^+ uncouples photophosphorylation completely, the actual activity of the enzyme at the low concentrations of NH_4^+ likely to exist inside the chloroplast *in vivo* would be minimal (Miflin and Lea 1976, 1977; Givan 1979). Similar reasoning can be used to show that it is unlikely that more than a small part of the ammonia generated during glycine decarboxylation in leaf mitochondria is assimilated by glutamate dehydrogenase. It follows that another system for assimilating NH_4^+ must exist inside chloroplasts. Indeed, these organelles contain a high activity of glutamine synthetase, which catalyses the reaction.

$$\text{L-glutamate} + NH_3 + ATP \xrightarrow{Mg^{2+}} \text{L-glutamine} + ADP + Pi \quad (10.5)$$

Glutamine synthetase has a K_m for ammonia of less than 10^{-4}M and its action is favoured under the conditions of high pH, Mg^{2+} ion concentration and ATP/ADP ratio that exist in the illuminated chloroplast (Miflin and Lea 1976, 1977; Givan 1979). Hence ammonia is rapidly converted to glutamine and prevented from accumulating. Glutamine synthetase is also present in the cytosol of the leaf and may serve to assimilate ammonia generated during glycine decarboxylation in mitochondria (Fig. 10.1), since little or no glutamine synthetase is present in leaf mitochondria (Keys, Bird, Cornelius, Lea, Wallsgrove, and Miflin 1978; Jackson, Dench, Morris, Lui, Hall, and Moore 1979). Glutamine formed in the cytosol might then enter the chloroplast for further metabolism, although it could also serve as a nitrogen donor for biosynthetic processes in the cytoplasm (Rhodes, Sims, and Folk 1980).

The first clue to the fate of glutamine in chloroplasts came in 1970, when Tempest, Meers, and Brown detected the presence in bacteria of a new enzyme: glutamine-2-oxoglutarate aminotransferase, often simply referred to as glutamate synthetase or 'GOGAT'. It catalyses the reaction

$$\text{L-glutamine} + \text{2-oxoglutarate} \xrightarrow{\text{reductant}} \text{2 L-glutamate} \qquad (10.6)$$

Subsequent work detected the enzyme in plant cell cultures, roots and cotyledons (Miflin and Lea 1976, 1977; Givan 1979). The reductant found to be used was NADH or NADPH. For example, the enzyme from soybean cell suspension cultures is active with either nucleotide (Chiu and Shargool 1979).

In contrast, no NAD(P)H-dependent glutamate synthetase could be found in chloroplasts or leaves. The solution to this problem came when chloroplasts were shown to contain a glutamate synthetase that uses reduced ferredoxin as electron donor (Lea and Miflin 1974). The enzymes purified from *Vicia faba*, spinach, and corn leaves were found to have high maximal velocities and affinities for their substrates (e.g. the K_m values of the *Vicia* enzyme were found to be for glutamine 0.33 mmol l^{-1}, for 2-oxoglutarate 0.15 mmol l^{-1} and for reduced ferredoxin 0.002 mmol l^{-1}) and so glutamine can be rapidly assimilated provided that 2-oxoglutarate is supplied at sufficient rates (Wallsgrove *et al.* 1977; Matoh, Suzuki, and Ida 1979; Tamura, Kanki, Hirasawa, and Oto 1980). Chloroplasts do contain a low activity of $NADP^+$-dependent isocitrate dehydrogenase (Elias and Givan 1977) but there would doubtless have to be import of some oxoglutarate from the cytoplasm via the dicarboxylate translocator (Chapter 6). Chloroplasts cannot synthesize isocitrate and this must also be imported for isocitrate dehydrogenase action. Presumably these imports occur in exchange for the passage of glutamate into the cytoplasm (Fig. 10.1).

The above consideration of the kinetics and activity of glutamine and glutamate synthetases strongly implies that the major pathway of assimilation for the ammonia produced in leaf tissues by nitrite reductase (and by glycine decarboxylase) is conversion into glutamine, which then forms glutamate. Extensive evidence supporting this proposal has come from isotope studies using $^{15}NO_3^-$ and from the use of azaserine, an inhibitor of glutamate synthetase, together with methionine sulphoxime, which inhibits glutamine synthetase. These compounds markedly reduce ammonia assimilation (Miflin and Lea 1976, 1977; Givan 1979; Rhodes *et al.* 1980). A mutant of the plant *Arabidopsis thaliana* with greatly reduced glutamate synthetase activity in the leaves rapidly accumulates NH_3 to an extent which inhibits carbon dioxide fixation when it is placed under conditions that favour photorespiration (Somerville and Ogren 1980). It is therefore difficult to attribute any role to the chloroplast and mitochondrial glutamate dehydrogenases, except perhaps that of ammonia assimilation at high ammonia concentrations, as, for example, when ammonia is supplied externally to the plant in large amounts. The glutamate dehydrogenase reaction is therefore designated as a minor pathway in Fig. 10.1.

This analysis need not apply to photosynthetic organisms other than higher plants, however. For example, chloroplast fractions from the marine alga *Caulerpa simpliciuscula* contain not only glutamine and glutamate synthetases but also a glutamate dehydrogenase with a K_m for NH_4^+ of only 0.4–0.7 mmol l^{-1} (McKenzie,Ch'ng, and Gayler 1979). Both pathways of ammonia assimilation could therefore be operative in this organism, a point which requires further investigation.

10.4 Nitrate assimilation in C_4 plants

In so far as is known, the enzymology of NO_3^- assimilation and conversion to amino acids in C_4 leaves is not very different from that in C_3 leaves. Most nitrate taken up by C_4 leaves is metabolized by the mesophyll cells, which contain the bulk of the nitrate reductase and nitrite reductase activities of the leaf together with substantial amounts of glutamine synthetase and glutamate synthetase (e.g. Harel, Lea, and Miflin 1977; Neyra and Hageman 1978; Moore and Black 1979). Nitrate reductase again seems to be the rate-limiting step in nitrogen assimilation. Reduction of NO_3^- and assimilation of ammonia may well be a major fate of the reducing power generated by the mesophyll chloroplasts in aspartate-forming C_4 leaves (Chapter 5). Glutamine synthetase activity is also found in the bundle-sheath cells, however, and it might perhaps serve to assimilate ammonia generated by glycine decarboxylation, since photorespiration is confined to these cells in C_4 leaves (Chapter 7).

10.5 Do chloroplasts synthesize amino acids?

When $^{14}CO_2$ is supplied to illuminated leaf tissues, the amino acids that become most rapidly labelled are glutamate, glutamine, aspartate, alanine, glycine, and serine. In C_3 plants, isoenzymes of glutamate-aspartate and glutamate-pyruvate

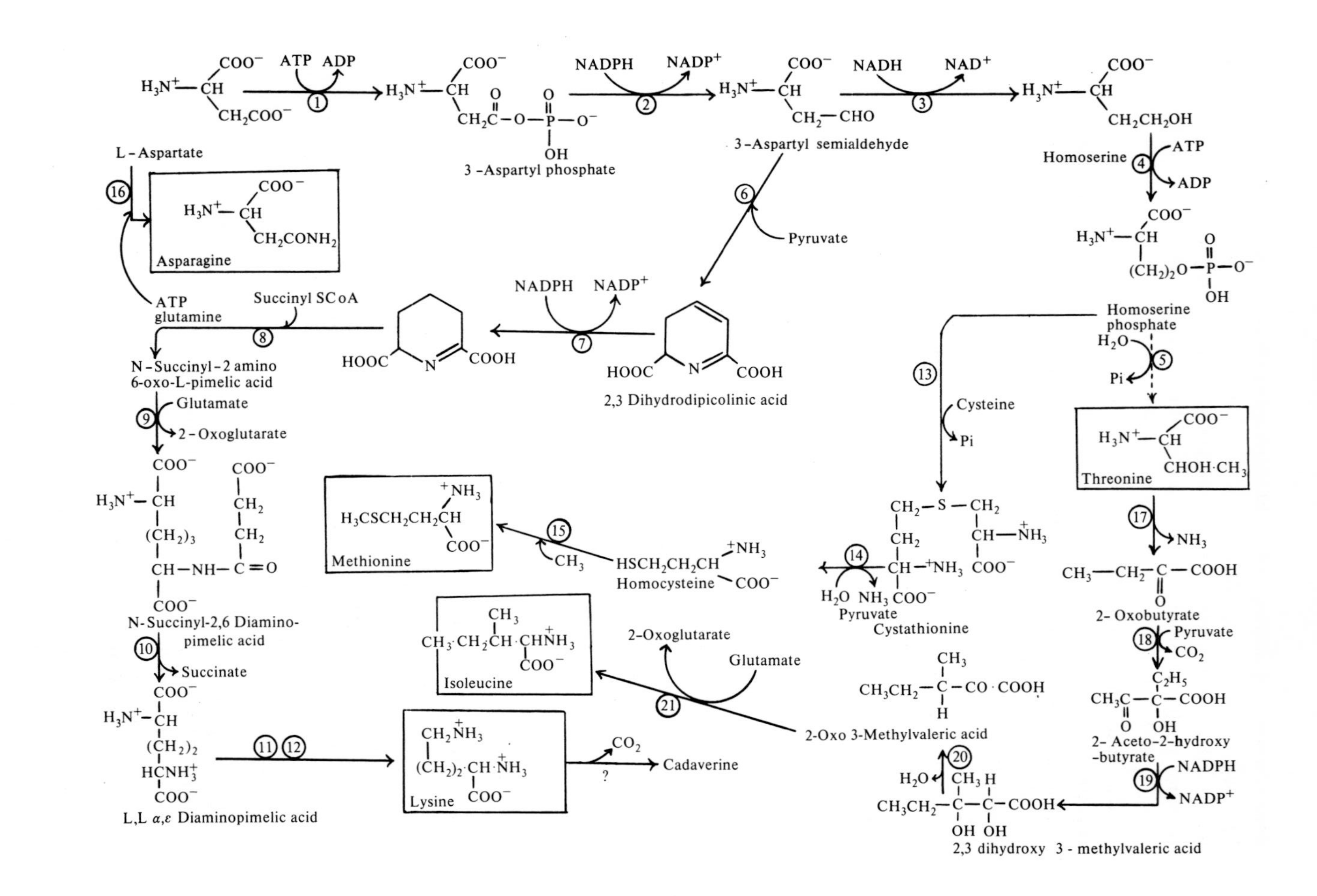
L-Aspartate
ATP
ADP
3-Aspartyl phosphate
NADPH
NADP+
3-Aspartyl semialdehyde
NADH
NAD+
Homoserine
ATP
ADP
Homoserine phosphate
H2O
Pi
Threonine
NH3
2-Oxobutyrate
Pyruvate
CO2
2-Aceto-2-hydroxy -butyrate
NADPH
NADP+
2,3 dihydroxy 3-methylvaleric acid
H2O
2-Oxo 3-Methylvaleric acid
Glutamate
2-Oxoglutarate
Isoleucine
Pyruvate
2,3 Dihydrodipicolinic acid
NADPH
NADP+
Succinyl SCoA
ATP
glutamine
Asparagine
N-Succinyl-2 amino 6-oxo-L-pimelic acid
Glutamate
2-Oxoglutarate
N-Succinyl-2,6 Diamino-pimelic acid
Succinate
L,L α,ε Diaminopimelic acid
Lysine
CO2
Cadaverine
?
Cysteine
Pi
Cystathionine
H2O
NH3
Pyruvate
Homocysteine
CH3
Methionine

aminotransferases are located in both chloroplasts and cytoplasm (e.g. Huang, Liu, and Youle 1976; Leech and Murphy 1976) and so it seems likely that at least some aspartate and alanine are made in chloroplasts *in vivo* by transamination, the oxaloacetate and pyruvate required being imported from the cytoplasm (Chapter 6; Larsson 1979; Mills and Joy 1980).

Many additional aminotransferases are present in chloroplast fractions. These can transfer amino groups from alanine, aspartate, or glutamate to a variety of keto-acid acceptors, and there have been occasional claims that chloroplasts are capable of the synthesis of several such keto-acids. Unfortunately, many studies of the capacity of chloroplasts to synthesize amino acids have used crude chloroplast fractions contaminated with cytosol and with other organelles. Even when enzymes of amino acid synthesis have been found to be associated with chloroplasts by density gradient centrifugation, the possibility of non-specific adsorption onto the envelope has rarely been ruled out by performing latency experiments of the type described in Chapter 1. For example, purified type A chloroplasts are *incapable* of synthesizing glycine or serine, although label is rapidly detected in these compounds if $^{14}CO_2$ is supplied to crude chloroplast fractions in the light (Chapter 7).

10.5.1 The aspartate family of amino acids

Asparagine, threonine, lysine, and isoleucine all originate from L-aspartic acid in plant tissues (Fig. 10.2). The major precursor of cystathionine is *O*-phosphohomoserine rather than the acylhomoserines used by other organisms

---

Fig. 10.2 (*opposite*). Amino acids derived from aspartate in plant tissues. Enzymes involved: (1) aspartate kinase; (2) aspartate semialdehyde dehydrogenase; (3) homoserine dehydrogenase; (4) homoserine kinase; (5) threonine synthetase; (6) dihydrodipicolinate synthetase; (7) dihydrodipicolinate dehydrogenase; (8) synthetase enzyme; (9) aminotransferase; (10) desuccinylase; (11) diaminopimelate epimerase; (12) diaminopimelate decarboxylase; (13) cystathionine synthetase; (14) cystathionine lyase; (15) methyltransferase, using N^5-methyl-tetrahydrofolate as donor; (16) asparagine synthetase (glutamine is preferred nitrogen donor); (17) threonine dehydratase; (18) acetolactate synthetase; (19) acetolactate mutase and reductase; (20) dihydroxyacid dehydratase; (21) valine transaminase. The major precursor of cystathionine in plant tissues is O-phosphohomoserine rather than the acylhomoserines used by other organisms (Giovanelli *et al.* 1974). Lupin chloroplast fractions have been reported to decarboxylate lysine into cadaverine (Wink *et al.* 1980). Control of the aspartate pathway differs from plant to plant (Miflin and Lea 1977). In pea chloroplast fractions threonine decreased the synthesis of homoserine from aspartic acid, as well as its own synthesis, probably by inhibiting aspartate kinase and homoserine dehydrogenase. Threonine did not strongly inhibit lysine formation, whereas lysine limited not only its own synthesis but also that of homoserine and threonine, perhaps by inhibiting a lysine-sensitive aspartate kinase (Mills *et al.* 1980). S-adenosyl-methionine potentiates this inhibition although itself it has only a small effect on aspartate kinase.

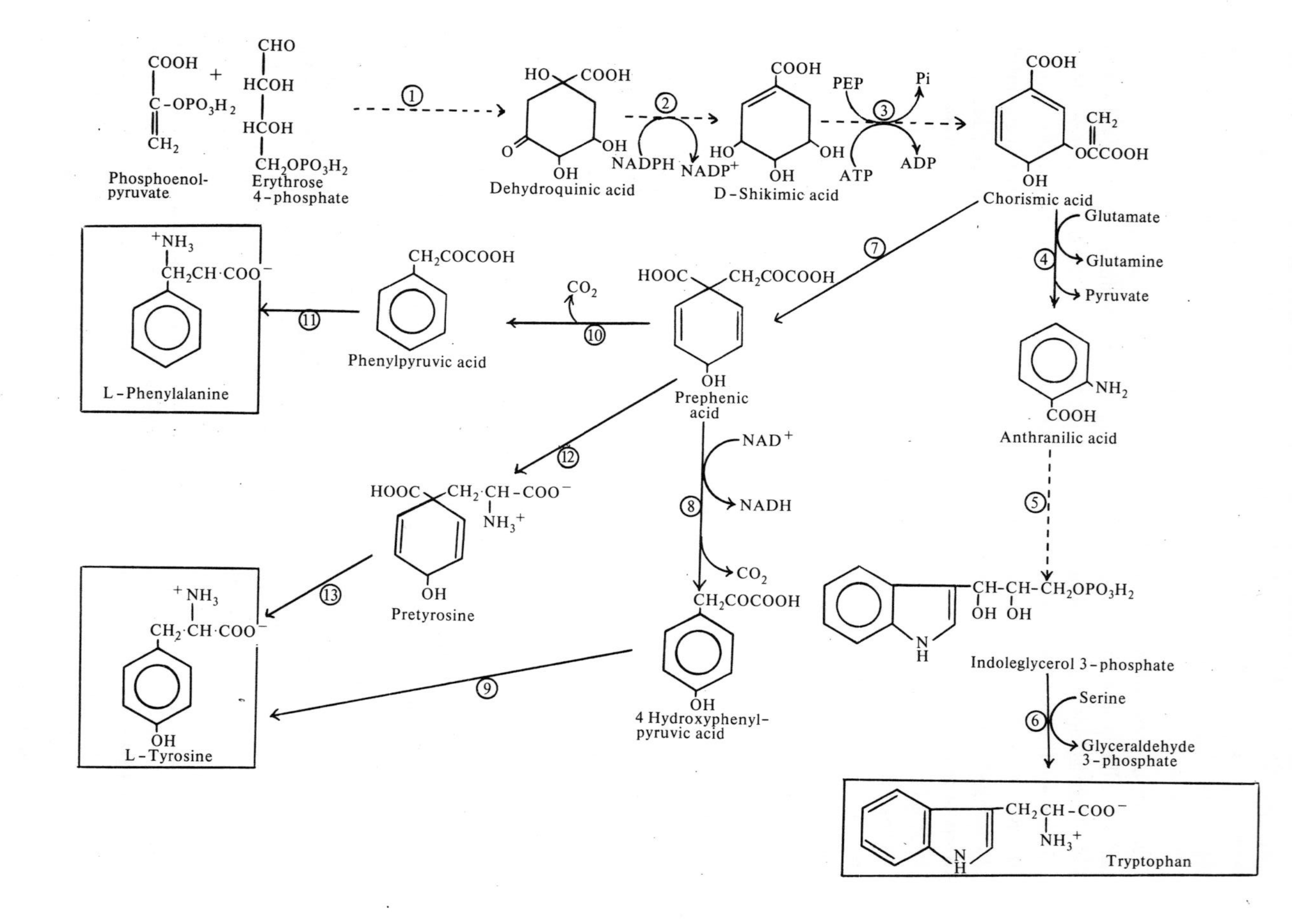

Phosphoenol-pyruvate
Erythrose 4-phosphate
Dehydroquinic acid
NADPH
$NADP^+$
D-Shikimic acid
PEP
Pi
ATP
ADP
Chorismic acid
Glutamate
Glutamine
Pyruvate
Anthranilic acid
Indoleglycerol 3-phosphate
Serine
Glyceraldehyde 3-phosphate
Tryptophan
Prephenic acid
NAD^+
NADH
CO_2
4 Hydroxyphenyl-pyruvic acid
Phenylpyruvic acid
L-Phenylalanine
Pretyrosine
L-Tyrosine

(Giovanelli, Mudd, and Datko 1974). Several experiments strongly suggest that at least some of the enzymes shown in Fig. 10.2 are present in purified chloroplasts, although latency experiments have not been done. Aspartate kinase has been detected in pea and in spinach chloroplasts (Wahnbaeck-Spencer, Mills, Henke, Burdge, and Wilson 1979), diaminopimelate decarboxylase in *Vicia faba* and pea chloroplasts (Mazelis, Miflin, and Pratt 1976), homoserine dehydrogenase in maize mesophyll chloroplasts (Bryan, Lissik, and Matthews 1977), dihydrodipicolinate synthetase in spinach chloroplasts (Wallsgrove and Mazelis 1980) and acetolactate synthetase also in spinach chloroplasts (Miflin 1974). Chloroplast fractions from young pea leaves have been reported to convert homocysteine into methionine, and aspartate into lysine, threonine, and isoleucine (Mills, Lea, and Miflin 1980). Although these chloroplast fractions had low cytoplasmic contamination further experiments need to be performed to check whether *all* the enzymes required are actually located within the chloroplast, in view of the complex organelle interactions already demonstrated for processes such as lipid biosynthesis (Chapter 9). Chloroplasts seem incapable of converting aspartate into asparagine, however (Lea and Fowden 1975).

Crude chloroplast fractions from leaves of lupin (*Lupinus polyphyllus*) have been reported to decarboxylate lysine and to convert the cadaverine so produced into the quinolizidine alkaloids sparteine and lupanine (Wink, Hartmann, and Witte 1980). Potato tubers exposed to light and cold develop a bitter taste, due to the alkaloid solanidine, and they also turn green as chlorophyll is synthesized. Chloroplast fractions isolated from greened potato peelings are able to fix carbon dioxide into the solanidine molecule. The Calvin cycle is absent from these chloroplasts and it has been proposed that carbon dioxide is fixed initially into formic acid, which is converted into glycine and then, via serine and pyruvate, into acetyl SCoA for mevalonate synthesis and conversion to solanidine (Ramaswamy, Behere, and Nair 1976). In general, however, our knowledge of the role that chloroplasts play in alkaloid metabolism is very poor

---

Fig. 10.3 (*opposite*). Pathway of biosynthesis of aromatic amino acids in plant tissues. The pathway is shown in outline only, since several of the enzymes involved have not yet been studied in plant tissues (Miflin and Lea 1977). Tryptophan may be a feedback inhibitor of glutamine-dependent anthranilate synthetase. Two routes of tyrosine synthesis have been detected in plant tissues: both use similar reactions but in reverse order (Rubin and Jensen 1979). Enzymes involved: (1) phospho-2-oxo-3-deoxy-heptonate aldolase and dehydroquinate synthetase; (2) dehydroquinate dehydratase and shikimate dehydrogenase; (3) Shikimate kinase, pyruvylshikimate-phosphate synthetase and chorismate synthetase; (4) anthranilate synthetase; (5) anthranilate phosphoribosyltransferase and indoleglycerolphosphate synthetase; (6) tryptophan synthetase; (7) chorismate mutase; (8) prephenate dehydrogenase; (9) transaminase; (10) prephenate dehydratase; (11) transaminase; (12) pretyrosine synthetase; (13) pretyrosine dehydrogenase. Homogentisic acid, a derivative of tyrosine, is a precursor of tocopherol and plastoquinone in chloroplasts (Chapter 9).

(for a review see Robinson 1974). If these molecules are made by chloroplasts *in vivo*, the process must be carefully controlled, since many alkaloids are powerful inhibitors of photophosphorylation (Vallejos and Andreo 1974).

10.5.2 Synthesis and metabolism of aromatic amino acids

Chloroplasts appear to contain some of the phenylalanine ammonia lyase activity present in green leaves (Chapter 9) and phenylalanine can cross the envelope at low rates (Chapter 6). Type A spinach chloroplast fractions have been reported to synthesize tyrosine, tryptophan, and phenylalanine from either assimilated $^{14}CO_2$ or from ^{14}C-shikimic acid (Bickel, Palm, and Schultz 1978). These results imply that chloroplasts contain the enzymes for biosynthesis of aromatic amino acids (Fig. 10.3). The rates of synthesis were extremely low, however, and more investigation is required.

10.6 Sulphate assimilation in plant tissues

Plants must take up sulphur from the soil for incorporation into amino acids, proteins, cofactors such as coenzyme A, glutathione, and other molecules such as the 'plant sulpholipid' found in chloroplast membranes. The major source of sulphur to the plant is the sulphate ion, SO_4^{2-}, in the soil. This sulphur in the oxidation state of +6 has to be reduced to sulphide (S^{2-}, oxidation state −2) before it is incorporated into amino acids and other molecules. The process by which such an eight-electron reduction is achieved in higher plants is known as 'assimilatory sulphate reduction' and takes place almost exclusively in the leaves. Here sulphate reduction occurs in the illuminated chloroplast (Schwenn and Trebst 1976; Schmidt 1979). Some of the reduced sulphur, if formed in excess, can then apparently be transported back to the roots, the major transport form being the tripeptide glutathione (Garsed and Read 1977; Rennenberg, Schmitz, and Bergmann 1979). Indeed, photosynthetic tobacco cells in suspension cultures release glutathione into the culture medium. If the sulphate supply in the medium becomes exhausted, the released glutathione is taken up again. These results suggest that glutathione in plant tissues may be a storage and transport form of reduced sulphur (Bergmann and Rennenberg 1978), although it also plays an important role in the metabolism of herbicides and in the protection of chloroplasts against toxic oxygen radicals (Chapter 8).

A minor source of sulphur to the plant is sulphur dioxide present in the atmosphere because of pollution. Sulphur dioxide enters the stomata and dissolves in the leaf water to produce sulphurous acid, which ionizes to give sulphite ions, SO_3^{2-}. These are eventually converted to sulphide by the processes described below (Ziegler 1975). Gaseous hydrogen sulphide can also act as a sulphur source to plants: indeed *Lemna minor* (duckweed) is capable of growing (slowly!) with H_2S or sulphur dioxide as sole sources of sulphur (Schmidt 1979).

This process is reversible, in that excess inorganic sulphur entering the leaves of plants can be released as H_2S in the light (Wilson, Bressan, and Filner 1978).

Sulphate taken into the leaves from the stem must enter the chloroplasts for reduction, yet little is known of the transport systems by which this doubly-charged ion penetrates the envelope, although an involvement of the phosphate translocator has been suggested (Hampp and Ziegler 1977). When so-called 'intact' (probably type B) spinach chloroplasts were supplied with $^{35}SO_4^{2-}$ in the light, radioactivity was detected in six compounds, viz. adenosine 5'-phosphosulphate (APS), 3'-phosphoadenosine 5'-phosphosulphate (PAPS), a bound form of sulphite, a bound form of sulphide and an unidentified product (P). No free sulphite, sulphide or thiosulphate ions were detected. On the basis of kinetic data the following series of reactions was proposed (Schwenn and Trebst 1976)

$$SO_4^{2-} \rightarrow APS \rightarrow (PAPS) \rightarrow \begin{matrix}\text{unknown}\\ \text{substance (P)}\end{matrix} \rightarrow \begin{matrix}\text{bound}\\ \text{sulphite}\end{matrix} \rightarrow \begin{matrix}\text{bound}\\ \text{sulphide}\end{matrix} \rightarrow \text{cysteine} \quad (10.7)$$

The bound sulphite can exchange label with any free sulphite added to the chloroplast fraction

$$\text{bound} - {}^{35}SO_3^{2-} + \text{free } SO_3^{2-} \rightleftarrows \text{free } {}^{35}SO_3^{2-} + \text{bound} - SO_3^{2-} \quad (10.8)$$

or with added thiosulphate ($S_2O_3^{2-}$)

$$\text{bound} - {}^{35}SO_3^{2-} + S_2O_3^{2-} \rightleftarrows \text{bound} - SO_3^{2-} + {}^{35}S\text{-}SO_3^{2-} \quad (10.9)$$

Bound sulphide can also exchange with added free sulphide

$$\text{bound} - {}^{35}S^{2-} + \text{free } S^{2-} \rightleftarrows \text{free } {}^{35}S^{2-} + \text{bound} - S^{2-} \quad (10.10)$$

or with thiol compounds (R-SH)

$$\text{bound} - {}^{35}S^{2-} + \text{R-SH} \rightleftarrows \text{bound} - \text{S-R} + {}^{35}S^{2-} \quad (10.11)$$

and such exchanges have frequently confused the results of isotope-labelling experiments.

The bound sulphate, sulphite, and sulphide of spinach chloroplasts are attached to a protein with a molecular weight of about 5000, about which little is known except that it probably binds these intermediates by means of a thiol (–SH) group.

10.6.1 Enzymes of sulphate assimilation

The first stage of sulphate assimilation in chloroplasts, conversion to APS (Fig. 10.4), is catalysed by the enzyme ATP-sulphurylase, which is present in chloroplast fractions (Schmidt 1979) although not yet rigorously proved by latency experiments of the type described in Chapter 1 to be located within the chloroplast. Such a location seems likely however. The purified spinach-leaf enzyme has a broad pH optimum centred around pH 8, a K_m for SO_4^{2-} of 3.1 mmol l^{-1} and a requirement for Mg^{2+} (Burnell and Anderson 1973a). It is thus adapted to function under the conditions of pH and Mg^{2+} concentration that exist in illuminated chloroplasts (Chapter 2). In leaves of C_4 plants, almost all ATP-sulphurylase activity is located in the bundle-sheath cells (Gerwick, Ku, and Black 1980). It therefore seems that whereas assimilation of carbon (CO_2) and of nitrogen (NO_3^-) by C_4 leaves begins in the mesophyll cells, assimilation of sulphate starts in the bundle-sheath.

ATP-sulphurylase catalyses an equilibrium reaction but the amount of APS present in the equilibrium mixture is very small. *In vivo* the reaction is pulled in the direction of APS synthesis because the pyrophosphate also produced is

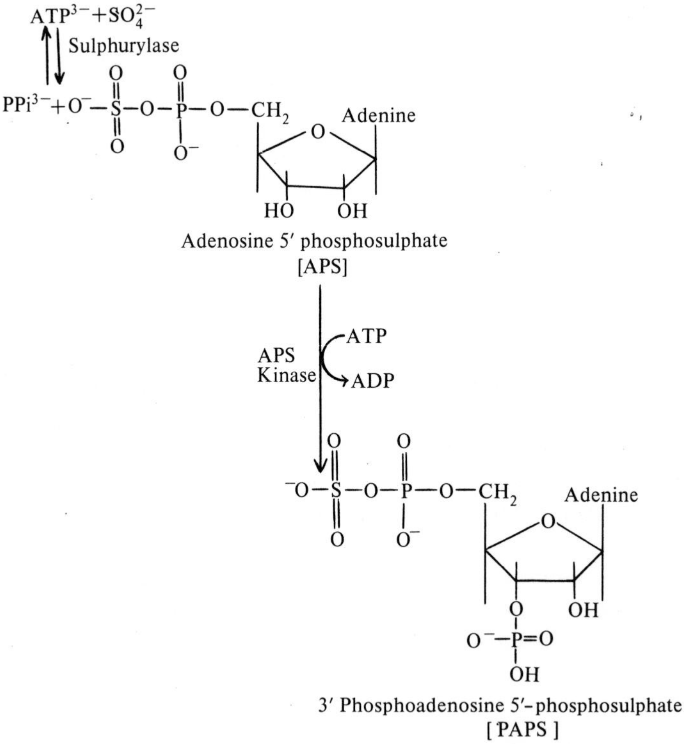

$ATP^{3-} + SO_4^{2-}$

Sulphurylase

PPi^{3-} + Adenosine 5′ phosphosulphate [APS]

APS Kinase (ATP → ADP)

3′ Phosphoadenosine 5′-phosphosulphate [PAPS]

Fig. 10.4. Formation of APS and PAPS in plant tissues.

immediately hydrolysed by chloroplast pyrophosphatase, a Mg^{2+}-dependent enzyme with an optimum pH of 8.2–8.6 (Chapter 4).

Spinach chloroplast fractions can convert APS into PAPS (Fig. 10.4), a compound detected during sulphate assimilation by chloroplasts, as mentioned above (Schwenn and Trebst 1976; Burnell and Anderson 1973b). An APS-kinase enzyme has been reported to be bound to the chloroplast thylakoids. However, there is considerable debate as to whether PAPS formation is an essential step in the conversion of sulphate to sulphide, or whether it is an alternative fate for APS, PAPS being used for the synthesis of sulphated lipids (Schmidt 1979; Schwenn and Trebst 1976). Certainly, the next stage in sulphate assimilation can operate with APS (Schmidt 1979). Differences between these data and earlier reports demonstrating activity with PAPS might have been due to the presence of phosphatases in some plant extracts that can convert PAPS into APS. Whether such phosphatases are present in chloroplasts remains to be established, however.

APS then reacts with the protein carrier molecule described above under the action of the enzyme APS-sulphotransferase. This enzyme catalyses the reaction shown below, in which *X* represents the carrier molecule bearing an —SH group.

$$\text{APS} + X\text{–SH} \rightarrow X\text{–S–SO}_3^- + \text{adenosine 5}'\text{phosphate} \qquad (10.12)$$

APS sulphotransferase is also capable of transferring sulphur from APS on to low-molecular-weight thiol compounds such as glutathione (eqn (10.13)) or dithioerythritol *in vitro*

$$\text{APS} + \text{GSH} \rightarrow \text{G–S–SO}_3^- + \text{AMP} \qquad (10.13)$$

Although glutathione is present in the chloroplast stroma at millimolar concentration (Foyer and Halliwell 1976) it is probably not a major substrate for the sulphotransferase enzyme *in vivo* because most labelled sulphur is found to be bound to the carrier when $^{35}SO_4^{2-}$ is supplied to illuminated chloroplasts (see above). The possibility that carrier *X* is a bound thioredoxin molecule deserves attention in view of the fact that chloroplasts contain several different thioredoxins (Chapter 3). However, the APS-sulphotransferase from cells of the green alga *Chlorella* will not accept *E. coli* thioredoxin as a substrate (Tsang and Schiff 1978). In spinach leaves, APS sulphotransferase is almost exclusively located in chloroplasts (Frankhauser and Brunold 1978) and the enzyme is inhibited by its product 5′AMP, which could represent a regulatory mechanism.

The sulphur in the product $X\text{–S–SO}_3^-$, which is formally equivalent to 'bound sulphite', is then reduced to sulphide whilst still attached to the carrier *X* (eqn (10.14)).

$$X - S - SO_3^- + 6e^- \rightarrow X - S - S^- \qquad (10.14)$$

Reduction is catalysed by a thylakoid-bound enzyme known as thiosulphonate reductase (Schwenn, Depka, and Hennies 1976). The electron donor is reduced ferredoxin and so the reaction proceeds only in the light. No intermediate valency states of sulphur have been detected.

Systems reducing sulphite in free solution to sulphide have been detected in several plant tissues, including spinach leaves (Asada, Tamura, and Bandurski 1969). These enzymes are usually assayed with methylviologen as electron donor; the natural donors are unknown although ferredoxin can sometimes be used in the presence of unidentified 'factors' (Schmidt 1979). The relationship of the 'free sulphite reductases' to the 'bound-sulphite reductase' of chloroplasts (thiosulphonate reductase) remains to be established. However, the 'free-sulphite reductases' do provide a mechanism by which SO_3^{2-} derived from sulphur dioxide can be assimilated.

The final step in assimilatory sulphate reduction is the formation of cysteine. Leaves contain an *O*-acetyl-L-serine sulphydrylase enzyme, which catalyses the reaction

$$O\text{-acetyl-L-serine} + H_2S \rightarrow \text{L-cysteine} + \text{acetate} \qquad (10.15)$$

About 20 per cent of the activity of this enzyme in spinach leaf extracts was present in the chloroplast band after density-gradient centrifugation (Frankhauser, Brunold, and Erisman 1976), although latency experiments were not carried out. Cysteine synthesis from SO_4^{2-} or from *O*-acetylserine has often been observed in illuminated chloroplast fractions (Schwenn and Trebst 1976; Ng and Anderson 1979).

A second cysteine-forming enzyme, serine sulphydrylase, is present in leaf tissues (Schmidt 1979) and catalyses the reaction

$$H_2S + \text{L-serine} \rightleftarrows \text{L-cysteine} + H_2O \qquad (10.16)$$

Whether this enzyme is involved in chloroplast cysteine biosynthesis remains to be established. It must be noted that chloroplasts cannot synthesise L-serine: they must import it from the cytoplasm (Chapter 7). The site of conversion of serine to *O*-acetyl serine remains to be established, assuming that, as seems likely, reaction (10.15) is responsible for at least part of cysteine synthesis in chloroplasts. Chloroplasts are known to have the capacity to convert acetate into acetyl SCoA (Chapter 9).

It must further be noted that the final product of assimilatory sulphate reduction in chloroplasts is bound sulphide, $X–S–S^-$. If sulphide is released for reaction (10.15) carrier *X* will be left in an oxidised form ('$X–S^+$') which must be reduced in a two-electron step to regenerate *X*–SH. How reduction is achieved remains to be established. Sulphide in excess of that required for cysteine synthesis may escape from the leaves in the form of H_2S (Wilson *et al.* 1978).

10.7 Other aspects of sulphur metabolism

Cysteine generated from sulphide must be used in part for the synthesis of chloroplast glutathione. Whether this synthesis occurs in chloroplast or cytoplasm has yet to be investigated. Another enzyme of sulphur metabolism present in some plant tissues is rhodanese, which catalyses the formation of thiocyanate from cyanide and thiosulphate (eqn (10.17))

$$S_2O_3^{2-} + CN^- \rightarrow CNS^- + SO_3^{2-} \tag{10.17}$$

CN^- can be replaced by dihydrolipoate or dihydrolipoamide

$$S_2O_3^{2-} + \text{lipoate (SH)}_2 \rightarrow \text{lipoate (S)}_2 + SO_3^{2-} + SH^- + H^+ \tag{10.18}$$

(Villarejo and Westley 1963; Silver and Kelly 1976). Rhodanese has recently been detected in crude chloroplast fractions from several plants (Tomati, Federici, and Cannella 1972) although this does not, of course, prove that it is located within the chloroplasts. It has been speculated that rhodanese could be involved in cysteine synthesis in chloroplasts (Schmidt 1979) or in the insertion of sulphur into the ferredoxin molecule (Finnazzi-Agro, Cannella, Graziani, and Cavallini 1971; Tomati *et al.* 1974).

References

Asada, K., Tamura, G., and Bandurski, R. S. (1969). *J. biol. Chem.* **244**, 4904–15.
Aslam, M., Huffaker, R. C., Rains, D. W., and Rao, K. P. (1979). *Pl. Physiol., Lancaster* **63**, 1205–9.
Bergmann, L. and Rennenberg, H. (1978). *Z. Pflanzenphysiol.* **88**, 175–85.
Bickel, H., Palme, L., and Schultz, G. (1978). *Phytochemistry* **17**, 119–24.
Bryan, J. K., Lissik, E. A., and Matthews, B. F. (1977). *Pl. Physiol., Lancaster* **59**, 673–9.
Burnell, J. N. and Anderson, J. W. (1973a). *Biochem. J.* **133**, 417–28.
—— (1973b). *Biochem. J.* **134**, 565–79.
Butt, V. S. and Beevers, H. (1961). *Biochem. J.* **80**, 21–7.
Canvin, D. T. and Atkins, C. A. (1974). *Planta* **116**, 207–24.
Chiu, J. Y. and Shargool, P. D. (1979). *Pl. Physiol., Lancaster* **63**, 409–15.
Cresswell, C. F., Tew, A. J., and Baxter, J. (1975). In *Proceedings of the third international congress on photosynthesis* (ed. M. Avron) pp. 1231–48. Elsevier, Amsterdam.
Dalling, M. J., Tolbert, N. E., and Hageman, R. H. (1972*a*). *Biochim. biophys. Acta* **283**, 505–12.

Dalling, M. J., Tolbert, N. E. and Hageman, R. H. (1972*b*). *Biochim. biophys. Acta* **283**, 513–19.
Elias, B. A. and Givan, C. V. (1977). *Pl. Physiol., Lancaster* **59**, 738–40.
Fair, P., Tew, J., and Cresswell, C. F. (1973). *Ann. Bot.* **37**, 831–44.
— — — (1974). *Ann. Bot.* **38**, 39–43.
Finnazzi-Agro, A., Cannella, C., Graziani, M. T., and Cavallini, D. (1971). *FEBS Lett.* **16**, 172–4.
Foyer, C. H. and Halliwell, B. (1976). *Planta* **133**, 21–5.
Frankhauser, H. and Brunold, C. (1978). *Planta* **143**, 285–9.
— — Erismann, K. H. (1976). *Experientia* **32**, 1494–6.
Garsed, S. G. and Read, D. J. (1977). *New Phytol.* **99**, 583–92.
Gayler, K. R. and Morgan, W. R. (1976). *Pl. Physiol., Lancaster* **58**, 283–7.
Gerwick, B. C., Ku, S. B., and Black, C. C. (1980). *Science, N.Y.* **209**, 513–15.
Giovanelli, J., Mudd, S. H., and Datko, A. H. (1974). *Pl. Physiol., Lancaster* **54**, 725–36.
Givan, C. V. (1979). *Phytochemistry* **18**, 375–82.
Hampp, R. and Ziegler, I. (1977). *Planta* **137**, 309–12.
Harel, E., Lea, P. J., and Miflin, B. J. (1977). *Planta* **134**, 195–200.
Haynes, R. H. and Goh, K. M. (1978). *Biol. Rev.* **53**, 465–510.
Hewitt, E. J. (1974). In *MTP international review of science,* Vol. II (ed. D. H. Northcote) pp. 199–245. Butterworths, London.
Huang, A. H. C., Liu, K. D. F., and Youle, R. J. (1976). *Pl. Physiol., Lancaster* **58**, 110–3.
Hucklesby, D. P., Cammack, R., and Hewitt, E. J. (1979). In *Nitrogen assimilation in plants* (eds. E. J. Hewitt and C. V. Cutting) pp. 245–54. Academic Press, London.
Jackson, C., Dench, J. E., Morris, P., Lui, S. C., Hall, D. O. and Moore, A. L. (1979). *Biochem. Soc. Trans.* **7**, 1122–4.
Kessler, E. (1964). *A. Rev. Pl. Physiol.* **15**, 57–72.
Keys, A. J., Bird, I. F., Cornelius, M. J., Lea, P. J., Wallsgrove, R. M. and Miflin, B. J. (1978). *Nature, Lond.* **275**, 741–3.
Larsson, C. (1979). *Physiol. Plant.* **46**, 221–6.
Lea, P. J. and Miflin, B. J. (1974). *Nature, Lond.* **251**, 614–16.
— and Fowden, L. (1975). *Biochem. Physiol. Pflanzen.* **168**, 3–14.
— and Thurman, D. A. (1972). *J. exp. Bot.* **23**, 440–9.
Lee, R. B. (1980). *Plant Cell Envir.* **3**, 65–90.
Leech, R. M. and Murphy, D. J. (1976). In *The intact chloroplast, Topics in photosynthesis,* Vol. 1 (ed. J. Barber) pp. 365–401. Elsevier, Amsterdam.
Matoh, T., Suzuki, F., and Ida, S. (1979). *Plant Cell Physiol.* **20**, 1329–40.
Mazelis, M., Miflin, B. J., and Pratt, H. M. (1976). *FEBS Lett.* **64**, 197–200.
McKenzie, G. H., Ch'ng, A. L., and Gayler, K. R. (1979). *Pl. Physiol., Lancaster* **63**, 578–82.
Miflin, B. J. (1974). *Pl. Physiol., Lancaster* **54**, 550–5.
— and Lea, P. J. (1976). *Phytochemistry* **15**, 873–85.
— — (1977). *A. Rev. Pl. Physiol.* **28**, 299–329.
Mills, W. R. and Joy, K. W. (1980). *Planta* **148**, 75–83.
— Lea, P. J., and Miflin, B. J. (1980). *Pl. Physiol., Lancaster* **65**, 1166–72.
Moore, R. and Black, C. C. (1979). *Pl. Physiol., Lancaster* **64**, 309–13.
Neyra, C. A. and Hageman, R. H. (1976). *Pl. Physiol., Lancaster* **58**, 726–30.
— — (1978). *Pl. Physiol., Lancaster* **62**, 618–21.
Ng, B. H. and Anderson, J. W. (1978). *Phytochemistry* **17**, 879–85.

Pate, J. S. (1980). *A. Rev. Pl. Physiol.* **31**, 313–40.
Ramaswamy, N. K., Behere, A. G. and Nair, P. M. (1976). *Eur. J. Biochem.* **67**, 275–82.
Rathnam, C. K. M. (1978). *Pl. Physiol., Lancaster* **62**, 220–3.
Rennenberg, H., Schmitz, K., and Bergmann, L. (1979). *Planta* **147**, 57–62.
Rhodes, D., Sims, A. P., and Folkes, B. F. (1980). *Phytochemistry* **19**, 357–65.
Robinson, T. (1974). *Science, N.Y.* **184**, 430–5.
Rogers, H. H., Campbell, J. C., and Volk, R. J. (1979). *Science, N.Y.* **206**, 333–5.
Rubin, J. L. and Jensen, R. A. (1979). *Pl. Physiol., Lancaster* **64**, 727–34.
Sawhney, S. K., Naik, M. S., and Nicholas, D. J. D. (1978). *Biochem. Biophys. Res. Commun.* **81**, 1209–16.
Schmidt, A. (1979). In *Photosynthesis II, Encyclopaedia of plant physiology*, Vol. 6 (eds. M. Gibbs and E. Latzko) pp. 481–96. Springer, Berlin.
Schwenn, J. D. and Trebst, A. (1976). In *The intact chloroplast, Topics in photosynthesis*, Vol. 1 (ed. J. Barber) pp. 315–34. Elsevier, Amsterdam.
— Depka, B. and Hennies, H. H. (1976). *Plant Cell Physiol.* **17**, 165–76.
Silver, M. and Kelly, D. P. (1976). *J. gen. Microbiol.* **97**, 277–84.
Somerville, C. R. and Ogren, W. L. (1980). *Nature, Lond.* **286**, 257–9.
Stewart, G. R. and Orejambo, T. O. (1979). *New Phytol.* **83**, 311–19.
Tamura, G., Kanki, M., Hirasawa, M., and Oto, M. (1980). *Agr. biol. Chem.* **44**, 925–7.
Tempest, D. W., Meers, J. L., and Brown, C. M. (1970). *Biochem. J.* **117**, 405–7.
Tomati, U., Federici, G., and Cannella, C. (1972). *Physiol. Chem. Phys.* **4**, 193–6.
Tsang, M. L. and Schiff, J. A. (1978). *Plant Sci. Lett.* **11**, 177–83.
Vallejos, R. H. and Andreo, C. S. (1974). *Biochim. biophys. Acta* **333**, 141–8.
Vennesland, B. and Guerrero, M. G. (1979). In *Photosynthesis II, Encyclopaedia of plant physiology*, Vol. 6 (eds. M. Gibbs and E. Latzko), pp, 425–44. Springer, Berlin.
Villarejo, M. and Westley, J. (1963). *J. biol. Chem.* **238**, 4016–20.
Wahnbaeck-Spencer, R., Mills, W. R., Henke, R. R., Burdge, E. L. and Wilson, K. G. (1979). *FEBS Lett.* **104**, 303–8.
Wallsgrove, R. M. and Mazelis, M. (1980). *FEBS Lett.* **116**, 189–92.
— Lea, P. J. and Miflin, B. J. (1979). *Pl. Physiol., Lancaster* **63**, 232–6.
— Harel, E., Lea, P. J., and Miflin, B. J. (1977). *J. exp. Bot.* **28**, 588–96.
Washitani, I. and Sato, S. (1977). *Plant Cell Physiol.* **18**, 1235–41.
Wells, G. N. and Hageman, R. H. (1974). *Pl. Physiol., Lancaster* **54**, 136–41.
Wilson, L. G., Bressan, R. A., and Filner, P. (1978). *Pl. Physiol., Lancaster* **61**, 184–9.
Wink, M., Hartmann, T., and Witte, L. (1980). *Z. Naturforsch.* **35C**, 93–7.
Woo, K. C., Jokinen, M., and Canvin, D. T. (1980). *Pl. Physiol., Lancaster* **65**, 433–6.
Ziegler, I. (1975). *Residue Rev.* **56**, 79–105.

APPENDIX 1

Spectrophotometric determination of chlorophyll

The total chlorophyll (a + b) content of a leaf extract in 80 per cent (v/v) acetone may be obtained by reading the absorbances at 645, 663, or 652 nm in a cell of 1 cm path length. The following formulae may then be used

$$\text{Chlorophyll conc. in acetone solution (mg l}^{-1}\text{)} = 20.2A_{645} + 8.02A_{663}$$

$$= \frac{10^3 A_{652}}{36}$$

Reference: Bruinsma, J. (1961). *Biochim. biophys. Acta* 52, 576–8.

APPENDIX 2

Botanical names of plants mentioned in the text

(alt. = alternative name)

Common name	*Botanical name*	*Plant family*
Alfalfa (alt. lucerne)	*Medicago sativa*	Leguminosae
Barley	*Hordeum vulgare*	Gramineae
Bean (broad)	*Vicia faba*	Leguminosae
Bean (French, alt. kidney)	*Phaseolus vulgaris*	Leguminosae
Bean (soy)	*Glycine max*	Leguminosae
Beet, spinach (alt. sugar)	*Beta vulgaris*	Chenopodiaceae
Cabbage	*Brassica oleracea*	Cruciferae
Carrot	*Daucus carota*	Umbelliferae
Clover, red	*Trifolium praetense*	Leguminosae
Coffee	*Coffea arabica*	Rubiaceae
Common millet	*Panicum miliaceum*	Gramineae
Corn (alt. maize)	*Zea mays*	Gramineae
Crabgrass	*Digitaria sanguinalis*	Gramineae
Cucumber	*Cucumnis sativus*	Cucurbitaceae
Duckweed	*Lemna minor*	Lemnaceae
Fig	*Ficus glabrata*	Moraceae
Geranium	*Pelargonium hortorum*	Geraniaceae
Hastate orache	*Atriplex hastata*	Chenopodiaceae
Horseradish	*Armoracia lapathifolia*	Cruciferae
Lettuce	*Lactuca sativa*	Compositae
Lucerne (alt. alfalfa)	*Medicago sativa*	Leguminosae
Lupin	*Lupinus polyphyllus*	Leguminosae
Maize (alt. corn)	*Zea mays*	Gramineae
Millet, common	*Panicum miliaceum*	Gramineae
Mung bean	*Phaseolus aureus*	Leguminosae
Mustard	*Sinapis alba*	Cruciferae
Oat	*Avena sativa*	Gramineae
Orache	*Atriplex hortensis*	Chenopodiaceae
Orange	*Citrus sinensis*	Rutaceae
Pea	*Pisum sativum*	Leguminosae
Pineapple	*Ananus sativus*	Bromeliaceae
Potato	*Solanum tuberosum*	Solanaceae
Purslane	*Portulaca oleracea*	Portulacaceae
Rape	*Brassica napus*	Cruciferae
Red orache	*Atriplex rosea*	Chenopodiaceae
Rye	*Secale cereale*	Gramineae
Safflower	*Carthamus tinctorius*	Compositae
Scots pine	*Pinus sylvestris*	Pinaceae
Sorghum	*Sorghum dochna*	Gramineae

Common name	*Botanical name*	*Plant family*
Soybean	*Glycine max*	Leguminosae
Spinach	*Spinacia oleracea*	Chenopodiaceae
Sugar-cane	*Saccharum officinarum*	Gramineae
Sycamore	*Acer pseudoplatanus*	Aceraceae
Tea	*Thea sinensis*	Theaceae
Tobacco	*Nicotiana tabacum*	Solanaceae
Tomato	*Lycopersicon esculentum*	Solanaceae
Wheat	*Triticum aestivum*	Gramineae

INDEX

Please note that plant species are indexed under both their systematic and their trivial names. Organic acids are indexed under the names of their anion, e.g. for acetic acid see 'acetate' and for malic acid see 'malate'. If a topic is mentioned in several places in the text, the pages containing the most extensive account of it are presented in **bold type**.